CONCEPTS IN DISCRETE MATHEMATICS
SECOND EDITION

SARTAJ SAHNI
University of Minnesota

THE CAMELOT PUBLISHING COMPANY
NORTH OAKS, MINNESOTA

The Camelot Publishing Co.
3 Hill Farm Circle
North Oaks
Minnesota 55127

Second Edition
Printing 9 8 7 6 5 4 3 2   87 88 89

Library of Congress Cataloging in Publication Data

Sahni, Sartaj
  Concepts in Discrete Mathematics

  Includes bibliographies and index.
  1. Discrete mathematics

ISBN 0-942450-00-0
Library of Congress: 83-104061

DEDICATION

to my parents: Dharam Nath and Santosh

      my wife: Neeta

and

  my children: Agam and Neha

PREFACE

This book contains a collection of mathematical topics that are of immense value to everyone who is pursuing a course of study in science or engineering. While a variety of mathematical tools are needed to successfully complete a course of study in these fields, most science and engineering curricula include mathematical courses in calculus and algebra only. Important concepts such as proof methods; difference equations; combinatorics; graph theory; etc. are often omitted and the student is expected to pick up these concepts along the way (somehow).

In this text, I have made an attempt to include those mathematical topics whose understanding is essential to science and engineering, but which are not covered in mathematical courses traditionally required of students in these disciplines. The topics covered are: logic; sets; relations; functions and computability; analysis of algorithms; recurrence equations; combinatorics; discrete probability; graphs; and algebra. While many of these topics are the subject of individual courses offered in traditional mathematical curricula, mathematics departments seldom have a one or two course sequence that covers all of them. The depth of coverage of each of the topics included in this text is about what is needed to successfully complete courses typically found in science and engineering curricula.

There is a bias towards computer science in this text. Such a bias can hardly be avoided today given the rapid growth in the use of computers and the permeation of computer science courses in virtually all curricula. The material in this text is illustrated by a large number of examples that have been carefully and completely worked out. There are over two hundred exercises that have been designed to enhance one's understanding of the material.

ACKNOWLEDGEMENTS

This text was prepared using the TROFF text formatting program. I am grateful to Terrie Christian for entering this text into the computer and also for doing the art work. Professor Bill Thompson provided invaluable assistance in using the TROFF program. The debugging of this text has been greatly facilitated by the able assistance of many of the students in my discrete structures course C.Sci. 3400. In this regard I would particularly like to acknowledge the assistance of Bill Larson and Richard Farrell. Special credit also goes to Jim Cohoon, Eliezer Dekel, Steve Lai, and Ragu Raghavan for carefully reading this manuscript and pointing out several bugs. The helpful comments of Professors Al Hevner, Kurt Maly, M.

Krishnamoorthy, Brian Leininger, and Lou Rosier are greatly appreciated. Finally, I wish to thank the University of Minnesota for encouraging, in every way, the writing of this book.

Sartaj Sahni
Minneapolis
June 1981

CONTENTS

CHAPTER 1 LOGIC

1.1 Propositions and Well Formed Formulas 1
1.2 Normal Forms and Boolean Algebra 8
1.3 Proof Methods 16
1.4 Tableau Method 26
1.5 The Davis-Putnam Procedure 30
1.6 First Order Logic 33
1.7 Prenex Normal Form 42
1.8 FOL Proofs 45
1.9 Other Logics 49
       References and Selected Readings 51
       Exercises 52

**CHAPTER 2 CONSTRUCTIVE PROOFS, MATHEMATICAL
                     INDUCTION AND PROGRAM CORRECTNESS**

2.1 Constructive Proofs 63
2.2 Mathematical Induction 67
2.3 Program Correctness 81
       2.3.1 Recursive Programs 81
       2.3.2 Iteratve Programs 85
       2.3.3 Loop Invariants 90
       2.3.4 The Predicate Transformer Method 93
       References and Selected Readings 118
       Exercises 119

CHAPTER 3 SETS

3.1 Sets, Multisets, and Subsets 129
3.2 Set Specification 130
       3.2.1 Explicit 130
       3.2.2 By Properties 131
       3.2.3 Grammars 132
3.3 Set Operations 139
       3.3.1 Definitions 139
       3.3.2 Properties 142
       3.3.3 Set Expressions 145
3.4 Correspondence, Countability, and Diagonalization 148
       Exercises 159

CHAPTER 4 RELATIONS
4.1 Introduction 163
4.2 Binary Relations 168
    4.2.1 Representations 168
    4.2.2 Properties 170
    4.2.3 Composition and Closure 172
    4.2.4 Equivalence Relations 175
    4.2.5 Partial Orders 178
4.3 Operations on k-ary Relations 186
4.4 Functional and Multivalued Dependencies 191
4.5 Normal Forms 195
    References and Selected Readings 199
    Exercises 201

CHAPTER 5 FUNCTIONS, RECURSION, AND COMPUTABILITY
5.1 Functions 205
    5.1.1 Terminology 205
    5.1.2 Properties 209
    5.1.3 The Pigeon Hole Principle 211
5.2 Recursion 213
5.3 Computability 238
    References and Selected Readings 245
    Exercises 247

CHAPTER 6 ANALYSIS OF ALGORITHMS
6.1 Complexity 251
6.2 Asymptotic Notation (O, Ω, Θ, o) 273
6.3 Practical Complexities 282
    References and Selected Readings 286
    Exercises 288

CHAPTER 7 RECURRENCE RELATIONS
7.1 Introduction 291
7.2 Substitution 295
7.3 Induction 301
7.4 Characteristic Roots 303
7.5 Generating Functions 318
    References and Selected Readings 332
    Exercises 333

CHAPTER 8 COMBINATORICS AND DISCRETE PROBABILITY

8.1 The Binomial Coefficient 336
8.2 Permutations and Combinations 340
8.3 The Multinomial Coefficient 343
8.4 Discrete Probability 346
    8.4.1 Events, Trials, and Probability 346
    8.4.2 Conditional Probability 349
    8.4.3 Bernoulli Trials 351
    8.4.4 Random Variables 353
    8.4.5 Chebychev's Inequality 357
    8.4.6 Law Of Large Numbers 358
    References and Selected Readings 361
    Exercises 361

CHAPTER 9 GRAPHS

9.1 The Basics 367
    9.1.1 Introduction 367
    9.1.2 Terminology 372
    9.1.3 Graph Representation 384
9.2 Spanning Trees and Connectivity 391
    9.2.1 Breadth First Search 391
    9.2.2 Depth First Search 395
    9.2.3 Minimum Cost Spanning Trees 398
    9.2.4 Cycle Basis 403
    9.2.5 Connectivity 405
9.3 Paths 411
    9.3.1 Euler Paths and Circuits 411
    9.3.2 Hamiltonian Paths and Circuits 415
    9.3.3 Shortest Paths and Transitive Closure 417
9.4 Rooted Trees 419
    9.4.1 Terminology 419
    9.4.2 Properties 425
    9.4.3 The Number of Binary Trees 427
9.5 Miscellaneous Topics 431
    9.5.1 Planar Graphs 431
    9.5.2 Matchings 435
    9.5.3 Cliques and Independent Sets 437
    9.5.4 Colorings and Chromatic Number 438
    References and Selected Readings 439
    Exercises 442

CHAPTER 10 MODERN ALGEBRA
10.1 Algebras 449
10.2 Binary Algebras 450
10.3 Rings and Fields 454
10.4 Homomorphisms 455
      References and Selected Readings 458
      Exercises 459

APPENDIX A: SYMBOLS 463

APPENDIX B: ALGORITHMIC LANGUAGE 465

INDEX 470

CHAPTER 1

LOGIC

1.1 PROPOSITIONS AND WELL FORMED FORMULAS

Logical reasoning is the stuff proofs are made of and proofs are what scientific and other knowledge rests upon. The importance of proofs and thus of logical reasoning and logic cannot be overstated. Our faith in the thousands of theories postulated by scientists, mathematicians, etc. would be considerably less if there did not exist strong (and often conclusive) logical arguments in favor of these theories. What would be the status of the following statements if it were not for the existence of mathematically acceptable proofs establishing their validity?

(a) If A is a right angled triangle with sides of length a, b, and c, then $a^2 + b^2 = c^2$ where c is the length of the hypotenuse.
(b) The sum of the angles in any triangle is 180 degrees.
(c) The derivative of x^3 is $3x^2$.
(d) $\sum_{i=1}^{n} i = n(n+1)/2$.
(e) The area of a square of side d is d^2.

Given the importance of logical reasoning to mathematics, science, engineering, etc., it is appropriate that we begin our study of mathematical concepts with the study of the principles of reasoning (i.e., logic). First, we introduce some terms.

A *declarative sentence* is any sentence that can possibly be true or false. Some examples of declarative sentences are:

(a) The voltage across a resistor is the product of the current and the resistance ($V = IR$).
(b) There exist intelligent life forms on planets other than earth.
(c) Tom dislikes the discrete structures course.
(d) This text is fantastically clear.
(e) Mary had a little lamb.

It is quite meaningful for us to consider whether each of the above declarative sentences is true or false. Every electrical engineer knows that

1

(a) is true for ideal resistors. Tom knows whether (c) is true or not, and most three year olds have reasons to believe that (e) is true. One could have considerable debate over the truth of (b) and (d).

A *proposition* is a declarative sentence that must be either true or false but not both. Each of the five declarative sentences listed above is a proposition. We already know that each is either true or false. It is not too difficult to see that none of these five sentences can be both true and false. For example, this text is either fantastically clear or it is not. It cannot be both fantastically clear and not fantastically clear.

The use of the word either, in English, is often ambiguous. For example, consider the sentence:

Tom is either guilty or innocent.

This sentence is readily seen to be equivalent to the sentence:

Tom is either guilty or innocent, but not both.

On the other hand, the sentence:

Good performance on either the exams or the assignments is sufficient to pass the course.

is not equivalent to:

Good performance on either the exams or the assignments, but not on both, is sufficient to pass the course.

Rather, it is equivalent to:

Good performance on either the exams or the assignments, or on both, is sufficient to pass the course.

Generally, the context in which 'either a or b' is used determines whether 'either a or b or both' or 'either a or b but not both' is meant. To avoid possible confusion resulting from the use of the word 'either', we shall usually state explicitly which interpretation is intended. When no interpretation is provided, we shall always mean 'either or both'(in this text).

Not all sentences are declarative sentences. For example:

(a)   Pass me the butter.
(b)   Has flight 201 from New York arrived?
(c)   Can't you do anything right?

Furthermore, not all declarative sentences are propositions. For example, the sentence:

> This statement is false.

can be neither true nor false. If the statement is true, then it asserts that it is false. If it is false, then it must be true. All propositions obey the following law:

> *Propositional Calculus Axiom:* Every proposition is either true or false (but not both).

Observe that the above axiom includes the famous *law of contradiction* which states that no proposition is both true and false.

In algebra, symbols are used to denote numbers. For example, in the arithmetic expression $x + y$, the symbols x and y denote variables and the expression $x + y$ has value 10 when x is assigned the value 8 and y the value 2 or when $x = 6$ and $y = 4$, etc. In logic, we use capital letters $(A, B, ..., Z)$ as variables (called propositional variables). These variables can be assigned propositions as values. For instance, P could denote any of the following propositions :

(a)   It rains in Minneapolis.
(b)   Stan is a democrat.
(c)   Minnesota does not have a computer science department.
etc.

The *truth value* of a propositional variable P is true if the proposition assigned to it is true. It is false otherwise. If P denotes either of propositions (a) and (b) above, it is true. If P denotes proposition (c), then the truth value of P is false. Clearly, any propositional variable (i.e., $A, B, ..., Z$) can have a *truth value* either true or false depending upon which proposition it denotes. Propositional variables can be combined together using logical operators to get well formed formulas (wffs). This is similar to the use of $+$, $-$, $/$, $*$, etc. to combine together arithmetic variables to obtain arithmetic expressions. The logical operators we shall be dealing with are: $\neg$ (not), $\vee$ (or), $\wedge$ (and), $\implies$ (implies), and $\iff$ (if and only if).

NOT ($\neg$)

The operator $\neg$ denotes negation. If P is a proposition then $\neg P$ (also written as $\bar{P}$) is its negation. The *negation* of a proposition P is another proposition that is true whenever P is false and is false whenever P is true. This can be stated in terms of a truth table (Figure 1.1(a)). In a truth table, the truth values true and false are abbreviated T and F respectively. The truth table

P	$\neg P$
T	F
F	T

P	Q	$P \lor Q$
T	T	T
T	F	T
F	T	T
F	F	F

P	Q	$P \land Q$
T	T	T
T	F	F
F	T	F
F	F	F

(a)$\neg P$ (b) $P \lor Q$ (c) $P \land Q$

P	Q	$P \Rightarrow Q$
T	T	T
T	F	F
F	T	T
F	F	T

P	Q	$P \Longleftrightarrow Q$
T	T	T
T	F	F
F	T	F
F	F	T

(d) $P \Rightarrow Q$ (e) $P \Longleftrightarrow Q$

Figure 1.1 Truth tables for logical operators.

for $\neg P$ has one column for P and one for $\neg P$. In the column for P we list the two possible truth values of P. The column for $\neg P$ gives the corresponding truth values for $\neg P$. Hence, from the truth value of P and the truth table of Figure 1.1(a) one can determine the value of $\neg P$. Consider the proposition:

This pie is good.

Its negation is:

This pie is not good.

Which is equivalent to:

It is not the case that this pie is good.

OR ($\lor$)

The operator $\lor$ obtains the disjunction of two propositions. The disjunction of the propositions P and Q is written $P \lor Q$ and read as "P or Q". Figure 1.1(b) gives the truth table for $P \lor Q$. The truth table for $P \lor Q$ has three columns. One for each of P, Q, and $P \lor Q$. There is one row for each combination of truth values of P and Q. The entry in the column for $P \lor Q$ in any row of the truth table gives the truth value of $P \lor Q$ when P and Q have the truth values given in that row. Note that $P \lor Q$ is true iff (if and only if) at least one of P and Q is true.

AND ($\land$)

The conjunction of two propositions P and Q is obtained by using the operator $\land$. It is denoted $P \land Q$ and read as "P and Q". Figure 1.1(c) gives the truth table for $P \land Q$. Observe that the truth value of $P \land Q$ is true iff both P and Q are true.

IMPLICATION ($\Longrightarrow$)

$P \Longrightarrow Q$ is read as "if P then Q" or as "P implies Q". P is the *antecedent* of "$\Longrightarrow$" and Q is its *consequent*. The truth table for $P \Longrightarrow Q$ is given in Figure 1.1(d). This truth table merits further discussion. The statement if P then Q essentially says that Q is true whenever P is true. It does not say anything about the truth value of Q when P is false. So, when P is false, Q can be either true or false. Hence the entries corresponding to $P = $ F and $Q = T$ or F are T. The only time the statement $P \Longrightarrow Q$ is false is when P is true and Q is false.

To understand the preceding discussion better, consider the proposition R:

If it rains, the ground will get wet.

Let P denote "it rains" and let Q denote "the ground will get wet". The proposition R is then equivalent to $P \Longrightarrow Q$. If it rains and the ground doesn't get wet, then R is false. So, the truth table entry for P true and Q false is F. Now, suppose it doesn't rain. It is still possible for the ground to get wet (someone may throw a bucket of dirty water on the ground). But, the fact that the ground has gotten wet despite the fact that it hasn't rained does not contradict R. This agrees with the truth table entry corresponding to P false and Q true. Similarly, if it doesn't rain and the ground isn't wet then P and Q are both false. Once again, this does not contradict the statement R and R remains true. The important point is that a statement of the type if P, then Q (written $P \Longrightarrow Q$) is false only if it is the case that Q is false when P is true. For $P \Longrightarrow Q$ to be true Q must be true whenever P is true. Q can take on any truth value when P is false.

IF AND ONLY IF ($\Longleftrightarrow$, iff)

$P \Longleftrightarrow Q$ (P iff Q) has the truth table given in Figure 1.1(e). $P \Longleftrightarrow Q$ is equivalent to the statement P implies Q and Q implies P. So, the truth table for $P \Longleftrightarrow Q$ must correspond to that for $(P \Longrightarrow Q) \wedge (Q \Longrightarrow P)$. One may easily verify that this is so.

Other logical operators such as exclusive or (XOR), not and (NAND), and not or (NOR) are defined in the exercises.

A *well formed formula* (wff) is defined recursively as below:

(a) All propositional variables and the constants true and false are wffs.
(b) If α and β are wffs, then $\neg \alpha$, $\bar{\alpha}$, (α), $[\alpha]$, $(\alpha \vee \beta)$, $(\alpha \wedge \beta)$, $(\alpha \Longrightarrow \beta)$, $(\alpha \Longleftrightarrow \beta)$, $[\alpha \vee \beta]$, $[\alpha \wedge \beta]$, $[\alpha \Longrightarrow \beta]$, and $[\alpha \Longleftrightarrow \beta]$ are all wffs.
(c) Nothing else is a wff.

Some examples of well formed formulas are:

(a) $(P \vee Q)$
(b) $(P \Longrightarrow Q)$
(c) $[P \wedge Q]$
(d) $(((P \wedge Q) \vee R) \Longrightarrow (A \wedge \bar{A}))$
(e) $(((P \Longleftrightarrow Q) \wedge (R \Longleftrightarrow S)) \vee (T \Longrightarrow S))$

We shall often eliminate many of the parentheses that arise in wffs. This, of course, will be done only when there is no confusion about the meaning of the wff. So, for example, the five wffs given above can also be written as:

(a) $P \vee Q$
(b) $P \Longrightarrow Q$
(c) $P \wedge Q$
(d) $(P \wedge Q) \vee R \Longrightarrow A \wedge \bar{A}$
(e) $((P \Longleftrightarrow Q) \wedge (R \Longleftrightarrow S)) \vee (T \Longrightarrow S)$

The logical operators may be assigned priorities, P, as below:

$$P(\neg) = 5; \ P(\wedge) = 4; \ P(\vee) = 3; \ P(\Longrightarrow) = 2; \text{ and } P(\Longleftrightarrow) = 1$$

These may be used to resolve ambiguities when parentheses have been dropped. Thus, if the sequence aQb appears in a wff (where a and b are logical operators), then Q is the right operand of a iff $P(a) \geqslant P(b)$. If Q is not the right operand of a, then it is the left operand of b.

Let us look at a few examples of translations of English statements into wffs. Consider the statement:

If Tom fails the discrete structures final, he will have to retake the final or be placed on probation.

Using the symbolism:

P: Tom fails the discrete structures final
Q: Tom will have to retake the final
R: Tom will be placed on probation

the above statement may be written as:

$P \Longrightarrow Q \lor R$

As another example, consider:

Mary can write her program in Pascal or Fortran or not write it at all. If she does not write her program she will get a zero and fail the course. If she fails the course she will be put on probation and if she gets a zero her boyfriend will desert her. If Mary writes her program in Fortran, she will fail the course but if she writes it in Pascal, she will pass.

Let us use the symbolism:

P: Mary writes her program in Pascal
Q: Mary writes her program in Fortran
R: Mary does not write her program
S: Mary gets a zero
T: Mary fails
U: Mary is put on probation
V: Mary's boyfriend deserts her

One might be tempted to write the first sentence concerning Mary as:

$P \lor Q \lor R$

Observe that it is not possible for P, Q, and R to all be simultaneously true. Assuming that it is possible for Mary to write her program in both Pascal and Fortran, the first sentence takes the symbolic form:

$(P \lor Q \lor R) \land (P \lor Q \Longrightarrow \bar{R})$

The wff corresponding to the set of statements about Mary is:

$$(P \lor Q \lor R) \land (P \lor Q \Longrightarrow \bar{R}) \land (R \Longrightarrow S \land T) \land (T \Longrightarrow U) \land (S$$
$$\Longrightarrow V) \land (Q \Longrightarrow T) \land (P \Longrightarrow \bar{T})$$

Given a truth value for each of the propositional variables appearing in a wff one can determine the truth value of the wff. A wff that evaluates to true for all possible truth assignments to its variables is a *tautology* (a tautology is also called a *theorem*). A *contradiction* is a wff that evaluates to false for all possible truth assignments to its variables. A wff that is not a contradiction is said to be *satisfiable*. Note that a wff is satisfiable iff there is at least one set of truth assignments to its variables under which the wff evaluates to true.

One way to determine if a wff is a tautology, is satisfiable, or is a contradiction is to use the truth table method used earlier. By examining the columns for $\neg P$, $P \lor Q$, $P \land Q$, $P \Longrightarrow Q$ and $P \Longleftrightarrow Q$ in Figure 1.1, we can conclude that neither of these wffs is a tautology. For example, $P \lor Q$ is false when both P and Q are false. Also, neither of these is a contradiction. Each of these wffs is satisfiable. Figure 1.2 gives truth tables for several other wffs. Each of the wffs considered in Figure 1.2 is important. $P \land \bar{P}$ is the negation of the law of contradiction. $P \lor \bar{P}$ is the law of the excluded middle. As expected, $P \land \bar{P}$ is a contradiction and $P \lor \bar{P}$ is a tautology. The wffs of Figures 1.2(c) to (e) are all tautologies. The tautology $(\bar{P} \lor Q)$ $\Longleftrightarrow (P \Longrightarrow Q)$ implies that we can do away with the operator " $\Longrightarrow$ ". As stated earlier, $P \Longleftrightarrow Q$ is equivalent to $(P \Longrightarrow Q) \land (Q \Longrightarrow P)$. So, the operator " $\Longleftrightarrow$ " can also be eliminated and we need only consider the three operators $\land$, $\lor$, and $\neg$. It is, however, often more convenient to use $\Longrightarrow$ and $\Longleftrightarrow$ in wffs rather than their equivalent forms. The tautologies of Figures 1.2(d) and (e) are known as DeMorgan's Laws.

Since each variable in a wff can be assigned one of two possible values (T or F), the number of rows in the truth table for a wff with r variables is 2^r. When r=3 the number of rows is 8 and when r=6, the number of rows is 64. The truth table method is therefore suitable only for wffs with a small number of variables. In subsequent sections, we shall examine alternate methods to determine if a wff is a tautology, is satisfiable or is a contradiction.

1.2 NORMAL FORMS AND BOOLEAN ALGEBRA

We have seen how to obtain a truth table for any given wff. Suppose we are given a truth table. How can we obtain a wff corresponding to this table?

P	$\bar{P}$	$P \wedge \bar{P}$
T	F	F
F	T	F

P	$\bar{P}$	$P \vee \bar{P}$
T	F	T
F	T	T

(a) $P \wedge \bar{P}$ (b) $P \vee \bar{P}$

P	Q	$\bar{P}$	$\bar{P} \vee Q$	$P \Rightarrow Q$	$(\bar{P} \vee Q) \Longleftrightarrow (P \Rightarrow Q)$
T	T	F	T	T	T
T	F	F	F	F	T
F	T	T	T	T	T
F	F	T	T	T	T

(c) $(\bar{P} \vee Q) \Longleftrightarrow (P \Rightarrow Q)$

P	Q	$P \wedge Q$	$\neg (P \wedge Q)$	$\bar{P}$	$\bar{Q}$	$\bar{P} \vee \bar{Q}$	$\neg (P \wedge Q) \Longleftrightarrow (\bar{P} \vee \bar{Q})$
T	T	T	F	F	F	F	T
T	F	F	T	F	T	T	T
F	T	F	T	T	F	T	T
F	F	F	T	T	T	T	T

(d) $\neg (P \wedge Q) \Longleftrightarrow (\bar{P} \vee \bar{Q})$

P	Q	$P \vee Q$	$\neg (P \vee Q)$	$\bar{P}$	$\bar{Q}$	$\bar{P} \wedge \bar{Q}$	$\neg (P \vee Q) \Longleftrightarrow (\bar{P} \wedge \bar{Q})$
T	T	T	F	F	F	F	T
T	F	T	F	F	T	F	T
F	T	T	F	T	F	F	T
F	F	F	T	T	T	T	T

(e) $\neg (P \vee Q) \Longleftrightarrow (\bar{P} \wedge \bar{Q})$

Figure 1.2 Truth table examples.

This problem is of special interest in the design of switching circuits. Figure 1.3 shows the schematic for an n input 1 output switching circuit. Each of the n inputs and the output can be either a 0 (corresponding to false) or a 1 (corresponding to true). Hence the mapping of input values to an output value can be given by a truth table as in Figure 1.4(a) (in keeping with switching circuit terminology we use 0 and 1 in place of F and T respectively). This mapping of inputs to an output is referred to as a *switching function*. A switching function need not specify output values for every possible input combination. Figure 1.4(b) gives a switching function for which output values are specified for only eight of the possible sixteen input combinations.

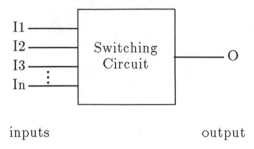

inputs   output

Figure 1.3 n/1 switching circuit.

INPUTS			OUT
P	Q	R	O
1	1	1	0
1	1	0	1
1	0	1	0
1	0	0	1
0	1	0	1
0	1	1	0
0	0	1	0
0	0	0	1

(a)

INPUTS				OUT
P	Q	R	S	O
0	0	1	0	1
1	1	1	0	0
0	1	0	1	1
1	1	1	1	0
1	1	0	1	0
1	0	1	0	1
0	0	0	0	1
1	0	0	1	0

(b)

Figure 1.4 Truth tables for a three and four input circuit.

A switching circuit *realizes* a given switching function iff for every input combination for which the switching function output is specified, the circuit output is the same as that given by the switching function.

One may verify that the circuit of Figure 1.5(a) realizes the function given in Figure 1.4(a). Boxes marked A denote and gates (i.e., a device with two inputs P and Q and output $P \wedge Q$), those marked O denote or gates and those marked N denote not gates. Figure 1.5(b) shows another circuit that realizes the switching function of Figure 1.4(a). This circuit is preferable to that of Figure 1.5(a) as it uses fewer gates and so is cheaper to build. The switching functions realized by the circuits of Figures 1.5(a) and (b) are easily seen to be equivalent to the wffs $(P \wedge \bar{R}) \vee (\bar{P} \wedge \bar{R})$ and $\bar{R}$ respectively. The equivalence of these two wffs may be established using Boolean algebra (as described later).

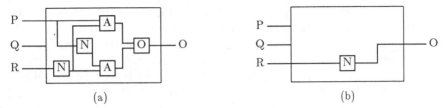

(a)   (b)

Figure 1.5 Two realizations of Figure 1.4(a).

The procedure to obtain a circuit realization starting from a truth table is to first obtain a wff corresponding to it; next minimize this wff (i.e. find an equivalent wff that will lead to a circuit with the fewest number of gates); finally obtain the circuit. The last of these steps is fairly easy as there is usually a one-to-one correspondence between the operators in the minimum wff and the gates available to build the circuit. In the example of Figure 1.5, $\vee$, $\wedge$, and $\neg$ were respectively implemented by or, and, and not gates.

In the remainder of this section we shall be concerned with only the first two steps of the circuit realization procedure outlined above. Further, in keeping with standard switching theory notation we shall use $\cdot$ to denote $\wedge$, $+$ to denote $\vee$, 0 to denote false and 1 to denote true. Also, the symbol $\cdot$ will usually be omitted as in ABC (this is equivalent to $A \cdot B \cdot C$). A propositional variable will also be referred to as a Boolean variable. Likewise, we shall use the terms wff and Boolean expression interchangeably. We shall expand our set of allowable variable names to include lower case letters (a-z) as well as subscripted letters (x_1, y_1, etc).

A *literal* is either a propositional (or Boolean) variable or its complement (e.g., X, $\bar{X}$, P, Q, $\bar{Q}$, $\bar{Z}$). A Boolean expression in the variables $x_1, x_2, ..., x_n$ is a *minterm* iff it is the conjunction $r_1 r_2 \cdots r_n$ where each r_i is either x_i or $\bar{x}_i$, $1 \leqslant i \leqslant n$. It is a *maxterm* iff it is the disjunction $r_1 + r_2 + ... + r_n$ where r_i is either x_i or $\bar{x}_i$, $1 \leqslant i \leqslant n$. One should note that there is exactly one truth assignment for which a minterm has truth value true and exactly one assignment for which a maxterm has truth value false.

Given a truth table, it is relatively easy to obtain a corresponding

Boolean expression (or wff) that is the disjunction of minterms. This is done by creating a minterm for each truth table row in which the function value is 1. As an example, consider the truth table of Figure 1.4(a). Row 2 has a function value of 1. This row corresponds to P and Q being true (i.e., having value 1) and R being false. The corresponding minterm is $PQ\bar{R}$. This term is true iff P and Q are true and R is false. Similarly, we obtain the minterms $P\bar{Q}R$, $\bar{P}QR$, and $\bar{P}\bar{Q}R$ corresponding, respectively, to rows 4, 5, and 8 of the truth table. A wff that corresponds to the Boolean function of Figure 1.4(a) can now be obtained by taking the disjunction of these four terms: $PQ\bar{R} + P\bar{Q}\bar{R} + \bar{P}Q\bar{R} + \bar{P}\bar{Q}\bar{R}$. It is easily verified that this wff is true whenever one or more of the minterms is true and it is false otherwise. Hence, it does indeed correspond to Figure 1.4(a). Using the procedure just outlined, we obtain the following Boolean expression for the function of Figure 1.4(b):

$$\bar{P}\,\bar{Q}R\bar{S} + \bar{P}Q\bar{R}S + P\bar{Q}R\bar{S} + \bar{P}\bar{Q}\bar{R}\bar{S}$$

A *clause* is either the disjunction or conjunction of literals. Examples are: $s_1 \vee \bar{x}_2 \vee x_3$, $x_4 \vee x_6 \vee x_7 \vee c_2$, $xy\bar{z}$, $x_1 y_1 \bar{a}b\bar{c}$, P, $\bar{Q}$, etc. A wff is in *conjunctive normal form* (CNF) iff it is the conjunction of clauses each of which is the disjunction of literals. A wff is in *disjunctive normal form* (DNF) iff it is the disjunction of clauses each of which is the conjunction of literals. Examples of CNF wffs are:

$$(x_1 + x_2)$$

$$(x_1 + \bar{x}_2 + x_3)(\bar{x}_4 + \bar{x}_6 + x_8 + \bar{x}_2)(x_1 + x_2)$$

$$(x_1 + x_9)(x_{10} + \bar{x}_3)$$

$$(X \vee Y \vee \bar{Z})(X \vee \bar{Y})$$

The following wffs are in DNF:

$$x_1 x_2 \bar{x}_3$$

$$x_1 x_2 + x_2 x_3 + \bar{x}_1 x_2 \bar{x}_3 x_4 \bar{x}_5$$

$$(X \wedge Y \wedge Z \wedge \bar{P}) \vee (\bar{X} \wedge \bar{Y}) \vee (X \wedge \bar{Y} \wedge \bar{Z} \wedge A \wedge B)$$

We have already seen how to obtain a DNF wff corresponding to a truth table. An equivalent CNF wff can be obtained just as easily. This time we create a maxterm for each truth table row having function value 0. Let $p_1, p_2, ..., p_n$ be the values of the variables $x_1, x_2, ..., x_n$ in some truth table row with function value 0. The maxterm corresponding to this row has $r_i = x_i$ iff $p_i = 0$. The maxterms corresponding to rows 1, 3, 6 and 7 of Figure 1.4(a) are $\bar{P} + \bar{Q} + \bar{R}$, $\bar{P} + Q + \bar{R}$, $P + \bar{Q} + \bar{R}$, and $P + Q + \bar{R}$. The resulting CNF wff is $(\bar{P} + \bar{Q} + \bar{R})(\bar{P} + Q + \bar{R})$ $(P + \bar{Q} + \bar{R})(P + Q + \bar{R})$. For Figure 1.4(b), $(\bar{P} + \bar{Q} + \bar{R} + S)$ $(\bar{P} + \bar{Q} + \bar{R} + \bar{S})$ $(\bar{P} + \bar{Q} + R + \bar{S})$ $(\bar{P} + Q + R + \bar{S})$ is the resulting CNF wff.

To see that a CNF wff obtained in this way does indeed correspond to the given truth table, observe that for a truth assignment corresponding to a row with zero function value exactly one of the maxterms will have value 0 (false). Hence, the whole wff has value 0 for this assignment. For any other assignment none of the maxterms will have value 0 and so the wff has value 1.

Boolean Algebra

It is useful at this point to introduce the notion of a Boolean algebra. Boolean algebra concepts are needed to obtain minimal forms for wffs. A *Boolean algebra* consists of a collection, K, of distinct elements (K is therefore a set, see Chapter 3) and two binary operators $\cdot$ and $+$ $\cdot$ $(K, \cdot, +)$ is a Boolean algebra iff the following laws (known as Huntington's Postulates) hold:

1. [*Closure under $\cdot$ and $+$*] For all elements a and b which are in K, $a \cdot b$ and $a + b$ are both elements in K.

2. [*Commutativity of $\cdot$ and $+$*] For all a and b in K, $a \cdot b = b \cdot a$ and $a + b = b + a$.

3. [*Distributivity of $\cdot$ and $+$*] For all a, b, and c in K, $a \cdot (b + c) = a \cdot b + a \cdot c$ and $a + (b \cdot c) = (a + b) \cdot (a + c)$

4. [*Identity and Zero elements*] K contains elements 1 and 0 with the properties $a \cdot 1 = a$ and $a + 0 = a$ for all elements a in K. 1 is called the *identity element* and 0 is the *zero element.*

5. [*Complement*] For every element a in K there exists another element $\bar{a}$ (also written $\neg a$), different from a such that $a \cdot \bar{a} = 0$ and $a + \bar{a} = 1$. $\bar{a}$ is the *complement* of a.

6. There are at least two distinct elements a and b, $a \neq b$, in K.

Let K be the set {false,true} and let $\cdot$ and $+$ respectively denote $\wedge$ (and) and $\vee$ (or). We see that ({false,true}, $\wedge$, $\vee$) satisfies postulate 1 because of the way $\wedge$ and $\vee$ were defined in Section 1.1. That postulates 2 and 3 hold may be verified using the truth tables for $\wedge$ and $\vee$. The requirements for the identity and zero elements are satisfied by true and false respectively. Also, true is the complement of false and false the complement of true. True and false are distinct elements. So, ({false,true}, $\wedge$, $\vee$) is a Boolean algebra.

Let α and β be two Boolean expressions. α is the *dual* of β iff α can be obtained from β by using the following mapping:

(a)   replace all occurrences of $\cdot$ by $+$ and of $+$ by $\cdot$
(b)   replace all occurrences of 0 by 1 and of 1 by 0

The expressions $a \cdot b$ and $a+b$ are duals of each other. The dual of $a+b+0 \cdot c$ is $a \cdot b \cdot (1+c)$ and the dual of $abc+cd+af$ is $(a+b+c) \cdot (c+d) \cdot (a+f)$.

An examination of Huntington's postulates reveals that each of the first five postulates actually contains a pair of requirements. For example, in postulate 2 we have the two requirements: $a \cdot b=b \cdot a$ and $a+b=b+a$. It is easily verified that for each pair of requirements corresponding to a given postulate, it is the case that one requirement can be obtained from the other by interchanging 0 and 1 and interchanging $\cdot$ and $+$. So, $a+b=b+a$ is obtained from $a \cdot b=b \cdot a$ by replacing $\cdot$ by $+$. Similarly, $a \cdot b=b \cdot a$ can be obtained from $a+b=b+a$ by replacing $+$ by $\cdot$. $a+0=a$ can be obtained from $a \cdot 1=a$ by replacing $\cdot$ by $+$ and 1 by 0. As a result of this, each requirement in a postulate is the dual of the other requirement. This duality in the postulates often enables one to prove the dual of a result by directly appealing to the proof of the original result.

Suppose we wish to establish that for every element a in K ($(K,\cdot,+)$ is a Boolean algebra) it is the case that $a \cdot a=a$. A proof of $a \cdot a=a$ is given below:

$$
\begin{aligned}
a \cdot a &= (a \cdot a)+0 & \text{Postulate 4} \\
&= (a \cdot a)+(a \cdot \bar{a}) & \text{Postulate 5} \\
&= a(a+\bar{a}) & \text{Postulate 3} \\
&= a \cdot 1 & \text{Postulate 5} \\
&= a & \text{Postulate 4}
\end{aligned}
$$

A proof of $a+a=a$ can be obtained from the above proof by simply replacing each $\cdot$ by a $+$, each $+$ by a $\cdot$, each 1 by a 0 and each 0 by a 1. The result is:

$$
\begin{aligned}
a+a &= (a+a)\cdot 1 & \text{Postulate 4} \\
&= (a+a)\cdot(a+\bar{a}) & \text{Postulate 5} \\
&= a+(a\cdot\bar{a}) & \text{Postulate 3} \\
&= a+0 & \text{Postulate 5} \\
&= a & \text{Postulate 4}
\end{aligned}
$$

In practice, if we have already proved that $a\cdot a=a$, then we shall conclude that $a+a=a$ follows from the duality of the Boolean algebra postulates.

Theorem 1.1 lists some of the important properties of a Boolean algebra. Each of these can be proved from Huntington's postulates. Again, it will be noticed that each property stated in the theorem has a dual. Both of these have been paired.

Theorem 1.1: [Boolean algebra theorem] The following statements are true for every Boolean algebra $(K,\cdot,+)$:
 (a) The identity and zero elements (1 and 0) are unique.
 (b) (Idempotence) For every a in K, $a\cdot a=a$ and $a+a=a$.
 (c) For every a in K, $a\cdot 0 = 0$ and $a+1=1$.
 (d) Every a in K has a unique complement $\bar{a}$.
 (e) For every a in K, $a = \neg\neg a$.
 (f) The identity and zero elements are distinct. Also, $\bar{1} = 0$ and $\bar{0} = 1$.
 (g) (De Morgan's law) For every a and b in K, $\overline{a\cdot b} = \bar{a} + \bar{b}$ and $\overline{a+b} = \bar{a}\cdot\bar{b}$.
 (h) (Associativity) For every a, b, and c in K, $a\cdot(b\cdot c)=(a\cdot b)\cdot c$ and $a+(b+c)=(a+b)+c$.
 (i) (Absorption) For every a and b in K, $a\cdot(a+b)=a$ and $a+(a\cdot b)=a$.
 (j) For every a, b in K, $a\cdot(\bar{a}+b)=a\cdot b$ and $a+(\bar{a}\cdot b)=a+b$.

Proof: Part (b) has already been proved. A proof of part (c) is given below:

$$
\begin{aligned}
a\cdot 0 &= (a\cdot 0)+0 & \text{Postulate 4} \\
&= (a\cdot 0)+(a\cdot\bar{a}) & \text{Postulate 5} \\
&= a\cdot(0+\bar{a}) & \text{Postulate 3} \\
&= a\cdot\bar{a} & \text{Postulates 4 and 2} \\
&= 0 & \text{Postulate 5}
\end{aligned}
$$

$a+1=1$ follows from $a\cdot 0=0$ by duality. The proofs for some of the other parts of the theorem can be found in the next section. The remaining proofs are left as an exercise. □

By using Huntington's postulates and Theorem 1.1 it is possible to simplify Boolean expressions. The following examples illustrate this.

Example 1.1: Let us simplify the Boolean expression $PQ\bar{R} + P\bar{Q}\bar{R} + \bar{P}Q\bar{R}$ $+ \bar{P}\bar{Q}\bar{R}$ corresponding to the truth table of Figure 1.4(a).

$$PQ\bar{R} + P\bar{Q}\bar{R} + \bar{P}Q\bar{R} + \bar{P}\bar{Q}\bar{R}$$
$$= P(Q\bar{R}+\bar{Q}\bar{R})+\bar{P}(Q\bar{R}+\bar{Q}\bar{R}) \text{ distributivity}$$
$$= P(Q+\bar{Q})\bar{R}+\bar{P}(Q+\bar{Q})\bar{R} \text{ comm., distr.}$$
$$= P(1\bar{R})+\bar{P}(1\bar{R}) \text{ complement}$$
$$= P\bar{R}+\bar{P}\bar{R} \qquad \text{1 is the identity element}$$
$$= (P+\bar{P})\bar{R} \qquad \text{Distributivity}$$
$$= 1\bar{R} \qquad \text{definition of complement}$$
$$= \bar{R} \qquad \text{1 is the identity element}$$

Example 1.2:
$$P\bar{Q}RS+\bar{P}RS+\bar{P}Q\bar{R}S+PQ\bar{R}S+PQRS$$
$$= P\bar{Q}RS+\bar{P}1RS+(\bar{P}+P)Q\bar{R}S+PQRS \quad \text{ident.,distr.}$$
$$= P\bar{Q}RS+\bar{P}(Q+\bar{Q})RS+1Q\bar{R}S+PQRS \quad \text{compl.}$$
$$= P\bar{Q}RS+\bar{P}QRS+\bar{P}\bar{Q}RS+Q\bar{R}S+PQRS \quad \text{distr.,ident.}$$
$$= (P+\bar{P})\bar{Q}RS+(\bar{P}+P)QRS+Q\bar{R}S \qquad \text{commut., distr.}$$
$$= 1\bar{Q}RS+1QRS+Q\bar{R}S \qquad \text{compl.}$$
$$= \bar{Q}RS+QRS+Q\bar{R}S \qquad \text{ident.}$$
$$= (\bar{Q}+Q)RS+Q\bar{R}S \qquad \text{distr.}$$
$$= 1RS+Q\bar{R}S \qquad \text{compl.}$$
$$= RS+Q\bar{R}S \qquad \text{ident. } \square$$

1.3 PROOF METHODS

In logic one often finds oneself in a situation where some conclusion or inference is to be drawn from a given set of premises. More precisely we would like to know if a particular wff, say Q, can be inferred from a given set of wffs. As an example, consider the following premises:

(P1) If Joe doesn't do the assignment or turns it in late, then he will get a zero.

(P2) If Joe gets a zero he will either withdraw from the course or fail.

From these two premises can we draw the conclusion:

(C) If Joe turns in the assignment late and does not withdraw from the course then he will fail.

Let's use the following symbols to denote the various components of (P1), (P2), and (C):

P: Joe doesn't do the assignment
Q: Joe turns the assignment in late
R: Joe gets a zero
S: Joe withdraws from the course
T: Joe fails.

P1, P2, and C respectively take the form:

$$P \lor Q \Longrightarrow R$$
$$R \Longrightarrow S \lor T$$
$$Q \land \bar{S} \Longrightarrow T$$

From the truth table for "$\Longrightarrow$", we see that the conclusion C follows from P1 and P2 iff

$$((P \lor Q \Longrightarrow R) \land (R \Longrightarrow S \lor T)) \Longrightarrow (Q \land \bar{S} \Longrightarrow T)$$

is a tautology. The truth table method can be used to determine whether or not the above wff is a tautology. As this wff contains 5 variables, the truth table corresponding to it will have 32 rows. Fortunately, the correctness of drawing conclusion C from the premises P1 and P2 can be established in a less cumbersome way. This requires us to make use of logical equivalences, rules of inference and proof methods.

Two wffs α and β are *equivalent* (written $\alpha \equiv \beta$) iff $\alpha \Longleftrightarrow \beta$ is a tautology. Note that if $\alpha \equiv \beta$, and both contain the same variables, then they have the same truth value for every assignment of truth values to these variables. In Section 1.1, we established some logical equivalences. For example, $P \Longrightarrow Q$ is identical to $\bar{P} \lor Q$ and $\neg(P \lor Q)$ is identical to $\bar{P} \land \bar{Q}$. Figure 1.6 lists some equivalences that will be useful in later proofs. α, β, and γ are wffs. The correctness of each of these can be established using the truth table method. Some of the equivalences listed in Figure 1.6 have names associated with them. These names together with the abbreviations we shall be using are given in the figure. Note that an equivalence can be used within a wff as well as on an entire wff. So, for example, the wff $P \land (Q \Longrightarrow R)$ can be replaced by the equivalent form $P \land (\bar{Q} \lor R)$ as a result of E16.

An *inference rule* allows us to infer the truth of a certain wff once we have established the truth of certain other wffs. Inference rules have the format:

$$\{\alpha_1, \alpha_2, ..., \alpha_k\} \models \beta$$

E1. $\alpha \wedge \bar{\alpha} \equiv$ false   contradiction
E2. $\alpha \vee \bar{\alpha} \equiv$ true   law of excluded middle
E3. $\alpha \wedge \alpha \equiv \alpha$
E4. $\alpha \vee \alpha \equiv \alpha$
E5. $\alpha \wedge$ true $\equiv \alpha$
E6. $\alpha \wedge$ false $\equiv$ *false*
E7. $\alpha \vee$ true $\equiv$ *true*
E8. $\alpha \vee$ false $\equiv \alpha$
E9. $\alpha \wedge \beta \equiv \beta \wedge \alpha$ commutativity (comm.)
E10. $\alpha \vee \beta \equiv \beta \vee \alpha$ commutativity (comm.)
E11. $\alpha \wedge (\beta \vee \gamma) \equiv (\alpha \wedge \beta) \vee (\alpha \wedge \gamma)$ distributivity (dist.)
E12. $\alpha \vee (\beta \wedge \gamma) \equiv (\alpha \vee \beta) \wedge (\alpha \vee \gamma)$ distributivity (dist.)
E13. $\alpha \equiv \neg(\neg \alpha)$
E14. $\neg(\alpha \wedge \beta) \equiv \bar{\alpha} \vee \bar{\beta}$ De Morgan's Law (De Morg.)
E15. $\neg(\alpha \vee \beta) \equiv \bar{\alpha} \wedge \bar{\beta}$ De Morgan's Law (De Morg.)
E16. $\alpha \implies \beta \equiv \bar{\alpha} \vee \beta$
E17. $\alpha \implies \beta \equiv \bar{\beta} \implies \bar{\alpha}$
E18. $\alpha \iff \beta \equiv (\alpha \implies \beta) \wedge (\beta \implies \alpha)$

Figure 1.6 *Logical equivalences.*

where the α_is and β are wffs. The interpretation of this format is: we may infer that β is true once we have shown that each of the α_is is true. Figure 1.7 gives some of the more commonly used inference rules. Consider the inference rule I4 ($\{\alpha, \alpha \implies \beta\} \models \beta$). This rule says that if we know that both α and $\alpha \implies \beta$ are true, then we can conclude that β is also true. As in the case of the equivalences, some of the inference rules also have names. These names are also given in Figure 1.7.

I1. $\{\alpha, \beta\} \models \alpha \wedge \beta$ conjunction (conj.)
I2.(a) $\{\alpha \wedge \beta\} \models \alpha$ simplification (simp.)
   (b) $\{\alpha \wedge \beta\} \models \beta$ simplification (simp.)
I3.(a) $\{\alpha\} \models \alpha \vee \beta$ addition (add.)
   (b) $\{\beta\} \models \alpha \vee \beta$ addition (add.)
I4. $\{\alpha, \alpha \implies \beta\} \models \beta$ modus ponens (MP)
I5. $\{\bar{\beta}, \alpha \implies \beta\} \models \bar{\alpha}$ modus tollens (MT)
I6. $\{\bar{\alpha}, \alpha \vee \beta\} \models \beta$ disjunctive syllogism (disj.syll.)
I7. $\{\alpha \implies \beta, \beta \implies \gamma\} \models \alpha \implies \gamma$
                        hypothetical syllogism (hypo. syll.)
I8. $\{\alpha \implies \beta, \gamma \implies \delta\} \models \alpha \wedge \gamma \implies \beta \wedge \delta$
I9. $\{\alpha \vee \beta, \bar{\alpha} \vee \beta\} \models \beta$

Figure 1.7 Inference rules.

With each inference rule $\{\alpha_1, \alpha_2, \ldots, \alpha_k\} \models \beta$, we may associate a wff $\alpha_1 \wedge \alpha_2 \wedge \ldots \wedge \alpha_k \Longrightarrow \beta$. From the truth table for "$\Longrightarrow$", we see that an inference rule is valid iff the associated wff is a tautology.

Figure 1.8 gives the truth table for the wff associated with the inference rule I7. Since this wff is a tautology, inference rule I7 is indeed a valid inference rule. The validity of each of the inference rules of Figure 1.7 may be established in this way.

α β γ	$\alpha \Longrightarrow \beta$	$\beta \Longrightarrow \gamma$	$(\alpha \Longrightarrow \beta) \wedge (\beta \Longrightarrow \gamma)$	$\alpha \Longrightarrow \gamma$	I7
T T T	T	T	T	T	T
T T F	T	F	F	F	T
T F T	F	T	F	T	T
T F F	F	T	F	F	T
F T T	T	T	T	T	T
F T F	T	F	F	T	T
F F T	T	T	T	T	T
F F F	T	T	T	T	T

Figure 1.8 *Truth table for inference rule I7.*

Inference rules I4 and I5 should be studied carefully. I4 says that from the truth of α and $\alpha \Longrightarrow \beta$, we can infer that β is true. I5 says that from the falsity of β and the truth of $\alpha \Longrightarrow \beta$, we can infer that α is false. This leaves us with two other possibilities: (i) α is false and (ii) β is true. It is a common mistake to infer $\bar{\beta}$ from $\bar{\alpha}$ and $\alpha \Longrightarrow \beta$. Such an inference is certainly not supported by the truth table for $\alpha \Longrightarrow \beta$ (Figure 1.1(d)) as this table has two rows corresponding to α being false and $\alpha \Longrightarrow \beta$ being true. One of these rows has β = true and the other has β = false. So, we cannot infer that β is false. Another common mistake is to infer α from β and $\alpha \Longrightarrow \beta$. Again, we see that the truth table for $\alpha \Longrightarrow \beta$ has a row with $\alpha = T, \beta = T$ and $\alpha \Longrightarrow \beta = T$ as well as one with $\alpha = F, \beta = T$ and $\alpha \Longrightarrow \beta = T$. So, α may be either true or false. These two mistakes are so common that logicians have given names to them. Mistakes of the first kind (i.e., infering $\bar{\beta}$ from $\bar{\alpha}$ and $\alpha \Longrightarrow \beta$) are called errors of *denying the antecedent* and mistakes of the second kind (i.e., infering α from β and $\alpha \Longrightarrow \beta$) are called errors of *affirming the consequent.*

There exist several valid proof methods (other than constructing a truth table) that may be used to show that a wff is a tautology. Let us look at six of these methods. The first four are applicable only when the wff α is of the form $\alpha_1 \Longrightarrow \alpha_2$. From the truth table for $\Longrightarrow$ (Figure 1.1(d)), it follows that there are four conditions under which $\alpha_1 \Longrightarrow \alpha_2$ is true. These are:

(a) α_2 is true. Now, only rows 1 and 3 of Figure 1.1(d) apply and $\alpha_1 \Longrightarrow \alpha_2$ is true for both rows.
(b) α_1 is false. In this case, only rows 3 and 4 of Figure 1.1(d) apply and $\alpha_1 \Longrightarrow \alpha_2$ is true for both.
(c) If α_1 is true then α_2 must be true in order for $\alpha_1 \Longrightarrow \alpha_2$ to be true.
(d) If α_2 is false then α_1 must also be false. Otherwise $\alpha_1 \Longrightarrow \alpha_2$ is false.

These four conditions lead to the first four proof methods. The fifth proof method is applicable when α is of the form $\alpha_1 \lor \alpha_2 \lor \alpha_3 \lor \ldots \lor \alpha_k \Longrightarrow \beta$ and the sixth can be used regardless of the form of α. The six proof methods are stated below as methods M1-M6.

M1: Trivial Proof (triv. pf.)

To show that $\alpha_1 \Longrightarrow \alpha_2$ is a tautology, using this method, we show that α_2 is a tautology. From (a) above it follows that this is sufficient to show that $\alpha_1 \Longrightarrow \alpha_2$ is a tautology. This proof method can be stated in the form of an inference rule as below:

$$\{\alpha_2\} \models \alpha_1 \Longrightarrow \alpha_2$$

M2: Vacuous Proof (vac. pf.)

This proof method follows from condition (b). In a vacuous proof of $\alpha_1 \Longrightarrow \alpha_2$ we establish that α_1 is a contradiction. The corresponding inference rule form is:

$$\{\neg \alpha_1\} \models \alpha_1 \Longrightarrow \alpha_2$$

M3: Direct Proof (dir. pf.)

Here, it is shown that from the truth of α_1 we can infer that α_2 is also true.

M4: Indirect Proof (ind. pf.)

An indirect proof of $\alpha_1 \Longrightarrow \alpha_2$ begins by assuming that α_2 is false (i.e., $\neg \alpha_2$ is true) and showing that this implies that α_1 is also false. I.e., $\neg \alpha_2 \Longrightarrow \neg \alpha_1$. From E17, it follows that this is equivalent to showing that $\alpha_1 \Longrightarrow \alpha_2$. It is interesting to note that an indirect proof of $\alpha_1 \Longrightarrow \alpha_2$ is, in fact, a direct proof of $\neg \alpha_2 \Longrightarrow \neg \alpha_1$.

M5: Proof by cases (by cases)

When α is of the form $\alpha_1 \vee \alpha_2 \vee ... \vee \alpha_k \Longrightarrow \beta$, the truth of α can be established by showing that each of the implications $\alpha_1 \Longrightarrow \beta$, $\alpha_2 \Longrightarrow \beta$, ..., $\alpha_k \Longrightarrow \beta$ is true. The corresponding inference rule form is:

$$\{\alpha_1 \Longrightarrow \beta, \alpha_2 \Longrightarrow \beta, ..., \alpha_k \Longrightarrow \beta\} \models \alpha_1 \vee \alpha_2 \vee ... \vee \alpha_k \Longrightarrow \beta$$

M6: Proof by contradiction (contr.)

In a proof by contradiction, we begin by assuming that α is false and then arrive at a contradiction. For example, we may be able to show that if $\neg \alpha$ is true then both P and $\bar{P}$ must be true for some variable P. However, E1 states that $P \wedge \bar{P}$ is false. So, the assumption that $\neg \alpha$ is true must be invalid and hence α must be true. The corresponding inference rule is:

$$\{\bar{\alpha} \Longrightarrow P \wedge \bar{P}\} \models \alpha$$

The six methods described above are adequate to establish that a wff (as defined in Section 1.1) is a tautology. We shall introduce additional proof methods in later sections and chapters as the need for them arises. Let us consider some simple examples to illustrate the proof methods M1 - M6.

Example 1.5: (Trivial proof) $R \Longrightarrow P \vee \bar{P}$ is a tautology. This can be proved using method M1 as below:

1. $P \vee \bar{P}$ E2
2. $R \Longrightarrow P \vee \bar{P}$ 1, M1

A proof consists of several lines of assertions (wffs that are true under the assumptions made so far). Each assertion is followed by a justification for its truth. So, in line 1 of the above proof $P \vee \bar{P}$ is the assertion and E2 (equivalence 2 of Figure 1.6) is the justification for the claim that $P \vee \bar{P}$ is true. In line 2, $R \Longrightarrow P \vee \bar{P}$ is the assertion and its justification is line 1 and proof method M1. Since, $R \Longrightarrow P \vee \bar{P}$ is what we had set out to prove, the proof is complete. Note that $R \Longrightarrow P \vee \bar{P}$ is true even though there is no relationship between R and P. $\square$

Example 1.6: (Vacuous proof) $P \wedge \bar{P} \Longrightarrow R$ can be proved using method M2. The proof is:

1. $P \vee \bar{P}$ E2
2. $\neg(P \wedge \bar{P})$ 1, De Morg., E13, comm.
3. $P \wedge \bar{P} \Longrightarrow R$ 2, Vac. pf.

When giving the justification for an assertion, we may list the equivalences, inferences rules and proof methods used symbolically as in E1, E6, I4, M1, M4, etc. or, we may use their names (if any). □

Example 1.7: (Direct proof) A proof of $(P \Longrightarrow Q) \Longrightarrow (P \Longrightarrow (P \wedge Q))$ is:

1. $P \Longrightarrow Q$ assumption 1 (or A1)
2. $P \Longrightarrow (P \wedge Q)$
   2.1 P A2
   2.2 Q 1, 2.1, MP
   2.3 $P \wedge Q$ 2.1, 2.2, conj.
   2.4 $P \Longrightarrow P \wedge Q$ A2, 2.3, dir.pf.
3. $(P \Longrightarrow Q) \Longrightarrow (P \Longrightarrow (P \wedge Q))$ A1, 2, dir. pf.

In the above proof, we set out to prove that $(P \Longrightarrow Q)$ implies that $P \Longrightarrow (P \wedge Q)$. As required by a direct proof, we began with the assumption (line 1 or A1) that $P \Longrightarrow Q$ was true and proceeded to establish the truth of $P \Longrightarrow (P \wedge Q)$ (line 2). The justification for line 2 is itself another proof. So, this was indented right and provided in lines 2.1 - 2.4. This sub-proof itself is a direct proof. In 2.1 the antecedent of line 2 is assumed true. This assumption leads (in line 2.3) to the truth of the consequent of line 2. MP and conj., respectively, refer to inference rules I4 and I1. □

Example 1.8: (Indirect proof) $(P \Longrightarrow \bar{P}) \Longrightarrow \bar{P}$ can be proved as follows:

1. P, A1
2. $\neg (P \Longrightarrow \bar{P})$
   2.1 $\neg (\bar{P} \vee \bar{P})$
      2.1.1 $P \wedge P$ 1,1, conj.
      2.1.2 $\neg (\bar{P} \vee \bar{P})$ 2.1.1, De Morg.
   2.2 $\neg (P \Longrightarrow \bar{P})$ 2.1, E16, E13
3. $(P \Longrightarrow \bar{P}) \Longrightarrow \bar{P}$ 1, 2, ind.pf.

The preceding proof can also be written as:

1. P A1
2. $P \wedge P$ 1, 1, conj.
3. $\neg (\bar{P} \vee \bar{P})$ 2, De Morg.
4. $\neg (P \Longrightarrow \bar{P})$ 3, E16, E13
5. $(P \Longrightarrow \bar{P}) \Longrightarrow \bar{P}$ 1,4, ind. pf.

There is an important stylistic difference between these two versions of the same proof. In the second version, the reasons for making the assertions of lines 2 and 3 does not become apparent until line 4. So, a person reading this proof might wonder what lines 2 and 3 have to do with the proof. This problem becomes more acute when the proof is long. In the first version of the proof the goals are stated first. So, in line 2 we state that we are going to prove that $\neg(P \Longrightarrow \bar{P})$ is true. It is clear at this point why we wish to do this (1. and 2. together prove the wff). The assertion of 2.1 is readily recognized as being identical to that of 2. So, the style adopted in the first proof is expected to lead to proofs that will be easier to read. In this style we first state what we are trying to prove (and if necessary, why) and then prove it. □

Example 1.9: (By cases and contradiction) $[(P \wedge (P \Longrightarrow \bar{Q})) \vee (Q \Longrightarrow \bar{Q})] \Longrightarrow \bar{Q}$ may be proved by cases:

1. $P \wedge (P \Longrightarrow \bar{Q}) \Longrightarrow \bar{Q}$
   1.1 $P \wedge (P \Longrightarrow \bar{Q})$ A1
   1.2 P 1.1, simp.
   1.3 $P \Longrightarrow \bar{Q}$ 1.1, simp.
   1.4 $\bar{Q}$ 1.2, 1.3, MP
   1.5 $P \wedge (P \Longrightarrow \bar{Q}) \Longrightarrow \bar{Q}$ 1.1, 1.4, dir. pf.
2. $(Q \Longrightarrow \bar{Q}) \Longrightarrow \bar{Q}$
   2.1 $Q \Longrightarrow \bar{Q}$ A2
   2.2 $\bar{Q}$
      2.2.1 Q A3
      2.2.2 $\bar{Q}$ 2.2.1, 2.1, MP
      2.2.3 $\bar{Q}$ 2.2.1, 2.2.2, contr.
   2.3 $(Q \Longrightarrow \bar{Q}) \Longrightarrow \bar{Q}$ 2.1, 2.2, dir.pf.
3. $[(P \wedge (P \Longrightarrow \bar{Q})) \vee (Q \Longrightarrow \bar{Q})] \Longrightarrow \bar{Q}$ 1, 2, by cases

In the above proof, one should observe that the assertions of lines 2.2.2 and 2.2.3 are identical. The proof of $\bar{Q}$ does not end at line 2.2.2 as here the truth of $\bar{Q}$ is conditional on the assumption A3. □

Example 1.10: $((P \Longrightarrow Q) \Longrightarrow Q) \Longrightarrow P \vee Q$ may be proved as follows:

1. $(P \Longrightarrow Q) \Longrightarrow Q$ A1
2. $P \vee Q$
   2.1 $\bar{P} \Longrightarrow Q$
      2.1.1 $\bar{P}$ A2
      2.1.2 $P \Longrightarrow Q$ 2.1.1, vac. pf.
      2.1.3 Q 2.1.2, 1, MP
      2.1.4 $\bar{P} \Longrightarrow Q$ 2.1.1, 2.1.3, dir. pf.
   2.2 $P \vee Q$ 2.1, E16
3. $((P \Longrightarrow Q) \Longrightarrow Q) \Longrightarrow P \vee Q$ 1, 2, dir. pf. □

Example 1.11: In this example we return to Joe's assignment dilemma introduced at the beginning of this section. We wish to show that the conclusion C follows from the premises P1 and P2. In symbolic form, we have:

$$P1 : P \vee Q \Longrightarrow R$$
$$P2 : R \Longrightarrow S \vee T$$
$$C : Q \wedge \bar{S} \Longrightarrow T$$

A proof of P1 $\wedge$ P2 $\Longrightarrow$ C is given below:

1. $Q \wedge \bar{S}$ A1
2. Q 1, simp.
3. $P \vee Q$ 2, add.
4. R 3, P1, MP
5. $S \vee T$ 4, P2, MP
6. $\bar{S}$ 1, simp.
7. T 6, 5, disj. syll.
8. $Q \wedge \bar{S} \Longrightarrow T$ 1, 7, dir. pf.

Note that P1 and P2 refer to the premises. □

Example 1.12: Consider the premises:

$$P1: Q \Longrightarrow P$$
$$P2: R \Longrightarrow S \wedge \bar{P}$$
$$P3: T \Longrightarrow R \vee Q$$

and the conclusion

$$C: T \wedge \bar{P} \Longrightarrow S$$

A proof of P1 $\wedge$ P2 $\wedge$ P3 $\Longrightarrow$ C is:

1. $T \wedge \bar{P}$ A1
2. T 1, simp.
3. $R \vee Q$ 2, P3, MP
4. $\bar{P}$ 1, simp.
5. $\bar{Q}$ 4, P1, MT
6. R 3, 5, disj. syll.
7. $S \wedge \bar{P}$ 6, P2, MP
8. S 7, simp.
9. $T \wedge \bar{P} \Longrightarrow S$ 1, 8, dir.pf. □

Example 1.13: Show that the conclusion:

C: $T \Longrightarrow (P \Longrightarrow U)$

follows from the premises:

P1: $P \Longrightarrow Q$
P2: $R \Longrightarrow \bar{Q} \lor S$
P3: $T \Longrightarrow R \lor U$
P4: $S \Longrightarrow \bar{Q}$

Proof:
1. T A1
2. $P \Longrightarrow U$
   2.1 P A2
   2.2 Q 2.1, P1, MP
   2.3 $R \lor U$ 1, P3, MP
   2.4 $\bar{S}$ 2.2, P4, MT
   2.5 $Q \land \bar{S}$ 2.2, 2.4, conj.
   2.6 $\neg(\bar{Q} \lor S)$ 2.5, De Morg.
   2.7 $\bar{R}$ 2.6, P2, MT
   2.8 U 2.3, 2.7, disj. syll.
   2.9 $P \Longrightarrow U$ 2.1, 2.8, dir.pf.
3. $T \Longrightarrow (P \Longrightarrow U)$ 1, 2, dir.pf.□

Example 1.14: Show that
   C: $U \Longrightarrow Q$

follows from:

P1: $P \lor \bar{T} \Longrightarrow \bar{U} \lor Q$
P2: $(S \Longrightarrow R) \Longrightarrow P$
P3: $(\bar{R} \lor T) \Longrightarrow \bar{S}$

Proof:
1. U A1
2. Q
   2.1 $\bar{Q}$ A2
   2.2 $U \land \bar{Q}$ 1, 2.1, conj.
   2.3 $\neg(\bar{U} \lor Q)$ 2.2, De Morg.
   2.4 $\neg(P \lor \bar{T})$ 2.3, P1, MT
   2.5 $\bar{P} \land T$ 2.4, De Morg.
   2.6 $\bar{P}$ 2.5, simp
   2.7 $\neg(S \Longrightarrow R)$ 2.6, P2, MT
   2.8 $S \land \bar{R}$ 2.7, E16, De morg.
   2.9 $\bar{R}$ 2.8, simp.

$$2.10 \ \bar{R} \vee T \qquad 2.9, \text{add.}$$
$$2.11 \ \bar{S} \qquad 2.10, \text{P3, MP}$$
$$2.12 \ S \qquad 2.8, \text{simp.}$$
$$2.13 \ Q \qquad \text{A2, 2.11, 2.12, contr.}$$
$$3. \ U \Longrightarrow Q \qquad 1,2, \text{dir.pr.}$$

The above proofs correspond quite closely to how one might write proofs in English. Consider the proof of Example 1.14. In English, this would take the form:

> **Proof:** We shall provide a direct proof of $P \Longrightarrow Q$. Assume that P is true. Call this assumption, A1. We must now show that Q is true. This will be established by contradiction. So, assume that $\bar{Q}$ is true. Call this assumption, A2. From A1 and A2, it follows that $\neg(\bar{U} \vee Q)$ is true. This, together with P1, leads to the truth of $\neg(P \vee \bar{T})$, etc.

Example 1.15: The utility of proof methods M1 - M6 is not limited to wffs. We may show that the identity and zero elements of Boolean algebra are unique by providing a proof by contradiction. Suppose that the identity element is not unique. Then, there exist two distinct identity elements 1_1 and 1_2. From postulate 4, it follows that $a \cdot 1_1 = a$ and $a \cdot 1_2 = a$ for every element a in K. Substituting 1_2 for a in the first equality and 1_1 for a in the second, we obtain $1_2 \cdot 1_1 = 1_2$ and $1_1 \cdot 1_2 = 1_1$. But, $1_2 \cdot 1_1 = 1_1 \cdot 1_2$ (postulate 2). So, $1_1 = 1_2$ which contradicts the assumption that 1_1 and 1_2 are different. The uniqueness of the zero element now follows by duality. □

1.4 A TABLEAU METHOD

Using the techniques developed in the preceding section, one can determine if any given wff α is a tautology. A wff α is a contradiction iff $\neg \ \alpha$ is a tautology. A wff α is satisfiable iff it is not a contradiction. Hence, by using the methods of the preceding section on $\neg \ \alpha$ we can determine whether α is a contradiction or is satisfiable. The proof methods discussed so far are not mechanical. A good deal of thought must be used to determine which methods (M1-M6) to use, what goals should be established, etc. In this section and the next, we shall examine two mechanical procedures to determine if a wff is a tautology, contradiction or satisfiable.

The mechanical proof method to be discussed in this section can be applied directly only to DNF wffs. The method of the next section requires the wff to be in CNF. So, let us first see how to transform an arbitrary wff into an equivalent DNF wff and an equivalent CNF wff. The transformation into DNF may be accomplished by proceeding in the manner described below:

(1) Eliminate all occurrences of $\Longrightarrow$ and $\Longleftrightarrow$ by using equivalences E16 ($\alpha \Longrightarrow \beta \equiv \bar{\alpha} \bigvee \beta$) and E18 ($\alpha \Longleftrightarrow \beta \equiv (\alpha \Longrightarrow \beta) \bigwedge (\beta \Longrightarrow \alpha)$).

(2) Repeatedly use De Morgan's laws and equivalence rule E13 ($\alpha \equiv \neg \neg \alpha$) so that the resulting wff contains only literals and the operators $\bigvee$ and $\bigwedge$ (parentheses and brackets may also exist).

Example 1.16: $\neg(\neg(\bar{P} \bigwedge Q) \bigvee \bar{R})$ is successively transformed into the equivalent formulas $\neg(\neg \neg P \bigvee \bar{Q} \bigvee \bar{R})$, $\neg(P \bigvee \bar{Q} \bigvee \bar{R})$, $\bar{P} \bigwedge \neg(\bar{Q} \bigvee \bar{R})$, $\bar{P} \bigwedge(\neg \neg Q \bigwedge \neg \neg R)$, $\bar{P} \bigwedge Q \bigwedge R$. $\square$

(3) If the wff α remaining at this point is not in DNF, then it may be transformed into DNF by the repeated use of equivalence E11 ($\alpha \bigwedge (\beta \bigvee \gamma) \equiv (\alpha \bigwedge \beta) \bigvee (\alpha \bigwedge \gamma)$) and equivalence E9 ($\alpha \bigwedge \beta \equiv \beta \bigwedge \alpha$).

Example 1.17: Consider the wff $\alpha = AB \bigvee C(D \bigvee EG) \bigvee (P \bigvee Q) \bigwedge (R \bigvee S)$. $C(D \bigvee EG)$ and $(P \bigvee Q) \bigwedge (R \bigvee S)$ can be replaced by the equivalent forms $CD \bigvee CEG$ and $PR \bigvee PS \bigvee QR \bigvee QS$ respectively. $\square$

(4) The DNF formula resulting from steps 1-3 may be simplified by eliminating clauses containing both a variable and its complement (as such clauses can never evaluate to true). If there are several occurrences of the same literal in a given clause then these occurences may be replaced by a single occurrence of that literal as $P.P \equiv P$ and $\bar{P}.\bar{P} \equiv \bar{P}$. Other simplifications may also be possible.

Example 1.18: Using the DNF transformation process described above, $\alpha = \neg[\neg(P \Longleftrightarrow Q) \Longrightarrow \neg(P \bigvee R)]$ may be transformed into DNF as below:

$\alpha = \neg[\neg(P \Longleftrightarrow Q) \Longrightarrow \neg(P \bigvee R)]$

$\quad = \neg[\neg((P \Longrightarrow Q) \bigwedge (Q \Longrightarrow P)) \Longrightarrow \neg(P \bigvee R)]$ E18

$\quad = \neg[\neg((\bar{P} \bigvee Q) \bigwedge (\bar{Q} \bigvee P)) \Longrightarrow \neg(P \bigvee R)]$ E16

$\quad = \neg[((\bar{P} \bigvee Q) \bigwedge (\bar{Q} \bigvee P)) \bigvee \neg(P \bigvee R)]$ E16, E13

$\quad = [\neg(\bar{P} \bigvee Q) \bigvee \neg(\bar{Q} \bigvee P)] \bigwedge (P \bigvee R)$ De Morg.

$\quad = [(\neg \neg P \bigwedge \bar{Q}) \bigvee (\neg \neg Q \bigwedge \bar{P})] \bigwedge (P \bigvee R)$ De Morg.

$\quad = [(P \bigwedge \bar{Q}) \bigvee (Q \bigwedge \bar{P})] \bigwedge (P \bigvee R)$ E13

$\quad = [P \bigwedge \bar{Q} \bigwedge (P \bigvee R)] \bigvee [Q \bigwedge \bar{P} \bigwedge (P \bigvee R)]$ E9, E11, E9

$\quad = P \bar{Q} P \bigvee P \bar{Q} R \bigvee Q \bar{P} P \bigvee Q \bar{P} R$ E11

$\quad = P \bar{Q} \bigvee P \bar{Q} R \bigvee Q \bar{P} R$ E3, E1, E6

Since $P \bar{Q} \bigvee P \bar{Q} R$ is easily seen to be equivalent to $P \bar{Q}$, the last expression for α may be simplified to obtain

$\alpha = P\overline{Q}\bigvee Q\overline{P}R.$ □

An equivalent CNF formula can be obtained by using rule E12 ($\alpha \bigvee (\beta \wedge \gamma) \equiv (\alpha \bigvee \beta) \wedge (\alpha \bigvee \gamma)$) in place of E11 in step 3 of the DNF procedure. Step 4 is unchanged as clauses containing both a literal and its complement may still be deleted (these clauses always evaluate to true). Multiple occurrences of a literal within a clause can be replaced by a single such occurrence.

Example 1.19: Let α be the wff $\neg(P \Longrightarrow Q) \Longrightarrow \neg(R\bigvee S)$. Using E16 and E13, we get $\alpha \equiv \neg\neg(P \Longrightarrow Q)\bigvee\neg(R\bigvee S) \equiv (P \Longrightarrow Q)\bigvee\neg(R\bigvee S)$. Another application of E16 yields $\alpha \equiv (\overline{P}\bigvee Q)\bigvee\neg(R\bigvee S)$. This completes our use of rule 1. De Morgan's law now gives us $\alpha \equiv (\overline{P}\bigvee Q)\bigvee(\overline{R}\wedge\overline{S})$. An application of E12 yields the CNF $\alpha \equiv (\overline{P}\bigvee Q\bigvee\overline{R})\wedge(\overline{P}\bigvee Q\bigvee\overline{S})$. □

Example 1.20: Consider the same α as in Example 1.18. The transformations due to steps 1 and 2 are unchanged. So, we may start with the form:

$$\alpha=[(P\wedge\overline{Q})\bigvee(Q\wedge\overline{P})]\wedge(P\bigvee R)$$

Using E12 and E9, we obtain:

$$\alpha=((P\wedge\overline{Q})\bigvee Q)\wedge((P\wedge\overline{Q})\bigvee\overline{P})\wedge(P\bigvee R)$$
$$= (P\bigvee Q)\wedge(\overline{Q}\bigvee Q)\wedge(P\bigvee\overline{P})\wedge(\overline{Q}\bigvee\overline{P})\wedge(P\bigvee R)$$

Using E2, we obtain:

$$\alpha=(P\bigvee Q)\wedge(\overline{Q}\bigvee\overline{P})\wedge(P\bigvee R).$$ □

A DNF wff is satisfiable iff any one of its clauses is satisfiable. Since each clause of a DNF wff is the conjunction of some number of literals, it is easy to determine if any of the clauses is satisfiable. Clearly, any clause that does not contain both a variable and its complement is satisfiable (e.g., $\overline{A}BC$ is true when A is false and both B and C are true. $A\overline{A}BC$ is not satisfiable). A DNF wff is satisfiable iff it contains at least one clause that does not contain both a variable and its complement. If a wff is not satisfiable then it is a contradiction. So, for DNF wffs we are left only with the problem of determining tautology. This problem can be solved using the tableau method described below.

Let α be a DNF wff. Let C be any clause in α. α can therefore be written as $C + \beta$ where β is a DNF wff. We make the following observations:

1. If C contains only one literal, z, then α is true at least whenever z is true. So, we need only be concerned with determining if α is true when z is false. For this, we set z to false and $\bar{z}$ to true in β and proceed to determine if the resulting DNF wff, β', is a tautology.

2. If C contains more than one literal then C may be written as $z\wedge D$ where z is a literal and D is a clause. Now, $\alpha = (z\wedge D)\vee\beta \equiv (z\vee\beta)\wedge(D\vee\beta)$. So, α is a tautology iff both $z\vee\beta$ and $D\vee\beta$ are tautologies.

Note that the DNF wffs β', $z\vee\beta$ and $D\vee\beta$ referred to in 1. and 2. are all simpler (i.e., have fewer literals) than α. So, we can transform the problem of determing if α is a tautology into one of determining if either one simpler formula (i.e., β' of observation 1) or two simpler formulas (i.e., $z\vee\beta$ and $D\vee\beta$ of observation 2) are tautologies. By repeatedly using the observations 1. and 2., we can reduce the problem of determing if α is a tautology to that of determining if a collection of much simpler DNF wffs are tautologies. The reduction process may be terminated as soon as one of the resulting simpler wffs is recognized as not being a tautology or when all the remaining wffs are recognized as being tautologies.

To illustrate the tautology verification procedure just described, consider $\alpha = P\vee \overline{P}QR\vee \overline{PQ}R\vee \overline{PR}\vee PR$. α is easily seen to be a tautology. In fact, using Boolean algebra, we get:

$$
\begin{aligned}
\alpha &= P\vee\overline{P}(QR\vee\overline{Q}R\vee\overline{R})\vee PR \\
&= P\vee\overline{P}((Q\vee\overline{Q})R\vee\overline{R})\vee PR \\
&= P\vee\overline{P}(R\vee\overline{R})\vee PR \\
&= P\vee\overline{P}\vee PR \\
&= \text{true}
\end{aligned}
$$

Using the method of this section, α can be written as $C\vee\beta$ with $C=P$ and $\beta = \overline{P}QR\vee\overline{PQ}R\vee\overline{PR}\vee PR$. Since C has only one literal, we need only be concerned with the wff obtained by setting $P = $ false in β. This gives us $\beta' = QR\vee\overline{Q}R\vee\overline{R}$. Now, β' can be written as $\beta' = C_1\vee\beta_1$ where $C_1 = \overline{R}$ and $\beta_1 = QR\vee\overline{Q}R$. Once again, C_1 contains only one literal and we need only show that $\beta'' = Q\vee\overline{Q}$ is a tautology. Equivalence E2 implies that β'' is a tautology and we need go no further. α is a tautology. This proof can be represented schematically as in Figure 1.9(a).

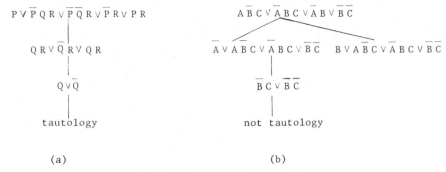

Figure 1.9 *Tableau proofs.*

As another example, consider $\alpha = \overline{A}\overline{B}C\vee\overline{A}BC\vee\overline{A}B\vee\overline{B}C$. We can use $C=\overline{A}B$ and $\beta = A\overline{B}C\vee\overline{A}BC\vee\overline{B}C$. With $z = \overline{A}$ and $D=B$, we need to determine if both $(z\vee\beta)$ and $(D\vee\beta)$ are tautologies. Let us continue with $z\vee\beta$. $z\vee\beta = \overline{A}\vee A\overline{B}C\vee\overline{A}BC\vee\overline{B}C$. Choosing $C=\overline{A}$, we are left with $\beta' = \overline{B}C\vee\overline{B}C$. β' is readily seen to be false when B = true. Hence, β' is not a tautology. Therefore, $z\vee\beta$ is not a tautology. So, α is not a tautology. Figure 1.9(b) shows this proof in diagram form.

1.5 THE DAVIS-PUTNAM PROCEDURE

The tautology problem for CNF wffs is easily solved as a CNF wff can be a tautology iff each of its clauses is one. Since each clause is the disjunction of literals, one may easily ascertain if a given clause is a tautology (e.g., $\overline{P}\vee P$ and $P\vee Q\vee\overline{P}\vee\overline{Q}$ are tautologies while $P\vee Q\vee R$ is not). A formula is a contradiction iff it is not satisfiable. So, for CNF wffs only the satisfiability problem merits further consideration. The Davis-Putnam procedure is a mechanical method to determine if a CNF formula is satisfiable.

Let α be a CNF wff. We can determine if α is satisfiable by repeatedly using the following rules to obtain new and simpler CNF wffs that are satisfiable iff α is.

(a)   If C_1 and C_2 are two clauses in α such that every literal in C_1 is also a literal in C_2 then eliminate C_2 from α and determine if the resulting wff is satisfiable. For example, if $C_1 = (\overline{P}\vee Q)$ and $C_2 = (\overline{P}\vee Q \vee R)$, then C_2 is eliminated from the wff.

(b)   If C is a one literal clause of α then set the literal in C to true and its complement to false in each of the remaining clauses of α. Delete C from α and proceed to determine if the resulting wff is satisfiable.

(c)   If a variable P appears only as P (or only as $\bar{P}$) throughout α then set P to true (or $\bar{P}$ to true if only $\bar{P}$ appears) throughout α. Eliminate the clauses that are now satisfied and continue with the remaining clauses.

(d)   If both P and $\bar{P}$ appear in α then separate the clauses of α into the three groups A, B, and F. A is the conjunction of all classes in α that contain P. B is the conjunction of all clauses containing $\bar{P}$ and F is the conjunction of the remaining clauses. Let $A' = D_1 D_2 \cdots D_k$ be A with all occurrence of P removed and let $B' = E_1 E_2 \cdots E_r$ be B with all occurrences of $\bar{P}$ removed. Replace α by

$$\bigwedge_{\substack{1 \leqslant i \leqslant k \\ 1 \leqslant j \leqslant r}} (D_i \vee E_j) \wedge F$$

and proceed to determine if this CNF wff is satisfiable.

Before attempting to justify each of the above rules, let us look at an example. Let $\alpha = \bar{P}(P \vee Q \vee R)(P \vee Q \vee R \vee S)(\bar{Q} \vee R \vee S)(Q \vee \bar{R})(Q \vee \bar{S})$. Since all the literals in clause 2 appear in clause 3, clause 3 may be eliminated by rule (a) to obtain $\alpha_1 = \bar{P}(P \vee Q \vee R)(\bar{Q} \vee R \vee S)(Q \vee \bar{R})(\bar{Q} \vee \bar{S})$. The first clause is a one literal clause and by rule (b) we may set $\bar{P}$ to true (i.e., P to false) throughout to obtain $\alpha_2 = (Q \vee R)(\bar{Q} \vee R \vee S)(Q \vee \bar{R})(\bar{Q} \vee \bar{S})$. Now, we may use rule (d) on Q. Separating the clauses into the groups A, B, and F, we get $A = (Q \vee R)(Q \vee \bar{R}), B = (\bar{Q} \vee R \vee S)(\bar{Q} \vee \bar{S})$, and F contains no clauses. $A' = R \wedge \bar{R}$ and $B' = (R \vee S) \wedge \bar{S}$. So, $D_1 = R$, $D_2 = \bar{R}$, $E_1 = (R \vee S)$, and $E_2 = \bar{S}$. α_2 is to be replaced by $\alpha_3 = (D_1 \vee E_1)(D_2 \vee E_1)(D_1 \vee E_2)(D_2 \vee E_2) = (R \vee R \vee S)(\bar{R} \vee R \vee S)(R \vee \bar{S})(\bar{R} \vee \bar{S})$. With some simplication, α_3 becomes $(R \vee S)(R \vee \bar{S})(\bar{R} \vee \bar{S})$.

Using rule (d) on R, we obtain $A = (R \vee S)(R \vee \bar{S})$, $B = (\bar{R} \vee \bar{S})$, and F contains no clauses. So, $A' = S \wedge \bar{S}$ and $B' = \bar{S}$. We must now determine if $\alpha_4 = (S \vee \bar{S})(\bar{S} \vee \bar{S}) = \bar{S}$ is satisfiable. It is easy to see that α_4 is true when S = false. So, the original wff α is satisfiable. A truth assignment for which α is true is obtained by working backwards from α_4 to α. For α_4 to be satisfied, S must be false. Now, for α_3 to also be true, we must have R = true. For α_2 to be true when S = false and R = true, we must have Q = true. For α_1 to now be true, we need P = false. So, the required truth assignment is P = false, Q = true, R = true, and S = false.

Theorem 1.2: The Davis-Putnam procedure is correct.

Proof: Let α be any CNF wff. Let β be the wff that results from an application of one of the rules (a)-(d). We shall show that α is satisfiable iff β is.

The correctness of the Davis-Putnam procedure is then an immediate consequence. The proof is by cases, one case for each of the four rules.

Case(i): If rule (a) is used then α contains two clauses C_1 and C_2 such that every literal in C_1 is also a literal in C_2. So, C_2 can be written as $C_1 \vee C'$ where C' is the disjunction of some literals. Since $C_1 \cdot C_2 \equiv C_1 \cdot (C_1 \vee C') = C_1 C_1 \vee C_1 C' = C_1 \vee C_1 C' = C_1 (\text{true} \vee C') = C_1$, the wff β obtained by deleting C_2 from α is satisfiable iff α is. Clearly, β is in CNF.

Case(ii): If C contains only one literal, z, then in every truth assignment for which α evaluates to true, z must be true. Hence, the β obtained by setting z to true in α is satisfiable iff α is. Note that β is in CNF.

Case(iii): If α contains a variable P whose complement, $\bar{P}$, appears in no clause, then there is no advantage to setting $P = $ false (as this assignment does not result in any of the clauses evaluating to true). So, we might as well set P to true. Similarly, if only $\bar{P}$ appears in α, we can set $\bar{P}$ to true. The wff left following an application of rule (c) is a CNF wff that is satisfiable iff the original wff is.

Case(iv): In this case, $\alpha \equiv A \cdot B \cdot F$ where $A = (P \vee D_1)(P \vee D_2)...(P \vee D_k)$ and $B = (\bar{P} \vee E_1)(\bar{P} \vee E_2)...(\bar{P} \vee E_r)$. From equivalence E12, it follows that $A = P \vee D_1 D_2...D_k = P \vee A'$ and $B = \bar{P} \vee E_1 E_2...E_r = \bar{P} \vee B'$. So, $\alpha = (P \vee A')(\bar{P} \vee B') \cdot F$ and the variable P does not appear in F. Since the clauses in A' and B' do not contain the variable P, $(P \vee A')(\bar{P} \vee B')$ is satisfiable iff $A' \vee B'$ is. Futhermore, since the variable P does not appear in F, α is satisfiable iff $(A' \vee B') \cdot F$ is.

$A' \vee B'$ is satisfiable iff at least one of A' and B' is satisfiable. This is the case iff

$$\beta = \bigwedge_{\substack{1 \le i \le k \\ 1 \le j \le r}} D_i \vee E_j)$$

is satisfiable. To see the truth of this latter statement, observe that if A' is satisfiable then there is a truth assignment for which each of $D_1, D_2, ..., D_k$ is true. Hence, for this assignment, each clause of β' is true. Similarly, if a truth assignment satisfies B', then it also satisfies β'. Hence if $A' \vee B'$ is satisfiable then so also is β'.

Now, suppose that β' is satisfiable. Consider a truth assignment that satisfies β'. If one of the D_is is false under this assignment then all the E_js must be true as each D_i is paired with every E_j. In this case the truth assignment satisfies B'. Similarly, if one the E_js is false then all the D_is must be true. So, it must be the case that any truth assignment that satisfies β' also satisfies A' or B' or both. Hence, $A'\bigvee B'$ is satisfiable iff β' is. Consequently, α is satisfiable iff $\beta = \beta'$. F is satisfiable. β is clearly a CNF wff. $\square$

Rules (a) and (b) reduce the number of clauses in α while rules (c) and (d) reduce the number of variables in α. Rule (d) does, however, increase the number of clauses. Since none of the rules introduces new variables into the wff, rules (c) and (d) can together be used only as many times as the number of variables originally in α. So, we are assured that if the rules are applied enough times, α will either run out of variables or clauses. Of course, the application of the rules will be halted as soon as we are left with a wff β which is sufficiently simple that we can easily determine whether or not it is satisfiable.

Example 1.21: Let $\alpha = (L\bigvee M\bigvee N)\bigwedge(L\bigvee\overline{M})\bigwedge(\overline{L}\bigvee M\bigvee\overline{N})\bigwedge(\overline{L}\bigvee\overline{M})$. Rules (a), (b), and (c) cannot be used at this time. Selecting variable L for elimination, we get $A = (L\bigvee M\bigvee N)(L\bigvee\overline{M})$ and $B = (\overline{L}\bigvee M\bigvee\overline{N})(\overline{L}\bigvee\overline{M})$. $D_1 = (M\bigvee N)$, $D_2 = \overline{M}$, $E_1 = (M\bigvee\overline{N})$, and $E_2 = \overline{M}$. So, $\beta = \beta'.F=$ $(M\bigvee N\bigvee M\bigvee\overline{N})(M\bigvee N\bigvee\overline{M})(\overline{M}\bigvee M\bigvee\overline{N})(\overline{M}\bigvee\overline{M})\equiv\overline{M}$. β is satisfied when $M =$ false. So, α is satisfiable. $\square$

Example 1.22: Let $\alpha=(X\bigvee Y)(\overline{X}\bigvee Y)(X\bigvee\overline{Y})(\overline{X}\bigvee\overline{Y})(Y\bigvee W)(Y\bigvee\overline{W})$. Using rule (d) and partitioning on the variable X, we get $A = (X\bigvee Y)(X\bigvee\overline{Y})$, $B=(\overline{X}\bigvee Y)\bigwedge(\overline{X}\bigvee\overline{Y})$, and $F=(Y\bigvee W)(Y\bigvee\overline{W})$. So, $A'=D_1 D_2 = Y\overline{Y}$ and $B'= E_1 E_2 = Y\overline{Y}$. The new formula to consider is $(Y\bigvee Y)(Y\bigvee\overline{Y})(\overline{Y}\bigvee Y)(\overline{Y}\bigvee\overline{Y})(Y\bigvee W)(Y\bigvee\overline{W}) = Y\overline{Y}(Y\bigvee W)(Y\bigvee\overline{W})$. Using rule (b) on the single literal clause Y, requires us to set Y to truc. The resulting formula is $[$false $\bigwedge W \bigwedge \overline{W}]$ which is false. So, α is not satisfiable. $\square$

1.6 FIRST ORDER LOGIC (FOL)/FIRST ORDER PREDI-CATE CALCULUS

The propositional logic discussed so far isn't powerful enough for us to make all the inferences that may be possible from a given set of premises. Consider the premises:

P1: Nothing intelligible puzzles me.
P2: Logic puzzles me.

and the conclusion:

C: Logic is unintelligible.

The validity of drawing the conclusion C from the premises P1 and P2 cannot be established in the propositional logic. First order logic provides us with additional tools that will enable us to show that P1 $\wedge$ P2 $\Longrightarrow$ C. Wffs of first order logic contain predicates and quantifiers in addition to the symbols permitted in wffs of propositional logic. First order logic is also called first order predicate calculus.

A *predicate* $P(x_1, x_2, \cdots, x_n)$, $n > 0$, is a mapping from $x_1, x_2, ..., x_n$ to the values true and false. n is the degree of the predicate. P is the predicate name and the x_is are *parameters* or *variables*. We shall use lower case letters from the end of the alphabet (with or without subscripts) to denote parameters (e.g., w, x, y, z, w_1, z_1, etc). Constants will be denoted by letters from the front of the alphabet (e.g. a_1, b, c_2, etc.). Parameters can be assigned only constants as values. Some examples are:

(a) $GREATER(x,y) = \begin{cases} true & \text{if } x > y \\ false & \text{if } x \leqslant y \end{cases}$

(b) $PRIME(x) = \begin{cases} true & \text{if } x \text{ is a prime number} \\ false & \text{otherwise} \end{cases}$

(c) $SUM(x,y,z) = \begin{cases} true & \text{if } x + y = z \\ false & \text{otherwise} \end{cases}$

(d) $PERSON(x) = \begin{cases} true & \text{if } x \text{ is a person} \\ false & \text{otherwise} \end{cases}$

(e) $I(x) = \begin{cases} true & \text{if } x \text{ is intelligible} \\ false & \text{otherwise} \end{cases}$

(f) $T(x) = \begin{cases} true & \text{if } x \text{ is a thing} \\ false & \text{otherwise} \end{cases}$

(g) $P(x) = \begin{cases} true & \text{if } x \text{ puzzles me} \\ false & \text{otherwise} \end{cases}$

(h) $LESS(x,y) = \begin{cases} true & \text{if } x < y \\ false & \text{otherwise} \end{cases}$

$= GREATER(y,x)$

(i) $GL(x,y,z) = \begin{cases} true & \text{if } x < y < z \\ false & \text{otherwise} \end{cases}$

$= LESS(x,y) \land LESS(y,z)$

For brevity, future predicate definitions will often be made by providing only the condition under which the predicate is true. For example, the predicates (a), (b) and (c) are equivalent to:

(a) $GREATER(x,y)$: $x > y$
(b) $PRIME(x)$: x is a prime number
(c) $SUM(x,y,z)$: $x + y = z$

Two types of quantifiers: universal ($\forall$) and existential ($\exists$) are permitted in first order logic. $\forall x P(x)$ (read: for all x $P(x)$) states that $P(x)$ is true for all x. $\exists x P(x)$ (read: there exists an x $P(x)$) states that $P(x)$ is true for at least one x.

Using quantifiers and predicates, it is possible to symbolize P1, P2, and C in such a way that P1 $\land$ P2 $\Longrightarrow$ C can be proved. Using the predicates I, T and P defined above, and the symbol a to denote logic, P1, P2 and C can be symbolized to get:

P1: $\forall x (T(x) \land I(x) \Longrightarrow \neg P(x))$
P2: $P(a)$
C: $\neg I(a)$

We also need the implied premise:

P3: $T(a)$

which asserts that logic is a thing.

C can be proved from P1, P2, and P3. Since P1 says $T(x) \land I(x) \Longrightarrow \neg P(x)$, for all x, we may substitute a for x to obtain $T(a) \land I(a) \Longrightarrow \neg P(a)$. This is equivalent to P(a) $\Longrightarrow \neg(T(a) \land I(a))$ (see E17). From

DeMorgan's law it follows that $P(a) \implies \overline{T}(a) \bigvee \overline{I}(a)$. From P2 we know that $P(a)$ is true. So, $\overline{T}(a) \bigvee \overline{I}(a)$ is true. P3 states that $T(a)$ is true. Therefore, $\neg I(a)$ must be true.

The *scope* of a quantifier is the wff to which it applies. For example in the wff $\forall x (P(x) \implies R(x))$ the scope of $\forall$ is $(P(x) \implies R(x))$; in $\forall x \exists y (P(x,y) \bigwedge R(x,y) \implies \exists z (G(x,z) \bigvee \overline{G}(y,z)))$ the scope of $\forall$ is $\exists y (P(x,y) \bigwedge R(x,y) \implies \exists z (G(x,z) \bigvee \overline{G}(y,z)))$ and the scope of the second $\exists$ is $(G(x,z) \bigvee \overline{G}(y,z))$. $\forall x$ and $\exists x$ *quantify* the predicate variable x. An occurrence of a predicate variable x is *free* iff that occurrence is not within the scope of a quantifier that quantifies x. An occurrence of a predicate variable is *bound* iff it is not free. With these definitions, we are in a position to define a wff of first order logic (or first order predicate calculus):

(1) All propositional variables, predicates and the constants true and false are first order logic wffs. A predicate with some of its parameters set to constants (as in $P(a)$) is also a FOL wff.
(2) If α and β are first order logic wffs, then $\neg \alpha$, $\overline{\alpha}$, $\neg(\alpha)$, $\neg[\alpha]$, $(\alpha \bigvee \beta)$, $[\alpha \bigvee \beta]$, $(\alpha \bigwedge \beta)$, $[\alpha \bigwedge \beta]$, $(\alpha \implies \beta)$, $[\alpha \implies \beta]$, $(\alpha \iff \beta)$, and $[\alpha \iff \beta]$ are FOL wffs provided that all occurrences of the same predicate name have the same number of parameters.
(3) $\forall x (\alpha)$ and $\exists x (\alpha)$ are first order logic wffs whenever α is a first order logic wff and x is not bound in α.
(4) Nothing else is a first order logic wff.

If α is an FOL wff with free variables $x_1, x_2, ..., x_k$, then we shall regard it as equivalent to the wff $\forall x_1 \forall x_2 ... \forall x_k \alpha$. Thus the wff $P(x) \implies Q(x)$ is equivalent to $\forall x [P(x) \implies Q(x)]$.

For simplicity, some of the brackets and parentheses will be omitted from wffs when such an omission introduces no ambiguity. Also, the sequences $\forall w \forall x ... \forall z$ and $\exists w \exists x \exists \cdots \exists z$ may, respectively, be abbreviated $\forall w,x,...,z$ and $\exists w,x,...,z$.

Let $\{\alpha_1, \alpha_2, ..., \alpha_n\}$ be a set of FOL wffs. Let $x_1, x_2, ..., x_k$ be the variables that appear in these wffs. By the term *universe of discourse* (or simply universe), we shall mean the set of all values assignable to the x_i's. Up to now the universe, U, has been implicitly assumed to consist of every conceivable object. This assumption on U often leads to unnecessarily complex wffs.

This is true for our earlier example concerning logic where it became necessary to introduce a new premise P3. If we restrict the universe so that it consists only of "things", then $T(x)$ is true for all x in the universe (recall that T is defined as $T(x)$: x is a thing). The symbolic versions when $U =$ 'all things' are:

P1: $\forall x (I(x) \implies \neg P(x))$
P2: $P(a)$
C: $\neg I(a)$

So, by choosing U appropriately, we can arrive at simpler wffs and can often also eliminate the need to explicitly introduce the implied premises.

As another example, consider the following sentences:

P1: No professors are ignorant.
P2: All ignorant people are vain.
C: Some vain persons are not professors.

Let U be the set of all people. Define the predicates:

$P(x)$: x is a professor
$V(x)$: x is vain
$I(x)$: x is ignorant

The wffs corresponding to P1, P2, and C are:

P1: $\neg \exists x [P(x) \wedge I(x)]$
P2: $\forall x [I(x) \implies V(x)]$
C: $\exists x [V(x) \wedge \bar{P}(x)]$

If the universe had not been restricted as above, then we would need the additional predicate:

$Q(x)$: x is a person

In this case the wffs corresponding to the given sentences are:

P1: $\neg \exists x [P(x) \wedge I(x)]$
P2: $\forall x [Q(x) \wedge I(x) \implies V(x)]$
C: $\exists x [Q(x) \wedge V(x) \wedge \bar{P}(x)]$

In addition, we need to add the implied premise that all professors are persons. This yields:

P4: $\forall x [P(x) \implies Q(x)]$

Example 1.23: [Lewis Carroll] Consider the premises and conclusion:

P1:   All philosophers are logical.
P2:   An illogical person is always obstinate.
C: Some obstinate persons are not philosophers.

Let U = "all persons". Define the following predicates:

$P(x)$: x is a philosopher
$Q(x)$: x is logical
$R(x)$: x is obstinate

We obtain the following symbolic forms:

P1: $\forall x(P(x) \Longrightarrow Q(x))$
P2: $\forall x(\bar{Q}(x) \Longrightarrow R(x))$
C: $\exists x(R(x) \wedge \bar{P}(x))$

The first two symbolic forms need no explanation. For C, however, it is very tempting to use:

C: $\exists x(R(x) \Longrightarrow \bar{P}(x))$

The two forms for C given above are not equivalent even though they may appear to be so. The second form says that there exists an x for which $R(x) \Longrightarrow \bar{P}(x)$ is true. The wff $R(x) \Longrightarrow \bar{P}(x)$ is true for all x for which $R(x)$ is false (regardless of whether $\bar{P}(x)$ is true or false). So, for $\exists x(R(x) \Longrightarrow \bar{P}(x))$ to be true there need only be an x such that x is not obstinate (i.e., $\bar{R}(x)$ is true). This is quite different from the assertion that there exists an obstinate person who is not a philosopher. □

Example 1.24: [Lewis Carroll]
P1:   All clear explanations are satisfactory.
P2:   Some excuses are unsatisfactory.
C: Some excuses are not clear explanations.

Define:

U = everything (i.e., the unrestricted universe)
$P(x)$: x is a clear explanation
$Q(x)$: x is satisfactory
$R(x)$: x is an excuse

The symbolic forms of P1, P2, and C are:

P1: $\forall x (P(x) \Longrightarrow Q(x))$
P2: $\exists x (R(x) \wedge \bar{Q}(x))$
C: $\exists x (R(x) \wedge \bar{P}(x))$

Again, note that the symbolisms:

P2: $\exists x (R(x) \Longrightarrow \bar{Q}(x))$

and

C: $\exists x (R(x) \Longrightarrow \bar{P}(x))$

are incorrect. □

Example 1.25: [Lewis Carroll] Consider the following sentences:

P1:   Some unauthorized reports are false.
P2:   All authorized reports are trustworthy.
P3:   Some false reports are not trustworthy.

Define:

U = all reports
$A(x)$: x is authorized
$T(x)$: x is trustworthy
$F(x)$: x is false

The corresponding wffs are:

P1: $\exists x [\bar{A}(x) \wedge F(x)]$
P2: $\forall x [A(x) \Longrightarrow T(x)]$
P3: $\exists x [F(x) \wedge \bar{T}(x)]$ □

Example 1.26: [Lewis Carroll] Let P1, P2, and C be as below:

P1:   No birds, except peacocks, are proud of their tails.
P2:   Some birds that are proud of their tails cannot sing.
C: Some peacocks cannot sing.

Define:

U = all birds
$P(x)$: x is a peacock
$T(x)$: x is proud of its tail
$S(x)$: x can sing

The corresponding wffs are:

P1: $\forall x[\bar{P}(x) \implies \bar{T}(x)]$
P2: $\exists x[T(x) \land \bar{S}(x)]$
C: $\exists x[P(x) \land \bar{S}(x)]$ □

Example 1.27: [Lewis Carroll]

P1:   There are no pencils of mine in this box.
P2:   No sugar plums of mine are cigars.
P3:   The whole of my property, that is not in this box, consists of cigars.
C: No pencils of mine are sugar plums.

Define:

U = all my things
$P(x)$: x is a pencil
$B(x)$: x is in this box
$C(x)$: x is a cigar
$S(x)$: x is a sugar plum

The corresponding wffs are:

P1: $\neg \exists x[P(x) \land B(x)]$
P2: $\neg \exists x[S(x) \land C(x)]$
P3: $\forall x[\bar{B}(x) \implies C(x)]$
C: $\neg \exists x[P(x) \land S(x)]$ □

Example 1.28:
P1:   Some students like every course they take.
P2:   All students get an A in at least one of the
        courses they take.
C: Some students get an A in a course they like.

Define:

$U =$ everything (i.e., the unrestricted universe)
$P(x)$: x is a student
$Q(x)$: x is a course

$R(x,y)$: x likes y
$S(x,y)$: x gets an A in y
$T(x,y)$: x takes y

The symbolic forms of P1, P2, and C are:

P1: $\exists x \forall y [P(x) \wedge (Q(y) \wedge T(x,y) \implies R(x,y))]$
P2: $\forall x \exists y [P(x) \implies Q(y) \wedge T(x,y) \wedge S(x,y)]$
C: $\exists x \exists y [P(x) \wedge Q(y) \wedge R(x,y) \wedge S(x,y)]$

P1 and P2 may also be written as:

P1: $\exists x [P(x) \wedge \forall y (Q(y) \wedge T(x,y) \implies R(x,y))]$
P2: $\forall x (P(x) \implies \exists y (Q(y) \wedge T(x,y) \wedge S(x,y)))$

Observe that the symbolisms:

P1: $\exists x \forall y [P(x) \wedge Q(y) \wedge T(x,y) \implies R(x,y)]$
P2: $\forall x \exists y [P(x) \wedge Q(y) \wedge T(x,y) \wedge S(x,y)]$
C: $\exists x \exists y [P(x) \implies Q(y) \wedge R(x,y) \wedge S(x,y)]$

are incorrect. □

The preceding examples lead us to the following cautions:
Caution 1: If x is bound by an existential quantifier, then be very cautious when using predicates involving x in the antecedent of an " $\implies$ ".
Caution 2: If x is bound by a universal quantifier, then be very cautious when using predicates involving x to form a conjunctive clause.

A common error made in the interpretation of first order logic wffs is concerned with the negation of the universal quantifier. The wff $\neg \forall x P(x)$ states that there is at least one x for which $P(x)$ is false. It does not state that $P(x)$ is false for all x. This latter statement is equivalent to $\forall x \overline{P}(x)$.

1.7 PRENEX NORMAL FORM

Most mechanical methods to determine if a propositional logic wff is a tautology or is satifiable begin with the wff in some standard form. The tableau method and the Davis-Putnam procedure, for example, begin with propositional logic wffs that are in DNF and CNF respectively. The starting point for mechanical proof methods applicable to FOL wffs is the prenex normal form (PNF). While these mechanical methods will not be discussed in this book, we shall nonetheless study the prenex normal form.

A wff in the first order logic is in *prenex normal form* (PNF) iff it is of the form:

$$Q_1 x_1 Q_2 x_2 \cdots Q_k x_k \alpha$$

where the Q_is are quantifiers ($\forall$ or $\exists$), the x_is are distinct predicate, parameters and α is an FOL wff containing no quantifiers.

Example 1.29: The wffs:

$$\forall x \, [P(x) \implies Q(x) \wedge R(x)]$$
$$\forall x \exists y \forall z \exists w \exists u \, [P(x,y,z) \wedge R(w,x,z) \implies Q(u) \vee S(u,w,x,y)]$$
$$\exists x \exists y \forall z [P(z) \wedge \overline{Q}(z,x) \iff Q(y,z) \wedge R(x)]$$

are in PNF. The formulas below are not in PNF:

$$\forall x \, P(x) \implies [\forall x \, Q(x) \wedge \forall x \, R(x)]$$
$$\exists x \forall y \, [P(x,y) \wedge \exists z \, R(z)] \iff \exists x \, T(x) \quad \square$$

For every FOL wff not in PNF there exists an equivalent FOL PNF wff. An equivalent FOL PNF wff can be obtained by making use of equivalences between certain wff forms. In Figure 1.10 some of the more useful FOL equivalences are listed. In this figure, Greek letters denote FOL wffs. The notation $\alpha(x)$ means that the wff α contains x as a predicate parameter. It may also contain other parameters. If $\alpha(x)$ and β are used in the same equivalence, it is implied that α contains x while β does not. If β also contains x then $\beta(x)$ will be used in place of β. In FE1 and FE2 the name change from x to y is permissible iff y does not appear in $\alpha(x)$. In FE11 and FE12 y is a parameter that does not appear in either $\alpha(x)$ or $\beta(x)$. One may easily verify that the equivalence and inference rules of Figures 1.6 and 1.7 hold even when α, β, δ, and γ are FOL wffs.

FE1. $\forall x\,\alpha\,(x) \equiv \forall y\,\alpha\,(y)$

FE2. $\exists x\,\alpha\,(x) \equiv \exists y\,\alpha\,(y)$

FE3. $\forall x\,\alpha\,(x) \vee \beta \equiv \forall x(\alpha\,(x) \vee \beta)$

FE4. $\exists x\,\alpha\,(x) \vee \beta \equiv \exists x(\alpha\,(x) \vee \beta)$

FE5. $\forall x\,\alpha\,(x) \wedge \beta \equiv \forall x(\alpha\,(x) \wedge \beta)$

FE6. $\exists x\,\alpha\,(x) \wedge \beta \equiv \exists x(\alpha\,(x) \wedge \beta)$

FE7. $\neg\,\forall x\,\alpha\,(x) \equiv \exists x(\neg\,\alpha\,(x))$

FE8. $\neg\,\exists x\,\alpha\,(x) \equiv \forall x(\neg\,\alpha\,(x))$

FE9. $\forall x\,\alpha\,(x) \wedge \forall x\,\beta\,(x) \equiv \forall x(\alpha\,(x) \wedge \beta\,(x))$

FE10. $\exists x\,\alpha\,(x) \vee \exists x\,\beta\,(x) \equiv \exists x(\alpha\,(x) \vee \beta\,(x))$

FE11. $\forall x\,\alpha\,(x) \vee \forall x\,\beta\,(x) \equiv \forall x\,\forall y(\alpha\,(x) \vee \beta\,(y))$ y is new

FE12. $\exists x\,\alpha\,(x) \wedge \exists x\,\beta\,(x) \equiv \exists x\,\exists y(\alpha\,(x) \wedge \beta\,(y))$ y is new

Figure 1.10 FOL equivalences

The validity of each of the equivalences FE1-FE12 is easily established. We shall consider only FE12 and leave the remaining as an exercise. FE12 can be established from FE2 and FE6 in the following way:

$\exists x\,\alpha\,(x) \wedge \exists x\,\beta\,(x)$

$\equiv \exists x\,\alpha\,(x) \wedge \exists y\,\beta\,(y)$ FE2

$\equiv \exists x(\alpha\,(x) \wedge \exists y\,\beta\,(y))$ FE6

$\equiv \exists x\,\exists y(\alpha\,(x) \wedge \beta\,(y))$ commutativity of $\wedge$ and FE6

It is important to note that

$$\forall x\,\alpha\,(x) \vee \forall x\,\beta\,(x) \not\equiv \forall x(\alpha\,(x) \vee \beta\,(x))$$

and

$\exists x \, \alpha(x) \wedge \exists x \beta(x) \not\equiv \exists x(\alpha(x) \wedge \beta(x))$.

Consider the second inequivalence. The left hand side says that there exists an x for which $\alpha(x)$ is true and there exists another (possibly different) x for which $\beta(x)$ is true. The right hand side, however, states that there is a single x for which both α and β are true. This is not implied by the left hand side. The following inference rules are however valid:

FI1: $\{\forall x \, \alpha(x) \vee \forall x \beta(x)\} \models \forall x(\alpha(x) \vee \beta(x))$

and

FI2: $\{\ \exists x(\alpha(x) \wedge \beta(x))\ \} \models \exists x \, \alpha(x) \wedge \exists x \beta(x)$

Using the FOL equivalences together with the propositional logic equivalences (all of these also hold for FOL wffs) an FOL wff may be transformed into an equivalent PNF wff by performing the three transformations listed below:

1. Use E16 and E18 to eliminate all occurrences of $\implies$ and $\iff$.

2. Move all occurrences of $\neg$ so that they appear immediately before the predicate symbols and propositional variables to which they apply. This may involve the elimination of double negations using E13 and also the application of equivalences E14 and E15 (De Morgan's Laws) as well as equivalences FE7 and FE8.

3. Use the FOL equivalences to move all the quantifiers to the left end of the wff.

Example 1.30:
$\exists x \, P(x) \implies \forall x \, Q(x)$
$\equiv \neg \exists x \, P(x) \vee \forall x \, Q(x)$ E16
$\equiv \forall x \, \bar{P}(x) \vee \forall x \, Q(x)$ FE8
$\equiv \forall x \forall y(\bar{P}(x) \vee Q(y))$FE11 $\square$

Example 1.31:
$\exists x \, P(x) \iff \exists x \, Q(x)$

$\equiv (\exists x P(x) \implies \exists x Q(x)) \wedge (\exists x \ Q(x) \implies \exists x P(x))$	E18
$\equiv (\neg \exists x P(x) \vee \exists x Q(x)) \wedge (\neg \exists x Q(x) \vee \exists x P(x))$	E16
$\equiv (\forall x \bar{P}(x) \vee \exists x Q(x)) \wedge (\forall x \bar{Q}(x) \vee \exists x P(x))$	FE8
$\equiv \forall x \, (\bar{P}(x) \vee \exists y Q(y)) \wedge \forall x \, (\bar{Q}(x) \vee \exists y P(y))$	FE2,FE3
$\equiv \forall x \exists y \, (\bar{P}(x) \vee Q(y)) \wedge \forall x \exists y(\bar{Q}(x) \vee P(y))$	FE4
$\equiv \forall x \exists y(\bar{P}(x) \vee Q(y)) \wedge \exists y(\bar{Q}(x) \vee P(y)))$	FE9
$\equiv \forall x \exists y \exists z((\bar{P}(x) \vee Q(y)) \wedge (\bar{Q}(x) \vee P(z)))$	FE12$\square$

Example 1.32:

$$\forall x(\overline{P}(x)\bigvee Q(x) \Longrightarrow \exists zR\,(x,z)\bigvee\forall yT(y))$$
$$\equiv \forall x(\neg(\overline{P}(x)\bigvee Q(x))\bigvee(\exists zR\,(x,z)\bigvee\forall yT(y))) \qquad\qquad \text{E16}$$
$$\equiv \forall x((\neg\neg P(x)\bigwedge \overline{Q}(x))\bigvee \exists zR\,(x,z)\bigvee\forall yT(y)) \qquad\qquad \text{DeMorg.}$$
$$\equiv \forall x \exists z((P(x)\bigwedge \overline{Q}(x))\bigvee R\,(x,z)\bigvee\forall yT(y)) \qquad\qquad \text{E13,FE4}$$
$$\equiv \forall x\ \exists z\forall y((P(x)\bigwedge \overline{Q}(x))\bigvee R\,(x,z)\bigvee T(y)) \qquad\qquad \text{FE3}\square$$

1.8 FOL PROOFS

The proof methods M1-M6 introduced in Section 1.3 can also be used for FOL wffs. In addition, three other proof methods are useful here:

M7: By example: To show that $\exists x\,\alpha(x)$, it is sufficient to show that $\alpha(x)$ is true for some x.

M8: By counter example: To show that $\forall x\,\alpha(x)$ is false, it is sufficient to exhibit an x for which $\alpha(x)$ is false.

M9: By generalization: To show that $\forall x\,\alpha(x)$ is true, it is sufficient to show that $\alpha(x)$ is true for *arbitrary* x. Note that this is not the same as showing that $\alpha(x)$ is true for some particular x.

Consider the statement:

There exists a red apple.

To prove the correctness of this statement, we need merely exhibit an apple that is red (i.e., a proof by example). Now, consider the statement:

All apples are red.

To prove that this statement is false, we need only exhibit an apple that is not red (i.e., a proof by counter example). Finally, consider the statement:

Every person has a name.

To establish the correctness of this statement, we could start by letting x be any arbitrary person and then proceed to show that x has a name. This would be a proof by generalization.

Example 1.33: The statement:

There exists an odd number.

is true. As proof, we need only show that 1 is odd. The wff $\forall x\, P(x)$ where $P(x)$ is the predicate:

$$P(x) = \begin{cases} true & \text{if } x \text{ is not a positive integer} \\ true & \text{if } x \text{ is a perfect square} \\ false & \text{otherwise} \end{cases}$$

is false. The proof is by counter example. The number 3 is a positive integer that is not a perfect square. □

Note that a proof by example can be used only to establish $\exists x\, \alpha(x)$. It cannot be used to establish $\forall x\, \alpha(x)$. Using proofs by example to establish $\forall x\, \alpha(x)$ is an all too common occurrence. This is the fallacy of generalizing from an instance. Simply because 1 is an odd number, we cannot conclude that all numbers are odd. Methods M7 and M8 are duals of each other. An example that shows $\alpha(x)$ is true is also an example that shows $\neg\, \alpha(x)$ is false. Hence, this example establishes $\exists x\, \alpha(x)$ (proof by example) and at the same time establishes $\neg\, \forall x\, \bar{\alpha}(x)$ (by counter example). Of course, by FE7, $\exists x\, \alpha(x) \equiv \neg\, \forall x\, \bar{\alpha}(x)$.

In addition to the equivalences and inference rules already discussed, the following inference rules are useful in FOL:

FI3: $\{\alpha(a)\} \models \exists x\, \alpha(x)$ Existential Quantification (EQ)
FI4: $\{\exists x\, \alpha(x)\} \models \alpha(a)$ Existential Instantiation (EI)
FI5: $\{\forall x\, \alpha(x)\} \models \alpha(a)$ Universal Instantiation (UI)

Whenever FI4 (Existential instantiation) is used, a must be a symbol different from any others in use at the time. In the case of FI5 (UI), however, a can be any symbol. In particular, it can be one that is already in use. FI3 is essentially a restatement of proof method M7 (by example).

An FOL wff is a *theorem* iff its truth can be established using the proof methods, inference rules, and equivalence rules listed so far. A wff is *valid* iff it is a theorem.

Example 1.34: Show that $\forall x\, (P(x) \implies Q(x)) \wedge \forall x\, (R(x) \implies S(x))$ $\implies \forall x\, [P(x) \wedge R(x) \implies Q(x) \wedge S(x)]$ is a valid wff FOL.

Proof:
1. $\forall x(P(x) \Longrightarrow Q(x)) \wedge \forall x(R(x) \Longrightarrow S(x)))$, A1
2. $\forall x((P(x) \Longrightarrow Q(x)) \wedge (R(x) \Longrightarrow S(x)))$, 1, FE9
3. $\forall x(P(x) \wedge R(x) \Longrightarrow Q(x) \wedge S(x))$
   3.1 $P(a) \wedge R(a) \Longrightarrow Q(a) \wedge S(a)$, a arbitrary
      3.1.1 $(P(a) \Longrightarrow Q(a)) \wedge (R(a) \Longrightarrow S(a))$, 2, UI
      3.1.2 $P(a) \wedge R(a) \Longrightarrow Q(a) \wedge S(a)$, 3.1.1, I8
   3.2 $\forall x(P(x) \wedge R(x) \Longrightarrow Q(x) \wedge S(x))$, 3.1, gen.
4. $\forall x(P(x) \Longrightarrow Q(x)) \wedge \forall x(R(x) \Longrightarrow S(x)) \Longrightarrow \forall x(P(x) \wedge R(x) \Longrightarrow Q(x) \wedge S(x))$, 1, 3, dir. pf. □

Example 1.35: Define P1, P2 and C to be the wffs:

   P1: $\forall x(P(x) \Longrightarrow Q(x))$
   P2: $\exists x(R(x) \wedge \forall y(Q(y) \Longrightarrow \neg S(x,y)))$
   C: $\exists x(R(x) \wedge \forall y(P(y) \Longrightarrow \neg S(x,y)))$

Show that C is a consequence of P1 and P2.

Proof:
1. $R(a) \wedge \forall y(Q(y) \Longrightarrow \neg S(a,y))$ P2, EI
2. $\forall y[P(y) \Longrightarrow \neg S(a,y)]$
   2.1 $P(b) \Longrightarrow \neg S(a,b)$ b arbitrary
      2.1.1 $P(b) \Longrightarrow Q(b)$ P1, UI
      2.1.2 $Q(b) \Longrightarrow \neg S(a,b)$ 1, simp., UI
      2.1.3 $P(b) \Longrightarrow \neg S(a,b)$ 2.1.1, 2.1.2, hypo. syll.
   2.2 $\forall y[P(y) \Longrightarrow \neg S(a,y)]$ 2.1, gen.
3. $R(a) \wedge \forall y(P(y) \Longrightarrow \neg S(a,y))$ 1, simp., 2, conj.
4. $\exists x(R(x) \wedge \forall y(P(y) \Longrightarrow \neg S(x,y)))$ 3, EQ □

Example 1.36: Let P1, P2, and C be as below:
   P1: $\forall x \exists y[P(x,y) \wedge S(x,y)]$
   P2: $\forall x \forall y[P(x,y) \Longrightarrow R(x,y)]$
   C: $\forall x \exists y[R(x,y) \wedge S(x,y)]$

Show that C is a consequence of P1 and P2.

Proof:
1. $\exists y[R(c,y) \wedge S(c,y)]$ c is arbitrary
   1.1 $\exists y[P(c,y) \wedge S(c,y)]$ P1,UI
   1.2 $P(c,d) \wedge S(c,d)$ 1.1, EI
   1.3 $\forall y[P(c,y) \Longrightarrow R(c,y)]$ P2,UI

$$1.4 \; P(c,d) \Longrightarrow R(c,d) \qquad \text{1.3, UI}$$
$$1.5 \; R(c,d) \qquad \text{1.2, simp., 1.4, MP}$$
$$1.6 \; R(c,d) \wedge S(c,d) \qquad \text{1.5, 1.2, simp., conj.}$$
$$1.7 \; \exists y [R(c,y) \wedge S(c,y)] \qquad \text{1.6, EQ}$$
$$2. \; \forall x \exists y [R(x,y) \wedge S(x,y)] \qquad \text{1, gen.} \; \square$$

Example 1.37: Let P1, P2 and C be defined as:

P1: $\exists x \forall y [P(x) \wedge (Q(y) \wedge R(x,y) \Longrightarrow S(x,y))]$
P2: $\forall x \exists y [P(x) \Longrightarrow Q(y) \wedge T(x,y) \wedge \bar{S}(x,y)]$
C: $\exists x \exists y [P(x) \wedge Q(y) \wedge T(x,y) \wedge \bar{R}(x,y)]$

Show that C is a consequence of P1 and P2.

Proof:

1. $\forall y [P(a) \wedge (Q(y) \wedge R(a,y) \Longrightarrow S(a,y))]$ P1, EI
2. $P(a) \wedge \forall y (Q(y) \wedge R(a,y) \Longrightarrow S(a,y))$ 1, FE5
3. $P(a)$ 2, simp.
4. $\exists y [P(a) \Longrightarrow Q(y) \wedge T(a,y) \wedge \bar{S}(a,y)]$ P2, UI
5. $P(a) \Longrightarrow Q(b) \wedge T(a,b) \wedge \bar{S}(a,b)$ 4, EI
6. $Q(b) \wedge T(a,b) \wedge \bar{S}(a,b)$ 3, 5, MP
7. $\forall y (Q(y) \wedge R(a,y) \Longrightarrow S(a,y))$ 2, simp.
8. $Q(b) \wedge R(a,b) \Longrightarrow S(a,b)$ 7, UI
9. $\bar{S}(a,b)$ 6, simp.
10. $\neg(Q(b) \wedge R(a,b))$ 8, 9, MT
11. $\bar{Q}(b) \vee \bar{R}(a,b)$ 10, De Morg.
12. $Q(b)$ 6, simp.
13. $\bar{R}(a,b)$ 11, 12, disj. syll.
14. $P(a) \wedge Q(b) \wedge T(a,b) \wedge \bar{R}(a,b)$ 3,12,6,simp., 13, conj.
15. $\exists x \exists y [P(x) \wedge Q(y) \wedge T(x,y) \wedge \bar{R}(x,y)]$ 14, EQ, EQ $\square$

Example 1.38: Let P1, P2 and C be:

P1: $\forall x (P(x) \Longrightarrow \exists y (Q(y) \wedge S(x,y))$
P2: $\neg \exists x \exists y (P(x) \wedge R(y) \wedge S(x,y))$
C: $\exists x \, P(x) \Longrightarrow \exists y (Q(y) \wedge \bar{R}(y))$

Show that C is a consequence of P1 and P2.

Proof:

1. $\exists x \, P(x)$ A1
2. $\exists y (Q(y) \wedge \bar{R}(y))$
   2.1 $\neg \exists y (Q(y) \wedge \bar{R}(y))$ A2

2.2 $\forall y(\overline{Q}(y) \lor R(y))$ 2.1, FE8, De Morg., E13

2.3 $\forall x \forall y (\overline{P}(x) \lor \overline{R}(y) \lor \overline{S}(x,y))$ P2,FE8,De Morg.

2.4 $P(a)$ 1, EI

2.5 $P(a) \Longrightarrow \exists y(Q(y) \land S(a,y))$ P1, UI

2.6 $\exists y(Q(y) \land S(a,y))$ 2.4, 2.5, MP

2.7 $Q(b) \land S(a,b)$ 2.6, EI

2.8 $\overline{Q}(b) \lor R(b)$ 2.2, UI

2.9 $R(b)$ 2.7, simp., 2.8, disj. syll.

2.10 $\overline{P}(a) \lor \overline{R}(b) \lor \overline{S}(a,b)$ 2.3, UI, UI

2.11 $\neg(\overline{P}(a) \lor \overline{R}(b))$ 2.4, 2.9, conj., De Morg.

2.12 $\overline{S}(a,b)$ 2.10, 2.11, disj. syll.

2.13 $S(a,b)$ 2.7, simp.

2.14 $\exists y(Q(y) \land \overline{R}(y))$ 2.12, 2.13, contradiction

3. $\exists x P(x) \Longrightarrow \exists y(Q(y) \land \overline{R}(y))$ 1, 2, dir. pf.

An alternate and shorter proof is provided below:

1. $\exists x P(x)$ A1
2. $P(a)$ 1, EI
3. $P(a) \Longrightarrow \exists y(Q(y) \land S(a,y))$ P1, UI
4. $\exists y(Q(y) \land S(a,y))$ 2, 3, MP
5. $Q(b) \land S(a,b)$ 4, EI
6. $\forall x \neg \exists y (P(x) \land R(y) \land S(x,y))]$ P2, FE8
7. $\forall x \forall y [\neg(P(x) \land R(y) \land S(x,y))]$ 6, FE8
8. $\neg(P(a) \land R(b) \land S(a,b))$ 7, UI, UI
9. $[\overline{P}(a) \land S(a,b)] \lor \overline{R}(b)$ 8, E9, De Morg.
10. $P(a) \land S(a,b)$ 2, 5, simp., conj.
11. $\overline{R}(b)$ 10, 9, disj. syll.
12. $Q(b) \land \overline{R}(b)$ 5, simp., 11, conj.
13. $\exists y(Q(y) \land \overline{R}(y))$ 12, EQ
14. $\exists x P(x) \Longrightarrow \exists y(Q(y) \land \overline{R}(y))$ 1, 13, dir. pf. □

1.9 OTHER LOGICS

Two popular extensions of first order logic are (i) first order logic with equality and (ii) first order logic with operators. In the first order logic with equality, one is permitted to make explicit use of the relational operators $=$ and $\neq$ (as in $x = y$ and $x \neq y$). This eliminates the need to define the predicates:

$E(x,y)$: $x = y$

$N(x,y)$: $x \neq y$

In FOL, the wff corresponding to the sentence:

If $x = y$, then $y = x$.

is $\forall x \forall y \, [E(x,y) \implies E(y,x)]$. In the first order logic with equality, we may also use the wff $\forall x \forall y [x = y \implies y = x]$. Some other examples of wffs in the first order logic with equality are:

(a) $\forall x \forall y [(x = y) \wedge P(x,y) \implies P(x,x)]$
(b) $a = a$
(c) $\forall x \forall y [x = y \implies (P(x) \iff \bar{P}(y))]$
(d) $P(a) \wedge \bar{P}(b) \implies a \neq b$
(e) $P(a) \wedge \bar{P}(b) \implies \exists x \exists y [x \neq y]$

Observe that while all FOL wffs are also wffs of the first order logic with equality, not all first order logic with equality wffs are FOL wffs.

In the first order logic with operators, one is permitted to make explicit use of the arithmetic and relational operators: $+, -, *, /, >, <, =, \neq$, etc. together with some unspecified operators such as op1, op2, etc. Some examples of wffs in the first order logic with operators are:

(a) $\forall x \forall y [x+y = y+x]$
(b) $\forall x \exists y [x*y \neq x*x]$
(c) $\forall x [P(x) \wedge P(x+1) \implies P(x+2)]$
(d) $\forall x \forall y \forall z [x < y \wedge y < z \implies x < z]$

Example 1.39: In this example, the universe of discourse is the set of all numbers. We list below some sentences and their corresponding wffs.

(a) sentence: $x+y = y+x$
     wff: $\forall x \forall y [x+y = y+x]$
(b) sentence: If $x > y$ and $y > z$, then $x > z$.
     wff: $\forall x \forall y \forall z [x > y \wedge y > z \implies x > z]$
(c) sentence: For every x there is a y such that $x < y$.
     wff: $\forall x \exists y [x < y]$
(d) sentence: If $x=y$ then $P(x)$ else $Q(x,y)$.
     wff: $\forall x \forall y [(x = y \implies P(x)) \wedge (x \neq y \implies Q(x,y))]$ □

Another important logic system is elementary arithmetic (EA). The wffs of EA are composed of the natural numbers $N = \{0, 1, 2, ...\}$; the arithmetic operators $+$ and $-$; the relational operator $=$; variables $a, b, c, ..., z$; logical operators $\neg, \wedge, \vee, \implies, \iff$; and the quantifiers $\forall$ and $\exists$. The domain of discourse is the natural numbers. Note that EA contains no predicate symbols. Some examples of EA wffs are:

(a) $\forall x \forall y [x+y = y+x]$
(b) $\exists x \exists y [\neg (x-y+y = x)]$
(c) $\exists x [\neg (x=3) \implies x=4]$

A *theory* is a set of true wffs. The true wffs of EA constitute the theory known as *elementary number theory* (*ENT*). A theory is said to be *complete* iff all its formulas can be proved true by the use of some algorithm. The theory consisting of all the true wffs of propositional calculus is a complete theory as we can prove that each true propositional calculus wff is true by simply computing its truth table.

Theorem 1.3: [Gödel] Every "sufficiently" complex theory is incomplete. □

ENT is an example of an incomplete theory.

REFERENCES AND SELECTED READINGS

The book

A course in mathematical logic, by John Bell and M. Machover, North-Holland Publishing Co., N.Y., 1977.

is a good general reference for propositional and first order logic. Many of our examples for first order logic are taken from Lewis Carroll's classic:

Symbolic Logic, by Lewis Carroll, 4th Edition, Berkeley Enterprises Inc., New York, 1955.

Further material on Boolean algebra and its applicability to switching circuit design can be found in:

Introduction to switching theory and logical design, second edition, by Fredrick Hill and Gerald Peterson, John Wiley, New York, 1974.

and

Logic design of digital systems, by Arthur Friedman, Computer Science Press Inc., Maryland, 1975.

The Quine-McCluskey procedure to obtain minimal equivalent Boolean expressions (together with other methods to do this) is discussed in both of these books. Tableau methods for wffs are discussed in the book:

First Order Logic, by R. M. Smullyan, Springer-Verlag, 1968.

The Davis-Putnam procedure originally appeared in the paper:

A computing procedure for quantification theory, by Martin Davis and Hilary Putnam, *JACM*, Vol. 7, 1960, pp.201-215.

The Davis-Putnam procedure did not originally contain rule (a). The inclusion of this rule was proposed by Robert Harris in his Ph.D. dissertation:

A polynomial bound on the complexity of the Davis-Putnam algorithm applied to symmetrizable propositions, by R. Harris, Cornell University, Technical Report 72-131, Aug. 1972.

Harris's dissertation also discusses the tableau method of Section 1.4. Mechanical theorem proving methods are discussed in the book:

Symbolic logic and mechanical theorem proving, by Chin-Liang Chang and Richard Lee, Academic Press, New York, 1973.

EXERCISES

1. Define the two operators NAND (not and) and NOR (not or) as follows:

$$\alpha \text{ NAND } \beta \equiv \neg(\alpha \wedge \beta)$$
$$\alpha \text{ NOR } \beta \equiv \neg(\alpha \vee \beta)$$

(a)   Obtain the truth tables for NAND and NOR.

(b)   Extend the definition of propositional wffs to include the operators NAND and NOR. Obtain wffs α_1, α_2 and α_3 such that the only operator contained in the α_is is NAND, and $\alpha_1 \equiv \bar{P}$, $\alpha_2 \equiv P \vee Q$, and $\alpha_3 \equiv P \wedge Q$. P and Q are propositional variables. The α_is are not permitted to contain the constants **true** and **false**. For example, $\alpha_1 = P \text{ NAND } P \equiv \bar{P}$. Prove the equivalences using truth tables.

(c)   Do part (b) with the α_is restricted to contain only NORs.

2. Obtain propositional logic wffs for each of the following old wives tales. In each case state what each propositional variable stands for.

(a) If you sing before you eat, you'll cry before you sleep.

(b) Coins thrown in a fountain bring good luck.

(c) If you enter a house through one door, you must leave through the same door, or you'll bring more company to the house.

(d) [Pregnancy blues] If a pregnant woman goes to an art museum, her child will be cultured. If she takes anything belonging to another, her child will be a thief. A pregnant woman should never walk over a grave, or her child will be born club-footed. If a pregnant woman steps over a broom, she will bear a hairy child and if she doesn't satisfy any craving, her child will have a birthmark in the shape of the food she was craving.

(e) [The itch] If your right hand itches, you're going to get money. Itching of the elbow means you'll soon sleep in a strange bed or you'll sleep with a stranger. If your nose itches, someone is soon to visit, or you'll kiss a fool. If your knee itches, you're jealous. If your left ear tingles, someone is speaking critically of you. If your right ear tingles, you are being praised.

3. Obtain CNF and DNF wffs corresponding to each of the truth tables of Figure 1.11. Use the postulates of Boolean algebra together with the results of Theorem 1.1 to simplify the resulting wffs.

P	Q	R	O
1	1	1	1
1	1	0	1
1	0	1	1
1	0	0	0
0	1	0	0
0	1	1	0
0	0	1	1
0	0	0	0

(i)

P	Q	R	O
1	1	1	0
1	1	0	1
1	0	1	0
1	0	0	1
0	1	0	0
0	1	1	0
0	0	1	1
0	0	0	1

(ii)

P	Q	R	O
1	1	1	1
1	1	0	1
1	0	1	0
1	0	0	1
0	1	0	0
0	1	1	0
0	0	1	1
0	0	0	0

(iii)

P	Q	R	S	O
1	1	1	1	0
1	0	1	0	1
1	1	0	0	1
0	1	1	0	1
0	0	0	0	1
0	1	0	1	1
1	1	0	1	0
0	1	1	1	0

(iv)

P	Q	R	S	O
0	0	0	0	0
1	1	1	1	0
0	1	0	1	0
0	0	1	0	0
1	1	0	0	1
1	0	1	0	1
1	1	0	1	1

(v)

P	Q	R	S	O
0	0	1	0	0
1	0	0	0	1
1	1	0	0	0
0	1	1	0	1
0	1	0	1	0
1	1	1	0	1

(vi)

Figure 1.11

4. Prove Theorem 1.1. Parts (a), (b), and (c) have already been proved. So, only (d)-(j) need to be proved.

5. Use the Boolean algebra postulates and Theorem 1.1 to prove the following equivalences:

(a) $\neg((\bar{P} \vee Q) \wedge (\bar{Q} \vee R)) \vee (\bar{P} \vee R) \equiv$ true
(b) $PQ \vee P\bar{Q}R \equiv PR \vee PQ\bar{R}$

6. Define the logical operator exclusive or (written as $\oplus$) such that $\alpha \oplus \beta \equiv (\alpha \vee \beta) \wedge \overline{(\alpha \wedge \beta)}$.

(i)   Obtain the truth table for $\alpha \oplus \beta$.

(ii) Using the Boolean algebra postulates and Theorem 1.1 show that (F and T respectively denote the constants false and true):

(a) $P \oplus P \equiv F$
(b) $P \oplus F \equiv P$
(c) $P \oplus T \equiv \bar{P}$
(d) $P \oplus Q \equiv Q \oplus P$
(e) $P \oplus Q \equiv (P \wedge \bar{Q}) \vee (\bar{P} \wedge Q)$
(f) $P \oplus (Q \oplus R) \equiv (P \oplus Q) \oplus R$
(g) $P \wedge (Q \oplus R) \equiv (P \wedge Q) \oplus (P \wedge R)$

7. Use the truth table method to establish the correctness of the equivalence rules of Figure 1.6.

8. Use the truth table method to establish the correctness of the inference rules of Figure 1.7.

9. Prove that the following wffs are tautologies. Use the method of Section 1.3. Do not use truth tables.

(a) $(P \wedge Q \Longrightarrow R) \Longleftrightarrow (P \wedge \bar{R} \Longrightarrow \bar{Q})$
(b) $(\bar{P} \vee \bar{Q}) \wedge (R \Longrightarrow Q) \wedge (S \Longrightarrow P) \Longrightarrow \neg(R \wedge S)$
(c) $(P \vee Q) \wedge (\bar{R} \vee \bar{S}) \wedge (Q \Longrightarrow S) \wedge (P \Longrightarrow R)$
$$\Longrightarrow (R \Longrightarrow P) \wedge (S \Longrightarrow Q)$$
(d) $\neg((\bar{P} \vee Q) \wedge (\bar{Q} \vee R)) \vee (\bar{P} \vee R)$
(e) $\neg(P \Longleftrightarrow Q) \Longleftrightarrow (P \wedge \bar{Q}) \vee (\bar{P} \wedge Q)$

10. Show that each of the following conclusions follows from the given premises.

(a) P1: $P \wedge Q \implies R$
  P2: $P \vee Q \implies U$
  P3: $R \implies \bar{U}$
  C: $P \implies \bar{Q}$

(b) P1: $P \implies Q \vee R$
  P2: $R \implies S \vee T$
  P3: $S \implies T$
  P4: $Q \implies \bar{P}$
  C: $P \implies T$

(c) P1: $P \vee Q \implies R \vee S$
  P2: $(U \implies T) \implies P$
  P3: $\bar{T} \vee \bar{Q} \implies \bar{U}$
  C: $\bar{R} \implies S$

(d) P1: $S \wedge T \implies \bar{P} \wedge Q$
  P2: $(U \wedge \bar{V}) \vee P$
  P3: $U \implies V \wedge \bar{Q}$
  C: $T \implies \bar{S}$

(e) P1: $Q \implies \bar{P}$
  P2: $\bar{Q} \wedge \bar{S} \implies \bar{R}$
  P3: $\bar{R} \wedge \bar{U} \implies \bar{T}$
  P4: $S \implies Q$
  C: $\bar{T} \vee (P \implies U)$

(f) P1: $\bar{Q} \implies \bar{P}$
  P2: $Q \wedge \bar{S} \implies \bar{R}$
  P3: $T \implies R \vee U$
  P4: $Q \implies \bar{S}$
  C: $\bar{T} \vee \bar{P} \vee U$

(g) P1: $\bar{Q} \vee P$
  P2: $\bar{R} \vee (S \wedge \bar{P})$
  P3: $\bar{T} \vee R \vee Q$
  C: $\bar{S} \implies \bar{T} \vee P$

(h) P1: $\bar{U} \vee V \vee \bar{Q}$
  P2: $(U \wedge \bar{V}) \vee P$
  P3: $S \wedge T \implies \bar{P} \wedge Q$
  C: $\bar{T} \vee \bar{S}$

11. Transform each of the following into equivalent DNF wffs using the

method proposed in Section 1.4. Simplify the resulting clauses by replacing multiple occurrences of a literal by a single occurrence and eliminating clauses containing both P and $\bar{P}$.

(a) $((P \Longrightarrow Q) \Longrightarrow R) \Longrightarrow Q$
(b) $((P \Longleftrightarrow Q) \Longrightarrow P) \Longrightarrow (P \wedge Q)$
(c) $(P \wedge Q) \vee R \Longrightarrow \bar{Q} \vee \bar{R}$
(d) $\neg(P \vee Q \Longrightarrow R) \wedge (P \wedge Q) \Longrightarrow S \vee R$
(e) $((P \Longrightarrow Q) \Longrightarrow R) \Longrightarrow (P \Longrightarrow R) \wedge (Q \Longrightarrow R)$

12. Transform each of the wffs of Exercise 11 into equivalent CNF wffs. Simplify the CNF wffs as described in Exercise 11.

13. Use the tableau method of Section 1.4 to determine if the following are tautologies:

(a) $\neg(P \vee Q) \Longleftrightarrow \bar{P} \wedge \bar{Q}$
(b) $\neg(P \wedge Q) \Longleftrightarrow \bar{P} \vee \bar{Q}$
(c) $(P \Longrightarrow Q) \Longrightarrow (\bar{Q} \Longrightarrow \bar{P})$
(d) $(P \vee Q) \wedge \bar{P} \Longrightarrow Q$
(e) $P \vee (Q \wedge R) \Longrightarrow (P \vee Q) \wedge (P \vee R)$
(f) $P \wedge Q \Longrightarrow P$
(g) $(P \Longrightarrow Q) \wedge (R \Longrightarrow S) \Longrightarrow (P \wedge R \Longrightarrow Q \wedge S$
(h) $(P \Longrightarrow \bar{P}) \Longrightarrow \bar{P}$

14. Use the Davis-Putnam procedure to determine if the following are satisfiable:

(a) $(P \vee Q \vee R)(\bar{P} \vee \bar{Q} \vee \bar{R})(\bar{P} \vee Q)(\bar{Q} \vee R)$
(b) $(P \wedge Q) \vee R \Longrightarrow \bar{Q} \vee \bar{R}$
(c) $\neg(P \vee Q) \Longrightarrow \bar{P} \vee Q$
(d) $\neg(P \wedge Q) \Longrightarrow \bar{P} \vee Q$
(e) $\neg(P \wedge Q) \Longrightarrow \bar{P} \wedge Q$
(f) $P \wedge Q \Longrightarrow \bar{Q}$
(g) $(P \vee \bar{Q})(\bar{P} \vee Q \vee R)(\bar{P} \vee \bar{Q} \vee \bar{R})(P \vee \bar{Q} \vee R)(P \vee Q \vee \bar{R})$

15. [Lewis Caroll] Obtain FOL wffs for each of the following English sentences. Carefully state what each predicate stands for. Also prove that the resulting wffs are true.

(a) If philosophers are not conceited and if some conceited persons are not gamblers then some people who are not gamblers are not philosophers.
(b) If exciting books suit feverish patients and if unexciting books

make one drowsy, then all books suit feverish patients unless they make one drowsy.

(c) If some holidays are rainy days and all rainy days are tiresome, then some holidays are tiresome.

16. Use the predicates:

$D(x)$: x is a dog
$C(x)$: x is a dog catcher
$T(x)$: x is a town
$L(x,y)$: x lives in y
$B(x,y)$: x has bitten y

and obtain FOL wffs corresponding to:

(a) Every town has a dog catcher who has been bitten by every dog in town.
(b) No town has a dog catcher who has been bitten by every dog in town.
(c) At least one town has a dog catcher who has been bitten by none of the dogs in town.

17. Use the predicates:

$P(x)$: x is a person
$T(x)$: x is a time
$F(x,y,t)$: x can fool y at time t

to obtain an FOL wff corresponding to:

Everyone can fool everyone some of the time; all persons can be fooled by everyone some of the time; but no one can be fooled by anyone all the time.

18. Prove that the equivalences of Figure 1.10 are correct.

19. Transform each of the following into Prenex Normal Form:

1. $\forall x \exists y\, R(x,y) \wedge \exists z\, S(z) \Longrightarrow \exists w\, P(w)$
2. $\forall x\, R(x) \wedge \forall y\, S(y) \vee \exists y[R(y) \Longrightarrow S(y)]$
3. $\exists x\, P(x) \wedge \forall y[Q(y) \Longrightarrow S(y)]$
4. $\exists x[P(x) \wedge \forall y(Q(x,y) \Longrightarrow S(y))]$
5. $\forall x \forall y[P(x) \vee Q(y) \Longrightarrow \exists z\, R(x,z) \wedge \exists z\, S(y,z)]$

6. $\exists x P(x) \wedge \forall y Q(y) \wedge \exists z R(z) \vee \exists w S(w)$
7. $\forall x P(x) \Longrightarrow \forall x Q(x) \wedge \forall x R(x)$
8. $\exists x \forall y [P(x,y) \wedge \exists z R(z)] \Longleftrightarrow \exists x T(x)$
9. $\exists x [\bar{P}(x) \wedge [\forall y \bar{Q}(y) \Longrightarrow \exists z \bar{S}(x,z)]]$
10. $\neg \forall x\ Q(x) \vee (\neg \exists x\ \bar{R}(x)) \wedge \exists x\ S(x) \Longrightarrow \forall x\ \bar{Q}(x) \vee (\neg \exists x\ T(x))$

20. Show that each of the following conclusions follows from the given premises. Note that in some cases there are no premises. Use the inference rule method.

(a)   P1: $\forall x [P(x) \Longrightarrow Q(x)]$
      P2: $\forall x [S(x) \Longrightarrow R(x)]$
      P3: $\neg \exists x [Q(x) \wedge R(x)]$
      C: $\forall x [S(x) \Longrightarrow \bar{P}(x)]$

(b)   P1: $\forall x [\bar{Q}(x) \Longrightarrow P(x)]$
      P2: $\neg \exists x [R(x) \wedge Q(x)]$
      P3: $\neg \exists x [S(x) \wedge P(x)]$
      C: $\neg \exists x [R(x) \wedge S(x)]$

(c)   P1: $P(d)$
      P2: $\neg \exists x [R(x) \wedge \bar{Q}(x)]$
      P3: $\neg \exists x [Q(x) \wedge P(x)]$
      C: $\bar{R}(d)$

(d)   P1: $\forall x [P(x) \Longrightarrow Q(x)]$
      P2: $\forall x [S(x) \Longrightarrow \bar{R}(x)]$
      P3: $\forall x [\bar{P}(x) \Longrightarrow S(x)]$
      C: $\forall x [R(x) \Longrightarrow Q(x)]$

(e)   C: $\forall x [P(x) \Longrightarrow Q] \Longleftrightarrow [\exists x P(x) \Longrightarrow Q]$

(f)   C: $\forall x \exists y [P(x) \vee Q(y)] \Longleftrightarrow (\forall x\ P(x) \vee \exists y\ Q(y))$

(g)   C: $\neg \exists x \forall y [P(y,x) \Longleftrightarrow \bar{P}(y,y)]$

(h)   P1: $\forall x [P(x) \Longrightarrow Q(x) \vee R(x)]$
      P2: $\forall x [P(x) \wedge S(x)]$
      P3: $\neg [Q(a) \wedge S(a)]$
      C: $P(a) \Longrightarrow R(a)$

(i)   P1: $\exists x [P(x) \wedge \forall y (P(y) \wedge R(y) \Longrightarrow \neg G(x,y))]$
      P2: $\forall x [P(x) \Longrightarrow R(x)]$
      C: $\exists x [P(x) \wedge \forall y (P(y) \Longrightarrow \neg G(x,y))]$

(j)   P1: $\forall x \exists y [P(x,y) \wedge S(x,y)]$
        P2: $\forall x \forall y [P(x,y) \Longrightarrow R(x,y)]$
        C: $\forall x \exists y [R(x,y) \wedge S(x,y)]$

(k)   P1: $\exists x [P(x) \wedge \forall y (R(y) \wedge S(x,y) \Longrightarrow Z(x,y))]$
        P2: $\forall x [P(x) \Longrightarrow \exists y (R(y) \wedge \bar{U}(x,y) \wedge T(x,y))]$
        P3: $\forall x \forall y [P(x) \wedge R(y) \wedge T(x,y) \Longrightarrow S(x,y)]$
        C: $\exists x \exists y [P(x) \wedge R(y) \wedge Z(x,y) \wedge \bar{U}(x,y)]$

(l)   P1: $\forall x \forall y [P(x) \wedge Q(y) \wedge S(y) \Longrightarrow \bar{R}(x,y)]$
        P2: $\exists x \forall y [P(x) \wedge (Q(y) \wedge \bar{U}(x,y) \Longrightarrow \bar{T}(x,y))]$
        P3: $\forall x \exists y [P(x) \Longrightarrow Q(y) \wedge S(y) \wedge T(x,y)]$
        C: $\exists x \exists y [P(x) \wedge Q(y) \wedge U(x,y) \wedge \bar{R}(x,y)]$

(m)   P1: $\forall x \forall y [P(x,y) \Longrightarrow Q(x,y)]$
        P2: $\exists x \forall y [R(x,y) \Longrightarrow P(x,y)]$
        C: $\exists x \forall y [R(x,y) \Longrightarrow Q(x,y)]$

(n)   P1: $\forall x \forall y [P(x,y) \Longrightarrow Q(x,y)]$
        P2: $\forall x \forall y [R(x,y) \Longrightarrow S(x,y)]$
        C: $\forall x \forall y [P(x,y) \wedge R(x,y) \Longrightarrow Q(x,y) \wedge S(x,y)]$

(o)   P1: $\exists x \forall y [P(x) \wedge (Q(y) \Longrightarrow R(x,y))]$
        P2: $\forall x \forall y [P(x) \wedge S(y) \Longrightarrow \bar{R}(x,y)]$
        C: $\forall x [Q(x) \Longrightarrow \bar{S}(x)]$

(p)   P1: $\forall x \forall y [P(x) \wedge Q(y) \Longrightarrow R(x,y)]$
        P2: $\exists x \forall y [P(x) \wedge S(x,y) \Longrightarrow Q(y)]$
        P3: $\forall x \exists y [P(x) \wedge \bar{R}(x,y) \wedge T(x,y)]$
        C: $\exists x \exists y [P(x) \wedge \bar{S}(x,y) \wedge T(x,y)]$

(q)   P1: $\forall x \forall y [P(x) \vee Q(y) \Longrightarrow R(x,y)]$
        P2: $\forall x \forall y [P(x) \vee S(y) \Longrightarrow \bar{R}(x,y)]$
        C: $\forall x [S(x) \Longrightarrow \bar{Q}(x)]$

(r)   P1: $\forall x \forall y [P(x,y) \Longrightarrow R(x,y)]$
        P2: $\forall x \forall y [S(x,y) \Longrightarrow T(x,y)]$
        P3: $\forall x \exists y [P(x,y) \wedge S(x,y)]$
        C: $\forall x \exists y [R(x,y) \wedge T(x,y)]$

(s)   P1: $\forall x [P(x) \Longrightarrow \forall y (Q(y) \Longrightarrow R(x,y))]$
        P2: $\forall x [P(x) \Longrightarrow \forall y [S(y) \Longrightarrow \bar{R}(x,y)]]$
        C: $\forall x [P(x) \wedge Q(x) \Longrightarrow \bar{S}(x)]$

21. [Lewis Caroll] Obtain FOL formulas corresponding to each of the following premises and conclusions. Clearly state what the universe of discourse is and what each of the predicates stands for. For each set of premises and conclusion, show that the conclusion follows from the given premises. Note that in some cases it may be necessary to explicitly include some of the implied premises. For example, in (a) the premise "all babies are persons" is implied. In (h), the premise "all computer romances are books" is implied.

(a)   P1: Babies are illogical.
      P2: Nobody who can manage a crocodile is despised.
      P3: Illogical persons are despised.
       C: Babies cannot manage crocodiles.

(b)   P1: Everyone who is sane can do logic.
      P2: No lunatics are fit to serve on a jury.
      P3: None of your sons can do logic.
       C: None of your sons is fit to serve on a jury.

(c)   P1: Showy talkers think too much of themselves.
      P2: No really well informed people are bad company.
      P3: People who think too much of themselves are not good company.
       C: Showy talkers are not really well informed.

(d)   P1: Colored flowers are always scented.
      P2: I dislike flowers that are not grown outdoors.
      P3: No flowers grown outdoors are colorless.
       C: No scentless flowers please me.

(e)   P1: All deserts are nice.
      P2: This dish is a desert.
      P3: No nice things are wholesome.
       C:This dish is not wholesome.

(f)   P1: No one who is uneducated buys the New York Times.
      P2: No monkeys can read.
      P3: Those who cannot read are uneducated.
       C: No monkeys buy the New York times.

(g)   P1: All humming birds are richly colored.
      P2: No large birds live on honey.
      P3: Birds that do not live on honey are dull in color.
       C: All humming birds are small.

(h)   P1: The only books I do not recommend reading are bad.
       P2: The bound books are all well written.
       P3: All the computer romances are good.
       P4: I do not recommend you to read any of the unbound books.
        C: All the computer romances are well written.

(i)   P1: No interesting poems are unpopular among people of real
           taste.
       P2: No modern poetry is free from affectation.
       P3: All your poems are on the subject of soap bubbles.
       P4: No affected poetry is popular among people of real taste.
       P5: No ancient poem is on the subject of soap bubbles.
        C: All your poems are uninteresting.

(j)   P1: All writers who understand human nature are clever.
       P2: No one is a true poet unless he can stir the hearts of men.
       P3: Shakespeare wrote Hamlet.
       P4: No writer who does not understand human nature can stir the
           hearts of men.
       P5: None but a true poet could have written Hamlet.
        C: Shakespeare was clever.

(k)   P1: When I work a logic example without grumbling, you may be
           sure that it is one that I understand.
       P2: These problems are not arranged in proper order, like the
           ones I am used to.
       P3: No easy problem makes my headache.
       P4: I can't understand problems that aren't arranged in proper
           order, like those I am used to.
       P5: I never grumble at a problem unless it gives me a headache.
        C: These problems are difficult.

(l)   P1: No husband who is always giving his wife new dresses can
           be an honest man.
       P2: A methodical husband always comes home for his supper.
       P3: No one who leaves his hat on the TV can be a man who is kept
           in proper order by his wife.
       P4: A good husband is always giving his wife new dresses.
       P5: No husband can fail to be honest if his wife does not keep him
           in proper order.
       P6: An unmethodical husband always leaves his hat on the TV.
        C: A good husband always comes home for supper.

(m)   P1: The only animals in this house are cats.

P2: Every animal that loves to gaze at the moon is suitable for a pet.

P3: When I detest an animal, I avoid it.

P4: No animals are carniverous unless they prowl at night.

P5: No cat fails to kill mice.

P6: No animals take to me except what are in the house.

P7: Kangaroos are not suitable as pets.

P8: None but carnivora kill mice.

P9: I detest animals that do not take to me.

P10: Animals that prowl at night always love to gaze at the moon.

C: I always avoid a kangaroo.

CHAPTER 2

CONSTRUCTIVE PROOFS
MATHEMATICAL INDUCTION
AND
PROGRAM CORRECTNESS

2.1 CONSTRUCTIVE PROOFS

In Chapter 1, we studied several important proof methods. This list of proof methods is by no means exhaustive. Two other proof methods will be studied in this chapter. The first of these is: proof by construction. Suppose we wish to show that there is a natural number greater than 2 that is a prime number. In a constructive proof of this statement we are required to actually produce a natural number greater than 2 and demonstrate that it is prime. A possible constructive proof of the existence of a prime number greater than 2 is:

> Consider the natural number 5. Its only natural number divisors are 1 and 5. So, 5 is a prime. Also, 5>2. So, there exists a natural number greater than 2 that is a prime number.

Several alternate constructive proofs can be obtained by using 3, 7, 11, 13, 17, etc. in place of 5 in the above proof. By contrast, in a non-constructive proof of the existence of a prime number greater than 2 we would not actually produce such a number. For example, consider the following proof:

> Every natural number can be represented as the product of prime numbers. This implies that if there is no prime number greater than 2, then all natural numbers are of the form 2^i for i a natural number. However, we know that there exist natural numbers that are not a power of 2 (for example, 81). Therefore there must exist a prime number greater than 2.

Observe that the second sentence in the above proof is of the form $P \implies Q$ and the third is $\overline{Q}$. The fourth sentence is arrived at using modus tollens. Another point to note is that this proof does not directly give us a prime number greater than 2 nor does it tell us how to find one. Nonetheless, it does establish that a prime number greater than 2 exists.

Example 2.1: In this example, we shall show that the predicate $P(n)$ defined as:

> $P(n)$: n cents of postage can be made using only 2 cent and 3 cent stamps

is true for all natural numbers n, $n \geq 2$. This can be stated as an FOL wff using two additional predicates:

> $Q(n)$: $n < 2$
> $R(n)$: n is a natural number.

We are interested in showing that the wff:

$$(2.1) \quad \forall n (\bar{R}(n) \lor Q(n) \lor P(n))$$

is true when P, Q and R are the predicates defined above. Unlike our previous proofs where we did not have to use the definition of any of the predicates, here we shall have to rely heavily on what P, Q, and R stand for. This is so because the wff (2.1) is not true for all predicate definitions P, Q and R. (2.1) can be proved by considering two cases:

(a) n is either not a natural number or it is less than 2.

and

(b) n is a natural number that is greater than or equal to 2.

     For case (a), one immediately sees that $\bar{R}(n) \lor Q(n)$ is true. Hence, $\bar{R}(n) \lor Q(n) \lor P(n)$ is true. So, we are left with case (b). For this case we shall show that $P(n)$ is true. This proof will be a proof by construction. We shall actually demonstrate how to obtain n cents postage using only 2 cent and 3 cent stamps.

     Divide n by 3. Let q be the quotient and r the remainder. I.e., $n = 3q + r$, and $q \geq 0$, and $0 \leq r < 3$. If $r = 0$, then we need use only q 3 cent stamps. If $r = 1$, then q must be greater than or equal to 1 (because if $q = 0$ then $n = 1$. But, $n \geq 2$). Now, we can use $q - 1$ 3 cent stamps and two 2 cent stamps. If $r = 2$ then q 3 cent and one 2 cent stamp can be used. This takes care of all possible values of r. Hence, $P(n)$ is true for all natural numbers n, $n \geq 2$.

     This proof is readily seen to be a constructive proof. We have actually demonstrated how to make n cents worth of postage, $n \geq 2$, using 2 cent and 3 cent stamps. □

Example 2.2: We are given m processors P_1, P_2, ..., P_m and n tasks T_1, T_2,

...,T_n. Task T_i requires t_i, $t_i \geq 0$, units of processing, $1 \leq i \leq n$. It is permissible to process a portion of a task on one processor, another portion on another processor, etc. However, at any given time, a task can be processed by at most one processor. A *schedule* is an assignment of tasks to processors.

For example, consider the five tasks with times $(t_1, t_2, \ldots, t_5) = (7, 3, 9, 4, 1)$. Figure 2.1 gives some possible schedules for the case when $m = 2$.

The *finish time* of a schedule is the earliest time at which the processing of all the tasks has been completed. The schedules of Figures 2.1(a) and (b) have a finish time of 12 while that of Figure 2.1(c) has a finish time of 16.

For this scheduling problem we wish to show that there exists a schedule with finish time such that

$$f = \max \{\max_{1 \leq i \leq n} \{t_i\}, \frac{1}{m} \sum_{1}^{n} t_i\}$$

It is not too difficult to see that there can be no schedule with a finish time less than f. In a constructive proof of the existence of a schedule with finish time equal to f, we shall have to actually show how such a schedule may be built.

The tasks to be scheduled are considered one by one. The first task is assigned to P_1 from time 0 to time t_1 (note that $t_1 \leq f$). Task T_2 is now assigned to P_1 from t_1 to $\min\{f, t_1 + t_2\}$. If all of T_2 is assigned to P_1, then we consider task T_3 etc. until we either come to a task T_i such that all of it does not get assigned to P_1 or we have exhausted all the tasks. In the latter case, we are done and a schedule of length f has been constructed. In the former case, the portion of T_i not assigned so far is assigned to P_2 starting at time zero. This assignment cannot overlap the assignment (if any) of T_i on P_1 as $t_i \leq f$. Now, we proceed to assign tasks T_{i+1}, T_{i+2}, ... to P_2 until we reach a task that cannot be completed on P_2 by time f or all tasks are exhausted. In the former case, we start with P_3 and in the latter we are done with building the schedule. It should be clear that by proceeding in this way, an m processor schedule with finish time f will be obtained. □

Example 2.3: The sum $P(x) = \sum_{i=0}^{n} a_i x^i$, $a_n \neq 0$ is a *polynomial* in the variable x. The *degree* of $P(x)$ is n and the a_is are called the *coefficients*. Let (x_i, y_i), $1 \leq i \leq n$, be n pairs of points such that $x_i \neq x_j$, $i \neq j$. In this example, we shall prove that there exists a unique polynomial $P(x)$ of degree at most $n-1$ such that $P(x_i) = y_i$, $1 \leq i \leq n$. This polynomial is called the *interpolating polynomial*.

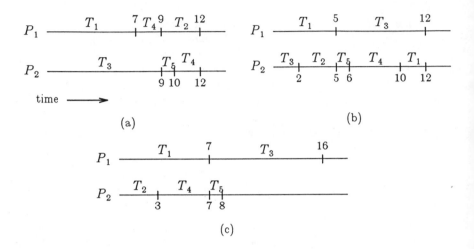

Figure 2.1 Possible schedules for Example 2.2.

First, let us prove that there is at most one polynomial that has the property given above. Let $P(x)$ and $Q(x)$ be two polynomials with the desired properties. It is easy to see that $R(x) = P(x)-Q(x)$ is a polynomial of degree at most $n-1$ and $R(x_i) = 0$, $1 \le i \le n$. It is well known that a polynomial of degree d, has at most d distinct roots (x is a *root* of $P(x)$ iff $P(x) = 0$) unless the polynomial identically equals zero. We have just seen that $R(x)$ has n distinct roots and has degree less than n. So, $R(x)$ must identically equal zero. I.e., $R(x) = 0$ for all x. Hence, $P(x) = Q(x)$ and there can be at most one interpolating polynomial.

Next, we provide a constructive proof that there exists at least one interpolating polynomial. This together with the proof of the preceding paragraph will imply that for every set of n pairs of points there exists a unique interpolating polynomial. Let $P(x)$ be as given below:

$$P(x) = \sum_{i \ne j} \left(\prod_{1 \le i \le n \atop 1 \le j \le n} \frac{x-x_j}{x_i-x_j} \right) y_i$$

One may readily verify that $P(x)$ is a polynomial of degree at most $n-1$ and that $P(x_i) = y_i$, $1 \le i \le n$. $\square$

2.2 MATHEMATICAL INDUCTION

Mathematical induction is an exceptionally powerful proof method. It can be used to establish that a predicate P is true whenever its parameters are assigned values from some given domains. Let us consider a single parameter predicate $P(n)$ and a domain (or universe) D such that D is the set of all values assignable to n. We shall assume that the members of D are labeled a, $a+1$, $a+2$, ... Henceforth, we shall simply refer to an element of D by its label (i.e., a or $a+1$ or $a+2$ or ...). In the next chapter, we shall see that not all domains can be labeled in this way. We shall use the notation

$$\forall x \in D \ P(x)$$

to mean "for all x in D, $P(x)$". Mathematical induction relies on the following inference rules:

MI1. $\{P(a), \ \forall x, x+1 \in D \ (P(x) \Rightarrow P(x+1))\} \models \forall x \in D \ P(x)$

MI2. $\{P(a) \bigwedge P(a+1), \ \forall x, x+1, x+2 \in D \ (P(x) \bigwedge P(x+1) \Rightarrow P(x+2))\} \models \forall x \in D \ P(x)$

MI3. $\{P(a) \bigwedge P(a+1) \bigwedge P(a+2), \ \forall x, x+1, x+2, x+3 \in D \ (P(x) \bigwedge P(x+1) \bigwedge P(x+2) \Rightarrow P(x+3))\} \models \forall x \in D \ P(x)$

.
.
.
.

Inference rules MI4, MI5, ..., etc. are defined in an analogous way. The statement of the above rules can be simplified somewhat by realizing that x, $x+1$, ..., $x+b$ are all in D iff x and $x+b$ are. So, for instance, $\forall x, x+1, x+2, x+3 \in D$ can also be written as $\forall x, x+3 \in D$. Before attempting to use the above inference rules to prove statements of the form $\forall x \in D \ P(x)$, let us establish their correctness.

Theorem 2.1: Inference rule MI1 is correct.

Proof: The wff associated with MI1 is:

$$[P(a) \bigwedge \forall x, x+1 \in D(P(x) \Rightarrow P(x+1))] \Rightarrow \forall x \in D \ P(x)$$

We need to show that this wff is a theorem. We shall provide a direct proof of this fact. Assume that $[P(a) \bigwedge \forall x, x+1 \in D(P(x) \Rightarrow P(x+1))]$ is true. Call this assumption, assumption 1 or A1.

Now, we need to show that $\forall x \in D \; P(x)$ follows from A1. The truth of $\forall x \in D \; P(x)$ will be established by contradiction. As in all proofs by contradiction, we begin with the assumption (A2) that $\forall x \in D \; P(x)$ is false. From equivalence FE7 (Figure 1.10), we see that A2 is equivalent to $\exists x \in D \; \bar{P}(x)$. So, $P(x)$ is false for some x in D. Because of the nature of D, there must therefore exist a least x in D for which $P(x)$ is false. Let this least x be c.

There are two possibilities for c. Either $c=a$ or $c>a$. If $c=a$, then from A1 (using simplification and replacing a by c) we obtain $P(c)$. But, by choice of c, $\bar{P}(c)$. So, we have a contradiction.

If $c>a$, then from the choice of c, it follows that $P(c-1)$ is true. Using simplification and universal instantiation on A1, we obtain $P(c-1) \Longrightarrow P(c)$. The inference rule modus ponens now yields $P(c)$. But, c was chosen such that $\bar{P}(c)$ is true. Once again, we have a contradiction.

Hence, there is no x, $x \in D$ such that $P(x)$ is false. I.e., $\forall x \in D \; P(x)$. This concludes the proof. $\square$

As we shall see in subsequent examples, MI1 is not adequate to prove the correctness of all wffs of the form $\forall x \in D \; P(x)$. MI2, MI3, ... will enable us to prove additional wffs of this form. Inference rules MI2, MI3, ... may be proved correct in essentially the same way as we proved MI1.

A proof by mathematical induction (abbreviated: proof by induction) can be divided into two distinct parts. If inference rule MIj is being used, then the first part of the proof establishes the truth of $P(a) \wedge P(a+1) \wedge \ldots \wedge P(a+j-1)$. This part of the proof is called the *induction base* (IB). In the second part of the proof it is shown that $\forall x, x+1, \ldots, x+j \in D \; (P(x) \wedge P(x+1) \wedge \ldots \wedge P(x+j-1) \Longrightarrow P(x+j))$. From parts one and two of the proof and the inference rule MIj, the truth of $\forall x \in D \; P(x)$ is inferred.

The second part of the proof is often a direct proof by generalization. We begin by assuming that $P(x) \wedge P(x+1) \wedge \ldots \wedge P(x+j-1)$ is true for an arbitrary x such that $x, x+1, \ldots, x+j$ are in D. This is called the *induction hypothesis* (IH). Next, it is shown that the truth of $P(x+j)$ follows from the induction hypothesis. This part of the proof is referred to as the *induction step* (IS).

Example 2.4: We shall show that $\sum_{i=0}^{n} i = n(n+1)/2$ for all natural numbers n. Inference rule MI1 will be used. The predicate P in this case is:

$$P(n): \sum_{i=0}^{n} i = n(n+1)/2$$

and $D = (0,1,2,...)$ is the ordered set of natural numbers. In the induction base, we have to prove that $P(0)$ is true. The induction hypothesis will be $\sum_{i=0}^{m} i = m(m+1)/2$ for some arbitrary natural number m. In the induction step we shall show that the induction hypothesis implies the truth of $P(m+1)$. I.e., $\sum_{i=0}^{m+1} i = (m+1)(m+2)/2$. The complete proof is given below. We use LHS (left hand side) to denote $\sum_{i=0}^{n} i$ and RHS (right hand side) to denote $n(n+1)/2$.

Proof:

Induction Base: When $n=0$, $\sum_{i=0}^{n} i = 0$ and $n(n+1)/2 = 0$. So, $P(0)$ is true.

Induction Hypothesis: Let m be any arbitrary natural number. Assume that $\sum_{i=0}^{m} i = m(m+1)/2$.

Induction Step: When $n = m+1$, we get

$$\text{LHS} = \sum_{i=0}^{m+1} i = m+1+ \sum_{i=0}^{m} i$$
$$= m+1+m(m+1)/2 \text{ (from the IH)}$$
$$= (m+1)(m+2)/2$$
$$= \text{RHS}$$

Note that since m is an arbitrary natural number, the induction hypothesis and induction step together provide a direct proof of $\forall\, m \in N\ (P(m) \implies P(m+1))$ where N denotes the set of all natural numbers. □

Example 2.5: Inference rule MI1 can also be used to show that $\sum_{i=0}^{n} a^i = (a^{n+1}-1)/(a-1)$ for all natural numbers n and $a \neq 1$. The proof is:

Induction Base: When $n = 0$, $\sum_{i=0}^{n} a^i = 1$ and $(a^{n+1}-1)/(a-1) = 1$.

Induction Hypothesis: Let m be any arbitrary natural number. Assume that $\sum_{i=0}^{m} a^i = (a^{m+1}-1)/(a-1)$.

Induction Step: When $n = m+1$,
$$\text{LHS} = \sum_{i=0}^{m+1} a^i = a^{m+1} + \sum_{i=0}^{m} a^i$$

$$= a^{m+1} + (a^{m+1}-1)/(a-1) \text{ (from the IH)}$$

$$= (a^{m+2}-a^{m+1}+a^{m+1}-1)/(a-1)$$

$$= (a^{m+2}-1)/(a-1)$$
$$= \text{RHS} \quad \square$$

Example 2.6: The Fibonacci numbers F_0, F_1, ... are defined by the equations:

$$F_0 = 0, F_1 = 1, \text{ and } F_{n+2} = F_{n+1}+F_n, n \geq 0$$

The first few Fibonacci numbers are therefore $F_0 = 0$, $F_1 = 1$, $F_2=1$, $F_3=2$. We wish to show that

$$F_n = \frac{1}{\sqrt{5}}\left(\frac{1+\sqrt{5}}{2}\right)^n - \frac{1}{\sqrt{5}}\left(\frac{1-\sqrt{5}}{2}\right)^n$$

for all natural numbers n, $n \geq 0$. This time we shall use the inference rule MI2.

Proof:
Induction Base: When $n = 0$, the right hand side of the formula for F_n evaluates to 0. Also, $F_0 = 0$. When $n = 1$, the formula evaluates to 1 which is equal to F_1. So, the formula for F_n is valid when $n = 0$ and $n = 1$.

Induction Hypothesis: Let m be any arbitrary natural number. Assume that the formula for F_n is valid when $n = m$ and when $n = m+1$.

Induction Step: (We shall show the validity of the formula for F_n when $n = m+2$). We shall use the following symbols:

$$a = \frac{1+\sqrt{5}}{2} \quad \text{and} \quad b = \frac{1-\sqrt{5}}{2}.$$

When $n = m+2$, we obtain:

(1) $F_{m+2} = F_{m+1} + F_m$ (from the definition of F)

$$(2) \qquad = \frac{1}{\sqrt{5}}(a^{m+1}-b^{m+1}+a^m-b^m) \text{ (from the IH)}$$

$$= \frac{1}{\sqrt{5}}(a^m(a+1)-b^m(b+1))$$

$$= \frac{1}{\sqrt{5}}(a^{m+2}-b^{m+2}) \quad (a^2 = a+1 \text{ and } b^2 = b+1)$$

= RHS

From the proof of the induction step it becomes apparent why MI2 had to be used. In going from equality (1) to equality (2), we used a value for F_{m+1} and another one for F_m. If MI1 were being used, then the induction hypothesis would contain an assumption for only one F_n. In this case the transition from (1) to (2) would be invalid. Of course, MI3, MI4, etc. could also be used. Using an MIj for higher j than necessary is of no advantage as more things have to be proved in the induction base. □

Example 2.7: Consider the sequence $N_0, N_1, N_2, \ldots$ defined as:

$$N_0 = 0, N_1 = 1, \text{ and } N_{h+2} = N_{h+1} + N_h + 1$$

for all natural numbers h, $h \geq 0$. We wish to show that $N_h = F_{h+2} - 1$, $h \geq 0$. The F_is are the Fibonacci numbers of Example 2.6. The validity of this relationship between the N_is and the F_is can be established using inference rule MI2.

Proof:
Induction Base: When $h = 0$, $N_h = 0$ and $F_{h+2} - 1 = F_1 + F_0 - 1 = 0$. When $h = 1$, $N_1 = 1$ and $F_3 - 1 = 2 - 1 = 1$.

Induction Hypothesis: Let m be any arbitrary natural number. Assume that $N_h = F_{h+2} - 1$ when $h = m$ as well as when $h = m+1$.

Induction Step: $N_{m+2} = N_{m+1} + N_m + 1$ (definition of N)
$$= F_{m+3} - 1 + F_{m+2} - 1 + 1 \text{ (from the IH)}$$
$$= F_{m+3} + F_{m+2} - 1$$
$$= F_{m+4} - 1 \text{ (definition of } F) \;\; \square$$

Once again, one should note that the induction step cannot be carried through using inference rule MI1[1]. □

Example 2.8: Suppose that $f(0) = 0$, $f(1) = 1$, and $f(n) = 5f(n-1) - 6f(n-2) + 4*3^n$, $n > 2$. We may show that $f(n) = 35*2^n - 35*3^n + 12n3^n$ for $n \geq 0$ by induction on n. Since $f(n)$ is defined in terms of $f(n-1)$ and $f(n-2)$, the inference rule MI2 will be used.

Proof:
Induction Base: When $n = 0$, $35*2^n - 35*3^n + 12n3^n = 0 = f(0)$. When $n = 1$, $35*2^n - 35*3^n + 12n3^n = 1 = f(1)$. So, the formula is correct for $n = 0$ and $n = 1$.

---

1. One should note that, strictly speaking, all forms of mathematical induction are the same as all the inference rules can be derived from MI1.

Induction Hypothesis: Let m be an arbitrary natural number. Assume that the formula is correct for $n = m$ and $n = m+1$.

Induction Step: When $n = m+2$, we get:

$$f(m+2) = 5f(m+1) - 6f(m) + 4*3^{m+2} \quad \text{(definition of } f)$$

$$= 175*2^{m+1} - 175*3^{m+1} + 60(m+1)3^{m+1}$$

$$-210*2^m + 210*3^m - 72m3^m$$

$$+4*3^{m+2} \quad \text{(from the IH)}$$

$$= 350*2^m - 525*3^m + 180m3^m + 180*3^m$$

$$-210*2^m + 210*3^m - 72m3^m$$

$$+36*3^m$$

$$= 140*2^m - 315*3^m + 108m3^m + 216*3^m$$

$$= 35*2^{m+2} - 35*3^{m+2} + 12(m+2)3^{m+2}$$

$$= \text{RHS of formula when } n = m+2 \quad \square$$

The inference rules for mathematical induction are readily generalized to permit different types of proofs by induction. Suppose that the domain D of the parameter x in the predicate $P(x)$ is $\{k, k-1, k-2, ...\}$. Then, we may obtain the following inference rules which are analogous to the rules MI1, MI2, etc.

MD1. $\{P(k), \forall x, x-1 \in D \ [P(x) \Longrightarrow P(x-1)]\} \models \forall x \in D \ P(x)$

MD2. $\{P(k) \wedge P(k-1), \forall x, x-1, x-2 \in D \ [P(x) \wedge P(x-1) \Longrightarrow P(x-2)]\} \models \forall x \in D \ P(x)$

MD3. $\{P(k) \wedge P(k-1) \wedge P(k-2), \forall x, x-1, x-2, x-3 \in D \ [P(x) \wedge P(x-1) \wedge P(x-2) \Longrightarrow P(x-3)]\} \models \forall x \in D \ P(x)$

.

.

.

Example 2.9: Consider the sequence $G_{100}, G_{99}, ..., G_0$ defined as below:

$$G_{100} = 9900, \ G_i = G_{i+1} - 2i, \ 0 \le i < 100$$

We wish to show that $G_i = i(i-1)$, $0 \le i \le 100$. As we shall see, this can be done using MD1.

Proof:

Induction Base: When $i = 100$, $i(i-1) = 9900 = G_{100}$.

Induction Hypothesis: Let m be an arbitrary natural number such that $0 < m \le 100$. Assume that $G_m = m(m-1)$.

Induction Step:
$$
\begin{aligned}
G_{m-1} &= G_m - 2(m-1) \quad \text{(definition of } G) \\
&= m(m-1) - 2(m-1) \quad \text{(IH)} \\
&= (m-1)(m-2) \\
&= \text{RHS of formula when } i = m-1 \quad \square
\end{aligned}
$$

The principle of mathematical induction can be extended to predicates with more than one parameter. Suppose that $P(x_1, x_2, \ldots, x_k)$ is a predicate and that D_1, D_2, ..., D_k are the domains of x_1, x_2, ..., x_k respectively. Assume that $D_i = \{a_i, a_i+1, \ldots\}$. We can prove that $P(x_1, x_2, \ldots, x_k)$ is true for all $x_i \in D_i$, $1 \le i \le k$ by using MI1 (e.g.) and regarding P as if it were only a one parameter predicate (say $Q(x_1)$). In this case the proof would take the form:

Induction Base: Show that $P(a_1, x_2, x_3, \ldots, x_k)$ is true for all x_i in D_i, $2 \le i \le k$.

Induction Hypothesis: Let m be any arbitrary element of D_1 such that $m+1$ is also in D_1. Assume that $P(m, x_2, \ldots, x_k)$ is true for all x_i in D_i, $2 \le i \le k$.

Induction Step: Show that $P(m+1, x_2, x_3, \ldots, x_k)$ is true for all $x_i \in D_i$, $2 \le i \le k$.

The general form of a proof by induction for a k parameter predicate (when the proof is based on MI1) takes the form:

Induction Base(1): Show that $P(a_1, x_2, x_3, \ldots, x_k)$ is true for all x_i in D_i, $2 \le i \le k$.

Induction Hypothesis(1): Let m_1 be any arbitrary element in D_1 such that m_1+1 is also in D_1. Assume that $P(m_1, x_2, \ldots, x_k)$ is true for all x_i in D_i, $2 \le i \le k$.

Induction Step(1): Show that $P(m_1+1, x_2, \ldots, x_k)$ is true for all x_i in D_i, $2 \le i \le k$.

Induction Base(2): Show that $P(m_1+1, a_2, x_3, \ldots, x_k)$ is true for all x_i in D_i, $3 \le i \le k$.

Induction Hypothesis(2): Let m_2 be any arbitrary element of D_2 such that m_2+1 is also in D_2. Assume that $P(m_1+1, m_2, \ldots, x_k)$ is true for all $x_i \in$

D_i, $3 \leq i \leq k$.

Induction Step(2): Show that $P(m_1+1, m_2+1, x_3, \ldots, x_k)$ is true for all $x_i \in D_i$, $3 \leq i \leq k$.

> **Induction Base(3):** Show that $P(m_1+1, m_2+1, a_3, x_4, \ldots, x_k)$ is true for all x_i in D_i, $4 \leq i \leq k$.
>
> **Induction Hypothesis(3):** Let m_3 be any arbitrary element of D_3 such that m_3+1 is also in D_3. Assume that $P(m_1+1, m_2+1, m_3, x_4, \ldots, x_k)$ is true for all x_i in D_i, $4 \leq i \leq k$.
>
> **Induction Step(3):** Show that $P(m_1+1, m_2+1, m_3+1, x_4, \ldots, x_k)$ is true for all x_i in D_i, $4 \leq i \leq k$.
>
> .
>
> .
>
> .
>
> > **Induction Base(k):** *Show that* $P(m_1+1, m_2+1, \ldots, m_{k-1}+1, a_k)$ is true.
> >
> > **Induction Hypothesis(k):** *Let* m_k *be any arbitrary element of* D_k *such that* m_k+1 *is also in* D_k. *Assume that* $P(m_1+1, m_2+1, \ldots, m_k)$ *is true.*
> >
> > **Induction Step(k):** *Show that* $P(m_1+1, m_2+1, \ldots, m_k+1)$ *is true.*
> > □

It is not essential for an induction proof of $P(x_1, x_2, \ldots, x_k)$ to contain all k levels of induction as given above. If it is possible to prove the induction step i for some i, $1 \leq i < k$ directly (say by generalization), then levels $i+1$, $i+2$, ..., k will not be present in the proof. If i levels of induction are present in a proof, the proof is an i level induction proof. When $i=2$, the proof is a double induction; when $i=3$ it is a triple induction; and so on.

Example 2.10: Let $S(n,m) = \sum\limits_{j=1}^{n} \sum\limits_{i=1}^{m} i * j$ for every pair of natural numbers m and n. We wish to show that $S(n,m) = n(n+1)m(m+1)/4$. A simple noninductive proof is:

$$S(n,m) = \sum_{j=1}^{n} \sum_{i=1}^{m} i * j$$

$$= \sum_{j=1}^{n} j * \sum_{i=1}^{m} i$$

$$= \sum_{j=1}^{n} jm(m+1)/2 \quad \text{(Example 2.4)}$$

$$= n(n+1)m(m+1)/4$$

We shall use this simple example to illustrate the mechanics of a two level inductive proof.

Proof: (by induction on n and m)

Induction Base (1): When $n = 0$, $S(n,m) = \sum_{j=1}^{0} \sum_{i=1}^{m} i*j = 0$ and $n(n+1)m(m+1)/4 = 0$.

Induction Hypothesis(1): Let p be an arbitrary natural number. Assume that $S(p,m) = p(p+1)m(m+1)/4$ for every natural number m.

Induction Step(1): We need to show that $S(p+1,m) = (p+1)(p+2)m(m+1)/4$ for every natural number m. We shall do this by induction on m.

Induction Base(2): When $m = 0$, $S(p+1,m) = 0$ and $(p+1)(p+2)m(m+1)/4 = 0$.

Induction Hypothesis(2): Let q be any arbitrary natural number. Assume that $S(p+1,q) = (p+1)(p+2)q(q+1)/4$.

Induction Step(2):

$$S(p+1,q+1) = \sum_{j=1}^{p+1} \sum_{i=1}^{q+1} i*j$$

$$= S(p+1,q) + \sum_{j=1}^{p+1} (q+1)j$$

$$= (p+1)(p+2)q(q+1)/4 + (p+1)(p+2)(q+1)/2$$

(From IH(2) and Example 2.4)

$$= (p+1)(p+2)(q(q+1) + 2(q+1))/4$$

$$= (p+1)(p+2)(q+1)(q+2)/4 \quad \square$$

Since the proof of IS(2) does not use IH(1), it is possible to carry out the proof by using generalization on n and induction only on m as below:

Proof: (by generalization on n and induction on m)

Induction Base: When $m = 0$, $S(n,m) = 0$ and $n(n+1)m(m+1)/4 = 0$.

Induction Hypothesis: Let q be an arbitrary natural number. Assume that $S(n,q) = n(n+1)q(q+1)/4$ for every natural number n.

Induction Step: $S(n,q+1) = \displaystyle\sum_{j=1}^{n} \sum_{i=1}^{q+1} i*j$

$$= S(n,q) + \sum_{j=1}^{n} (q+1)*j$$

$$= n(n+1)q(q+1)/4 + n(n+1)(q+1)/2$$

(From IH and Example 2.4)

$$= n(n+1)(q+1)(q+2)/4 \quad \square$$

Example 2.11: The Ackermann's function, $A(i,j)$, is defined as:

$$A(i,j) = \begin{cases} 2j & i=0 \\ 0 & i\geq 1 \text{ and } j=0 \\ 2 & i\geq 1 \text{ and } j=1 \\ A(i-1,A(i,j-1)) & i\geq 1 \text{ and } j\geq 2 \end{cases}$$

We wish to show that $A(i,j+1) > A(i,j)$ for all natural numbers i and j.

Proof (By induction on i and j):
Induction Base(1): When $i = 0$, $A(i,j) = 2j$. So, $A(0,j+1) > A(0,j)$ for all natural numbers j.

Induction Hypothesis(1): Let m be any arbitrary natural number. Assume that $A(m,j+1) > A(m,j)$ for all natural numbers j.

Induction Step(1): We need to show that $A(m+1,j+1) > A(m+1,j)$ for all j. We shall do this by induction on j.

> **Induction Base(2):** When $j = 0$, $A(m+1,j+1) = 2$ and $A(m+1,0) = 0$. So, $A(m+1,j+1) > A(m+1,j)$.

> **Induction Hypothesis(2):** Let r be any arbitrary natural number. Assume that $A(m+1,r+1) > A(m+1,r)$.

Induction Step(2): We shall show that $A(m+1,r+2) > A(m+1,r+1)$ by considering the two cases: (i) $r = 0$ and (ii) $r>0$. When $r = 0$, we need to show that $A(m+1,2) > A(m+1,1)$. From the definition of A, we see that $A(m+1,2) = A(m,A(m+1,1)) = A(m,2) = 4$ (see Exercise 6(a)). But, $A(m+1,1) = 2$.

Now, let's consider the case $r>0$. Once again, from the definition of A, we obtain the equality:

$$A(m+1,r+2) = A(m,A(m+1,r+1)).$$

From induction hypothesis (1), it follows that $A(m,x) > A(m,y)$ whenever $x>y$. From this and induction hypothesis (2), we conclude that $A(m,A(m+1,r+1)) > A(m,A(m+1,r))$. Since $r>0$, the definition of A yields the equality:

$$A(m+1,r+1) = A(m,A(m+1,r)).$$

Combining these equalities and inequalities, we obtain (for $r > 0$)

$$A(m+1, r+2) > A(m+1, r+1).$$

The proofs of cases (i) and (ii) complete the proof of induction step (2). This in turn completes the proof of induction step (1). $\square$

Weak Mathematical Induction[1]

Weak mathematical induction differs from the mathematical induction discussed so far (called strong mathematical induction) only in the nature of the induction hypothesis. In the induction hypotheses used so far we assumed that the predicate $P(x)$ is true for some fixed number of x's. For example, when using MI1, we assume that $P(x)$ is true for some arbitrary value $x = m$. We do not assume that $P(x)$ is also true for $x = a, a+1, \ldots, m-1$. Similarly, when MI2 is used, $P(x)$ is assumed true (in the induction hypothesis) for only two values of x i.e., m and $m+1$. When weak mathematical induction is used, $P(x)$ is assumed true for all values of x, $a \leq x \leq m$. In the induction step, it is proved that $P(m+1)$ is true. In the induction base, at least $P(a)$ is shown true. Depending upon the nature of the proof provided in the induction step, it may also be necessary to prove $P(a+1), P(a+2)$, etc. in the induction base.

Theorem 2.2: Strong mathematical induction is a valid proof method.

---

1. A word of caution regarding the terminology used is in order. In some texts, the use of the terms weak and strong mathematical induction is reversed from our use here. So, what we call weak induction here is called strong induction in some other texts.

Proof: In weak mathematical induction, the truth of $P(x)$ for all x in $D = \{a, a+1, a+2, ...\}$ is established by showing that:

(1) $P(a) \wedge \forall m, m+1 \in D \ (\forall x, a \leq x \leq m \ P(x) \Rightarrow P(m+1))$

is true. We shall proceed to show that if (1) is true, then there is no $j \in D$ for which $P(j)$ is false. The proof is by contradiction. Suppose there exists a $j \in D$ for which $P(j)$ is false. Since the elements of D are a, $a+1$, ..., the existence of a j for which $P(j)$ is false, implies the existence of a least j (say r) for which $P(r)$ is false. If $r = a$, then we have a contradiction with (1) as the truth of (1) implies that $P(a)$ is true. So, r must be greater than a and $P(a)$, ..., $P(r-1)$ are all true. Hence, $\forall x, a \leq x \leq r-1 \ P(x)$ is true. From (1) we can now conclude that $P(r)$ is also true. This contradicts the assumption that $P(r)$ is false. Hence, if (1) is true, there can be no $j \in D$ for which $P(j)$ is false. $\square$

The proof of Example 2.4 becomes a proof by weak mathematical induction if the induction hypothesis is changed to:

Assume that $\sum_{i=0}^{n} i = n(n+1)/2$ for $0 \leq n \leq m$ for some arbitrary natural number m.

No other changes in the proof are needed. Similarly, the proof of Example 2.7 becomes a proof by weak mathematical induction if we change the induction hypothesis to:

Assume that $N_h = F_{h+2} - 1$, $0 \leq h \leq m+1$ for some arbitrary m.

Note that it is still necessary to prove $N_0 = F_2 - 1$ and $N_1 = F_3 - 1$ in the induction base. If we had proved only $N_0 = F_2 - 1$ in the induction base, then we could not conclude $N_h = F_{h+2} - 1$ for all $h \geq 0$ as the case $h = 1$ does not follow from $N_0 = F_2 - 1$ and $\forall m (\forall h, 0 \leq h \leq m+1 \ (N_h = F_{h+2} - 1) \Rightarrow N_{m+2} = F_{m+4} - 1)$. To see this, observe that when $m = 0$ we can infer $N_2 = F_4 - 1$ from $N_0 = F_2 - 1$ and $N_1 = F_3 - 1$. But there is no natural number m that allows us to infer $N_1 = F_3 - 1$.

As pointed out in the proof of Theorem 2.2, the inference rule on which weak mathematical induction is based is:

WM1: $\{P(a), \forall m, m+1 \in D [\forall x, a \leq x \leq m \ P(x) \Rightarrow P(m+1)]\}$
$\models \forall x \in D \ P(x)$

Using a proof similar to that used in Theorem 2.2, one can obtain the following additional inference rules for weak mathematical induction on single parameter predicates:

WM2: $\{P(a) \wedge P(a+1), \forall m, m+1, m+2 \in D[\forall x, a \le x \le m+1 \; P(x) \implies P(m+2)]\} \models \forall x \in D \; P(x)$

WM3: $\{P(a) \wedge P(a+1) \wedge P(a+2), \forall m, m+1, m+2, m+3 \in D \; [\forall x, a \le x \le m+2 \; P(x) \implies P(m+3)]\} \models \forall x \in D \; P(x)$

.
.
.

While every proof by mathematical induction is easily changed to a proof by weak mathematical induction, the reverse is not true.

Example 2.12: Assume that $T(n)$ is defined as:

$$T(n) = \begin{cases} 0 & n=0 \\ T(\lfloor n/3 \rfloor) + T(\lfloor n/5 \rfloor) + 2T(\lfloor n/7 \rfloor) + n & n>0 \end{cases}$$

We shall use WM1 to show that $T(n) \le 6n$, $n \ge 0$.

Induction Base: When $n=0$, $T(n) = 0 \le 6n$.

Induction. Hypothesis: Let m be an arbitrary natural number. Assume that $T(n) \le 6n$ for $0 \le n \le m$.

Induction Step: When $n=m+1$, the definition of $T(n)$ yields:

$$T(m+1)=T(\lfloor \frac{m+1}{3} \rfloor) + T(\lfloor \frac{m+1}{5} \rfloor) + 2T(\lfloor \frac{m+1}{7} \rfloor) + m + 1$$

Since, $\lfloor \frac{m+1}{3} \rfloor, \lfloor \frac{m+1}{5} \rfloor$, and $\lfloor \frac{m+1}{7} \rfloor$ are all less than or equal to m, the induction hypothesis may be used to obtain:

$$T(m+1) \le 6 \lfloor \frac{m+1}{3} \rfloor + 6 \lfloor \frac{m+1}{5} \rfloor + 12 \lfloor \frac{m+1}{7} \rfloor + m + 1$$

$$\le 2(m+1) + 6(m+1)/5 + 12(m+1)/7 + m + 1$$

$$= (2+6/5+12/7+1)(m+1)$$

$$= 207(m+1)/35$$

$$< 6(m+1) \quad \square$$

Example 2.13: Suppose that:

$$T(n) \leq \begin{cases} b & 0 \leq n \leq 1 \\ cn + \dfrac{2}{n} \displaystyle\sum_{j=0}^{n-1} T(j) & n > 1 \end{cases}$$

We wish to show that $T(n) \leq 2(b+c)n\log_e n$, $n \geq 2$. This can be done using WM1 as below:

Induction Base: When $n=2$, we see that $T(2) \leq 2c + T(0) + T(1) \leq 2c + 2b \leq 2(b+c)*2*\log_e 2$.

Induction Hypothesis: Let m be an arbitrary natural number greater than 1. Assume that $T(n) \leq 2(b+c)n\log_e n$ for $2 \leq n \leq m$.

Induction Step: When $n = m+1$, we get:

$T(m+1)$

$$\leq c(m+1) + \frac{2}{m+1} \sum_{j=0}^{m} T(j)$$

$$\leq c(m+1) + \frac{4b}{m+1} + \frac{2}{m+1} \sum_{j=2}^{m} T(j)$$

$$\leq c(m+1) + \frac{4b}{m+1} + \frac{4(b+c)}{m+1} \sum_{j=2}^{m} j\log_e j \quad \text{(from IH)}$$

$$\leq c(m+1) + \frac{4b}{m+1} + \frac{4(b+c)}{m+1} \int_{2}^{m+1} x\log_e x \; dx$$

$$\leq c(m+1) + \frac{4b}{m+1}$$

$$+ \frac{4(b+c)}{m+1} \left[\frac{(m+1)^2\log_e(m+1)}{2} - \frac{(m+1)^2}{4} \right]$$

$$= c(m+1) + \frac{4b}{m+1} + 2(b+c)(m+1)\log_e(m+1) - (b+c)(m+1)$$

$$\leq 2(b+c)(m+1)\log_e(m+1) \quad \square$$

Like ordinary (or strong) mathematical induction, weak mathematical induction can also be used to prove predicates with several parameters. The extension to multiple parameter predicates is analogous to the corresponding extension of ordinary mathematical induction. Whether strong or weak mathematical induction is used, the proof is usually just referred to as a proof by induction.

2.3 PROGRAM CORRECTNESS

In this section, we shall apply what we have learnt about proof methods to the problem of establishing the correctness of computer programs. The programs we shall consider are written in an algorithmic language. The meaning of the control structures should be clear to anyone familiar with programming in structured languages such as Pascal, Algol, and PL/1. For those of you who are unsure about the interpretation of some of the control structures being used, an explanation is provided in Appendix B.

Our discussion of program correctness is divided into four parts. These four parts, respectively, deal with inductive proofs for recursive programs, inductive proofs for iterative programs that resemble recursive ones, inductive proofs using assertions or loop invariants, and the predicate transformer method.

Let π be any program. In order to prove that π is correct, we need a specification of what a correct version of π is supposed to do. For the most part, we shall provide this specification in an informal manner using English. Later when we get to the predicate transformer method of program correctness, we shall become more formal about this and also about what we mean by correctness.

The term *program state* is used to refer to the values of all variables in a program. A program begins in an *initial* state and is to terminate in a *result* (or *final*) state that describes the solution to the problem the program was designed to solve. A program is *correct* iff it behaves in this way for every legal initial state.

In establishing correctness, we assume that the programs do not meet any abnormal conditions such as: word overflow, undefined variables, ran out of time, etc. during execution.

2.3.1 Recursive Programs

Induction can often be used to provide simple correctness proofs for recursive programs. No ideas beyond those introduced in Section 2.2 are needed to obtain these proofs. So, we shall directly proceed to some examples. The example programs considered here correspond to problems that are normally not solved by recursive programs. However, we intend to use these same problems as examples in subsequent sections in which we shall examine the iterative programs for these problems. The techniques used here do, however, apply to recursive programs in general. In the remainder of this chapter, we use $A(i:j)$, $i \leq j$, as an abbreviation for $A(i)$, $A(i+1)$, ..., $A(j)$.

Example 2.14: [Sequential Search] Let n, x, and $A(1:n)$ be $n+2$ numbers. We wish to determine an i such that $i = 0$ iff X is not any of the $A(j)$s. If X is one of the $A(j)$s, then i is such that $A(i) = X$. We shall refer to i as the answer to the search problem. So, if $n=4$ and $A(1:4) = (10, 8, 12, 14)$, and $X = 8$, then the answer to the search problem is 2. If $X = 7$, then the answer is 0.

In a sequential search, one compares X with $A(n)$. If X is equal to $A(n)$, then the search terminates successfully with n as the answer. Otherwise, we search $A(1:n-1)$ for X. A recursive algorithm that returns the desired value of i is given in Algorithm 2.1.

line	
	procedure *SEARCH*(A, n, X)
	//Search $A(1:n)$ for X//
1	**declare** $A(n)$, n, X
2	**case**
3	:$n \leq 0$: **return**(0)
4	:$A(n)=X$: **return**(n)
5	:**else**: **return**$(SEARCH(A, n-1, X))$
6	**end case**
7	**end** *SEARCH*

Algorithm 2.1 Recursive sequential search

The correctness of procedure *SEARCH* may be established by generalization on the values $A(1:n)$, and X, and by induction on n. In the induction base, we shall show that procedure *SEARCH* works correctly when $n = 0$. The induction hypothesis will be the assumption that *SEARCH* works correctly for all arrays of size m and for all X. Here m is some arbitrary natural number. In the induction step it will be shown that the algorithm works correctly when $n = m+1$. Now, the correctness of the algorithm for all natural numbers n, all A, and all X follows from MI1.

Proof:
Induction Base: When $n = 0$, there are no numbers in the sequence $A(1:n)$ and so regardless of the value of X, the answer to the search problem is 0. We see that procedure *SEARCH* always returns 0 when $n = 0$ (line 3). So, procedure *SEARCH* works correctly for all A, and all X, when $n = 0$.

Inductions Hypothesis: Let m be an arbitrary natural number. Assume that procedure *SEARCH* works correctly for all $A(1:n)$ and all X when $n = m$.

Induction Step: We need to show that procedure *SEARCH* works correctly for all $A(1:n)$, and all X when $n = m+1$. Let $A(1:m+1)$ and X be arbitrary. If $A(m+1) = X$, then $m+1$ is an answer to the search problem (Note that X may occur several times in $A(1:m+1)$. So, the answer to the search problem may not be unique.). As can be seen from line 4, the procedure works correctly in this case. If $A(m+1) \neq X$, then the answer to the search problem with $n = m+1$ is the same as that to the search problem defined by $A(1:m)$ and X. From the induction hypothesis, this answer is found in line 5 of procedure *SEARCH*. This completes the proof of the induction step.

Observe that *SEARCH* works correctly even when $n<0$. Our proof, of course, does not establish this. Our proof will remain unchanged if the condition of line 3 is changed to $n=0$. $\square$

Example 2.15: [Insertion] Let $A(1:n)$ be a sorted array of n numbers. So, $A(1) \leq A(2) \leq \cdots \leq A(n)$. In the insertion problem, we wish to insert the number X into $A(1:n)$ in such a way that following the insertion the resulting array $A(1:n+1)$ is also sorted. For example, if $n = 4$, $A(1:4) = (3, 5, 8, 9)$, and $X = 6$, then following the insertion we have $A(1:5) = (3, 5, 6, 8, 9)$.

```
line  procedure INSERT(A, n, X)
         //Insert X into the sorted array A(1:n)//
1        declare A(n+1), n, X
2        case
3          :n=0:  A(1) ← X
4          :A(n)>X:  A(n+1) ← A(n); INSERT(A, n−1, X)
5          :else:  A(n+1) ← X
6        end case
7     end INSERT
```

Algorithm 2.2 Recursive procedure to insert X

A recursive procedure to accomplish this insertion is given in Algorithm 2.2. Notice that if n is a negative integer (say -5), then *INSERT* terminates abnormally as $A(-5)$ is undefined. The correctness of this procedure for all natural numbers n and all sorted sequences $A(1:n)$ may be established by using generalization on A and X, and induction on n as below.

Proof:
Induction Base: When $n = 0$, the insertion should simply result in X occupying position 1 of the array A. This is precisely what happens in line 3 of procedure *INSERT*. So, this procedure works correctly when $n = 0$.

Induction Hypothesis: Let m be an arbitrary natural number. Assume that

INSERT works correctly for all $A(1..n)$ and all X when $n = m$.

Induction Step: We need to show that procedure *INSERT* works correctly when $n = m+1$. Let $A(1:m+1)$ and X be arbitrary. If $A(m+1) > X$, then $A(m+1)$ will be in position $m+2$ following the insert. In addition, the position of the remaining numbers is correctly obtained by inserting X into $A(1:m)$. When $A(m+1) > X$, only line 4 of *INSERT* is executed. In this line, $A(m+1)$ is moved to position $m+2$ and a recursive call to *INSERT* made. From the induction hypothesis, it follows that this recursive call correctly inserts X into $A(1:m)$. So, the procedure works correctly for the case $A(m+1) > X$.

The case that remains is $A(m+1) \leq X$. A valid result of the insertion for this case has all the $A(i)$s in their original positions for $1 \leq i \leq m+1$ and $A(m+2) = X$. The procedure handles this case in line 5 and as can be seen exactly this configuration is produced. □

Example 2.16: [Binary Search] Binary search is a fast search method to determine if a given number X is present in a sequence $A(i:n)$, $i > 0$, of numbers that is sorted in nondecreasing order. For example, the sequence $A(1:5) = (3, 5, 5, 6, 8)$ is in nondecreasing order.

The basic idea is to compare X with the middle number $A(mid)$ where $mid = \lceil (i+n)/2 \rceil$. If $X = A(mid)$ then X has been found. If $X < A(mid)$ then only the sequence $A(i:mid-1)$ needs to be searched and if $X > A(mid)$, then only $A(mid+1:n)$ needs to be searched. Procedure *BINSRCH* (Algorithm 2.3) is the resulting algorithm. If X is in $A(i:n)$, then *BINSRCH* returns an index j such that $A(j) = X$. It returns 0 iff X is not in $A(i:n)$.

```
line   procedure BINSRCH(A, i, n, X)
          //search for X in A(i:n), i>0//
          //A(i)≤A(i+1)≤...A(n). Return 0//
          // if X is not in the sequence. Otherwise//
          // return a value j such that i≤j≤n and A(j)=X.//
  1       declare A(n), X, i, n, mid
  2       mid ← ⌈(i+n)/2⌉
  3       case
  4          :i>n: return(0)  //not found//
  5          :X=A(mid): return(mid)  //found//
  6          :X<A(mid): return(BINSRCH(A, i, mid−1, X)
  7          else: return(BINSRCH(A, mid+1, n, X)
  8       end case
  9     end BINSRCH
```

Algorithm 2.3 Recursive binary search

The correctness of *BINSRCH* may be established using weak mathematical induction on the number of elements $n-i+1$ in $A(i:n)$.

Proof:
Induction Base: When the number of elements in $A(i:n)$ is 0, the answer to the search problem should be 0. In this case, $n = i-1$ and the procedure executes only lines 2 and 4. The value 0 is the result.

Induction Hypothesis: Let m be an arbitrary natural number. Assume that procedure *BINSRCH* works correctly for all sorted $A(i:n)$ and all X whenever $i>0$ and $i-1 \le n \le i+m-1$ (i.e., whenever the number of elements is no more than m).

Induction Step: Let $A(i:n)$ be an arbitrary sorted array with $m+1$ elements. So, $n = i+m$. Assume that $i>0$. Let X be arbitrary. In addition to executing the statement of line 2, procedure *BINSRCH* will now execute the statements corresponding to exactly one of the cases of lines 5, 6, and 7. It is clear that if $X = A(mid)$, then the correct answer is produced. If $X<A(mid)$, then X is in $A(i:i+m)$ iff X is in $A(i:mid-1)$. Since the number of elements in $A(i:mid-1)$ is less than $m+1$, the induction hypothesis guarantees that the result from line 6 is correct. The proof for the case $X>A(mid)$ (line 7) is similar. $\square$

2.3.2 Iterative Programs

Many iterative programs are actually based on recursive algorithms. For example, the programs given in Algorithms 2.1 - 2.3 are seldom written recursively. Instead, they are written iteratively as in this section. We shall now reexamine the problems of Examples 2.14 - 2.16 and prove the correctness of the iterative versions of these programs. In addition, a new example is also introduced.

All the proofs of this section will explicitly use the fact that following the first iteration of the loop in each of the programs, the program behaves exactly as it would if started with a smaller problem instance. So, in effect, we use the fact that these programs are restatements of recursive programs!

Example 2.17: [Sequential Search] The problem being solved is the same as that described in Example 2.14. The iterative procedure for sequential search is given in Algorithm 2.4. This procedure avoids an explicit test for X not being in $A(1:n)$ by setting $A(0)$ to X. As a result, X is always one of $A(0:n)$. The answer 0 is, however, returned only if X is not one of $A(1:n)$. This procedure may be proved correct for all natural numbers n using induction on n as below.

Proof:
Induction Base: When $n = 0$, i is set to 0 and $A(0)$ to X in line 2. Since $A(0) = X$ and $i = 0$ the first time line 3 is reached, the **while** loop is not entered

```
line  procedure SEQSRCH(A,n,X)
      //Search A(1:n) for X//
 1       declare A(0:n),n,i,X
 2       i ← n; A(0) ← X
 3       while A(i) ≠ X do
 4          i ← i−1
 5       end while
 6       return(i)
 7    end SEQSRCH
```

Algorithm 2.4 Iterative sequential search

and the procedure terminates in line 6 with the answer $i = 0$. This is the correct answer for the case $n = 0$.

Induction Hypothesis: Let m be an arbitrary natural number. Assume that *SEQSRCH* works correctly for all $A(1:n)$ and all X when $n = m$.

Induction Step: Let $n = m+1$. Let $A(1:m+1)$ and X be arbitrary. The first time line 3 is reached, $i = m+1$. If $A(m+1) = X$, then the answer $i = m+1$ is returned from line 6. This is correct. Otherwise, the **while** loop is entered and i is decremented by 1. So, its value becomes m. From this point on, the procedure behaves exactly as it would if started with $n = m$. From the induction hypothesis, therefore, subsequent iterations of the **while** loop will correctly search $A(1:m)$ for X. □

Example 2.18: [Insertion] An iterative version of Algorithm 2.2 (Example 2.15) is given in Algorithm 2.5. We shall establish the correctness of this procedure by induction on the size, n, of the sorted sequence A.

Proof:
Induction Base: When $n = 0$, X is to be inserted into position $A(1)$. In line 2 of *INSERT*, i is set to 0 and $A(0)$ is set to X. The conditional (or predicate) $A(i)>X$ of line 3 is false (as $A(i) = A(0) = X$) and so lines 4 and 5 do not get executed. The next line to be executed is line 7 and here, $A(1)$ is set to X. So, the algorithm works correctly when $n = 0$.

Induction Hypothesis: Let m be any arbitrary natural number. Assume that the algorithm works correctly for all nondecreasing sequences of size m and all X.

Induction Step: Let $A(1:m+1)$ be any nondecreasing sequence of $m+1$ numbers. Let X be an arbitrary number. In line 2, i is set to $m+1$ and $A(0)$ is set to X. If $A(m+1)\leq X$ then the **while** loop (lines 3-6) is not entered and $A(m+2)$ is set to X in line 7. The resulting sequence $A(1:m+2)$ is clearly in nondecreasing order.

Hence, *INSERT* works correctly when $A(m+1) \leq X$. If $A(m+1) > X$, then the **while** loop is entered and $A(m+2)$ is set to $A(m+1)$ and i decreased to m. From this point on the algorithm behaves exactly as it would if started with the sequence $A(1:m)$ and with $n = m$. From the induction hypothesis, it follows that when line 7 is reached, $A(1:m+1)$ will be in nondecreasing order. Since $A(m+2) \geq A(j)$, $1 \leq j \leq m+1$, the sequence $A(1:m+2)$ is also in nondecreasing order. So, *INSERT* works correctly when $n = m+1$. $\square$

line   **procedure** *INSERT* (A,n,X)
        //insert X into the sorted array $A(1:n)$//
1   **declare** $A(n+1)$, n, X
2       $i \leftarrow n$; $A(0) \leftarrow X$
3   **while** $A(i) > X$ **do**
4           $A(i+1) \leftarrow A(i)$
5           $i \leftarrow i-1$
6   **end while**
7       $A(i+1) \leftarrow X$
8   **end** *INSERT*

Algorithm 2.5 Iterative insertion

Example 2.19: [Binary search] The binary search method was described in Example 2.16. An iterative procedure corresponding to this search method is given in Algorithm 2.6. Its correctness may be established by weak induction on the number of elements in A.

line   **procedure** *BINSRCH* (A,i,n,X)
            //search for X in $A(i:n)$, $i > 0$//
            //$A(i) \leq A(i+1) \leq \ldots A(n)$.//
1   **declare** $A(n),i,X,n,d,u,mid$
2           $d \leftarrow i$; $u \leftarrow n$
3   **while** $d \leq u$ **do**
4               $mid \leftarrow \lfloor (d+u)/2 \rfloor$
5   **case**
6   :$X < A(mid)$: $u \leftarrow mid-1$ //look in lower half//
7   :$X = A(mid)$: **return**(mid) //found X//
8   :else: $d \leftarrow mid+1$ //look in upper half//
9   **end case**
10   **end while**
11   **return**(0) //X not found//
12   **end** *BINSRCH*

Algorithm 2.6: Iterative binary search.

Proof:
Induction Base: When the number of elements in A is zero, $n = i-1$. In this case, d is set to i and u to $i-1$ in line 2. Since $d>u$, the conditional $(d \leq u)$ of line 3 has value **false** and the **while** loop is not entered. The value 0 is returned in line 11. So, the algorithm works correctly when A contains zero elements.

Induction Hypothesis: Let m be any arbitrary natural number. Assume the algorithm works correctly for all X, all $i>0$, and all nondecreasing sequences $A(i:n)$, whenever $i-1 \leq n \leq i+m-1$. I.e., whenever the number of elements in A is no more that m.

Induction Step: Let $A(i:i+m)$ be any nondecreasing sequence of $m+1$ numbers. Let X be any number. *BINSRCH* begins, in line 2, by setting $d = i$ and $u = i+$m. Since $m \geq 0$, $d \leq u$, and the **while** loop is entered. *mid* is set to $\lfloor (d+u)/2 \rfloor = \lfloor (2i+m)/2 \rfloor$. If $X<A(mid)$, then X cannot be one of $A(mid)$, $A(mid+1)$, ..., $A(i+m)$. After having set u to $mid-1$, the algorithm behaves exactly as it would if started with the input $A(i:mid-1)$. Since the number of elements in this sequence is fewer than $m+1$, the induction hypothesis can be used. It follows that the algorithm will work correctly on $A(i:mid-1)$ and so the correct value is returned.

If $X = A(mid)$, then the correct is returned in line 7. When $X>A(mid)$, d is set to $mid+1$ in line 8. From this point on the algorithm behaves exactly as it would if started with input $A(mid+1:i+m)$. The number of elements in this sequence is fewer than $m+1$ and so from the induction hypothesis it follows that the correct value is returned in this case too. This completes the proof of the induction step. $\square$

Example 2.20: Let us determine the maximum number of times the **while** loop of procedure *BINSRCH* can be iterated when we are searching for X in any nondecreasing sequence of p numbers. We shall use the inference rule WM1 to show that this number never exceeds $\lceil \log_2(p+1) \rceil$.

Induction Base: When $p = 0$, d is set to i and u to $i-1$ in line 2. Since $d>u$, the **while** loop isn't iterated even once. Also, $\lceil \log_2(p+1) \rceil$ equals 0 when $p = 0$.

Induction Hypothesis: Let m be any arbitrary natural number. Assume that the **while** loop is iterated at most $\lceil \log_2(p+1) \rceil$ times for all nondecreasing sequences $A(i:i+p-1)$, all X, and all p, $0 \leq p \leq m$.

Induction Step: Let $A(i:i+m)$ be any nondecreasing sequence of size $m+1$. Let X be arbitrary. The variables d and u are, respectively, initialized to i and $i+m$ in line 2 and the **while** loop entered. If $X = A(mid)$, then the algorithm terminates and the number of iterations of the loop is 1. This agrees with the bound as $\lceil \log_2(m+2) \rceil \geq 1$, $m \geq 0$.

If $X<A(mid)$, then u is updated to $mid-1$ in line 6. The number of elements in $A(i:mid-1)$ is $mid-i = \lfloor(d+u)/2\rfloor-i = \lfloor(2i+m)/2\rfloor-i = \lfloor m/2\rfloor$. Since the algorithm now behaves as it would if started with $A(i:mid-1)$, the induction hypothesis can be used. So, no more than $\lceil\log_2(\lfloor m/2\rfloor+1)\rceil$ additional iterations of the **while** loop will be made. So, when $X<A(mid)$, the total number of iterations of the **while** loop does not exceed $\lceil\log_2(\lfloor m/2\rfloor+1)\rceil+1 \le \lceil\log_2(m/2+1)\rceil+1 = \lceil\log_2(m+2)\text{-}1\rceil+1 = \lceil\log_2(m+2)\rceil$.

If $X>A(mid)$, then d is updated to $mid+1$ in line 8 and the number of elements in $A(mid+1:u)$ is $i+m-\lfloor(2i+m)/2\rfloor = m-\lfloor m/2\rfloor = \lfloor m/2\rfloor$. From the induction hypothesis, we see that no more than $\lceil\log_2(\lfloor m/2\rfloor+1)\rceil$ additional iterations of the **while** loop will be made. Hence, when $X>A(mid)$, the total number of iterations of the **while** loop is no more than $\lceil\log_2(\lfloor m/2\rfloor+1)\rceil + 1$. Since, $\lceil\log_2(\lfloor m/2\rfloor+1)\rceil$ cannot exceed $\lceil\log_2(m/2+1)\rceil$ regardless of whether m is odd or even, the number of iterations of the **while** loop is bounded by $\lceil\log_2(m/2+1)\rceil + 1 \le \lceil\log_2(m+2) \text{ - } 1\rceil + 1 \le \lceil\log_2(m+2)\rceil$.

So, the bound is valid in all three cases. □

Example 2.21: [Euclid's GCD algorithm] Let m and n be two positive integers. The *greatest common divisor* (gcd) of m and n (written as $\gcd(m,n)$) is the largest natural number that divides both m and n. For example, $\gcd(21,14) = 7$; $\gcd(18,9) = 9$; $\gcd(33,44) = 11$; $\gcd(75,30) = 15$; etc.

Euclid's algorithm to compute the gcd of m and n is given as procedure *EUCLID* (Algorithm 2.7). It assumes that $m\ge n>0$. In this algorithm, $\text{rem}(m,n)$ is a function that computes the remainder of m divided by n.

```
line  procedure EUCLID(m,n)
         //m ≥ n > 0//
  1       integer m,n,r
  2       loop
  3          r ← rem(m,n)
  4          if r = 0 then return(n) endif
  5          m ← n; n ← r
  6       repeat
  7    end EUCLID
```

Algorithm 2.7 Euclid's gcd algorithm.

The correctness of procedure *EUCLID* may be established by induction on m. Let *EUCLID*(m,n) be the value computed by procedure *EUCLID*. We wish to show that *EUCLID*(m,n) = gcd (m,n) for all m and n such that $m \ge n > 0$.

Induction Base: When $m=1$, there is only one possible value for n, i.e., $n=1$. It is easily seen that $EUCLID(1,1) = 1$ and that $\gcd(1,1) = 1$.

Induction Hypothesis: Let m' be an arbitrary natural number such that $m' \geq 1$. Assume that $EUCLID(m,n) = \gcd(m,n)$, $1 \leq n \leq m$, $1 \leq m \leq m'$.

Induction Step: We need to show that $EUCLID(m'+1,n) = \gcd(m'+1,n)$, $1 \leq n \leq m'+1$. We shall do this by generalization on n. Let b be an arbitrary integer value for n, $1 \leq b \leq m'+1$. If b divides $m'+1$, then $\text{rem}(m'+1, b) = 0$ and $EUCLID(m'+1, b) = b = \gcd(m'+1, b)$. If b does not divide $m'+1$, then $r \neq 0$ and following the first iteration of the loop, the algorithm proceeds to compute $EUCLID(b, \text{rem}(m'+1, b))$. Since $b < m'+1$ (as if $b = m'+1$ then b divides $m'+1$) and $0 < \text{rem}(m'+1,b) < b$, it follows from the induction hypothesis that $EUCLID(b, \text{rem}(m'+1, b)) = \gcd(b, \text{rem}(m'+1, b))$. In the exercises, it is shown that $\gcd(m'+1,b) = \gcd(b, \text{rem}(m'+1, b))$. So, $EUCLID(m'+1, b) = \gcd(m'+1,b)$. $\square$

2.3.3 Loop Invariants

Often, simpler correctness proofs for programs that contain loops can be obtained by performing induction on the number of iterations of these loops. When this approach is taken, one associates a predicate with a certain point in the loop (generally the start or end of the loop) and shows that this predicate is true each time program execution reaches this point of the loop. If the program has already been shown to terminate, then we know that this loop invariant is true the last time the loop is executed (there must be a last iteration as the program terminates). This fact is used to establish the correctness of the program.

A predicate whose truth value does not change over successive iterations of a loop is called a *loop invariant*. The following examples illustrate the method of loop invariants and also demonstrate that this approach results in simpler proofs.

Example 2.22: In this example, we shall present an alternate correctness proof for *BINSRCH* (Algorithm 2.6). Consider any arbitrary nondecreasing sequence $A(i:n)$, arbitrary integer n, $n \geq i-1$, and arbitrary X. Since the number of elements in $A(d:u)$ decreases following each iteration of the **while** loop, it follows that *BINSRCH* will terminate after at most $n-i+1$ iterations of this loop (From Example 2.20, it follows that actually at most $\lceil \log_2(n-i+2) \rceil$ iterations of this loop can be made).

If the algorithm terminates in line 7, then we see that the right answer is obtained. So, we need only show that line 11 is reached iff X is not one of $A(i:n)$. This is established by showing that each time the conditional of line 3 is tested, the predicate P:

P: X is one of $A(i:n)$ iff X is one of $A(d:u)$

is true. MI1 can be used to show that P is true each time line 3 is reached. The proof is:

Induction Base: The first time line 3 is reached, $d = i$ and $u = n$. The predicate P is trivially true.

Induction Hypothesis: Let m be an arbitrary positive integer. Assume that if line 3 is reached an mth time, then P is true.

Induction Step: We shall show that if line 3 is reached for the $m+1$st time, then also P is true. Assume that line 3 is reached for the $m+1$st time. Then, it must also have been reached an mth time. At this time, $d \le u$ (as line 3 is reached at least one more time) and P is true (from the IH). Only the cases of line 6 and 8 are possible for this iteration of the **while** loop. No matter which is executed, the fact that the numbers in A are sorted ensures that P will be true when line 3 is reached next.

So, the predicate P is a loop invariant for the **while** loop of procedure *BINSRCH*.

If line 11 is reached, then the last time line 3 is reached, $d > u$. From the truth of P, and the observation that $A(d:u)$ is an empty sequence when $d > u$, it follows that line 11 is reached iff X is not one of $A(i:i+n-1)$. □

Example 2.23: As in the case of *BINSRCH*, an alternate correctness proof (using a loop invariant) can be provided for procedure *EUCLID* (Algorithm 2.7). Let m' and n' be two arbitrary positive integers such that $m' \ge n'$. Let these, respectively, be the initial values for m and n. We observe that if $r = 0$, then the algorithm terminates in line 4. If $r \ne 0$, then $1 \le r < n'$. Hence, the next time line 3 is reached, the new value of n is smaller than the old one. Consequently, line 3 can be reached at most n' times (as by then n would have dropped to 1 and r will become 0). So, procedure *EUCLID* terminates for all m and n such that $m \ge n > 0$.

We are left with the problem of proving that the value returned in line 4 is in fact $\gcd(m',n')$. Consider the predicate P:

P: $\gcd(m',n') = \gcd(m,n)$

Using MI1 and the proof provided by Exercise 19, we may show that P is true whenever line 3 is reached. Hence, P is true the last time line 3 is reached. At this time, $r = 0$ and $n = \gcd(m,n) = \gcd(m',n')$. □

Example 2.24: [Gries] Let $A(1{:}m,1{:}n)$ be an array of integers such that:

(i) The sequence of numbers in every row of A is an increasing sequence left to right.

(ii) The sequence of numbers in every column of A is an increasing sequence top to bottom.

The procedure *ARRAY* (Algorithm 2.8) searches A for X. It returns the tuple (i,j) if $A(i,j) = X$ for some i,j such that $1 \leq i \leq m$ and $1 \leq j \leq n$. It returns the tuple $(0,0)$ iff no such i and j exist (i.e., X is not one of the numbers in A). We shall establish the correctness of this procedure by the method of loop invariants. With line 3, we associate the follwing predicate P:

P: X is an element of the array $A(1{:}m, 1{:}n)$ iff X is an element of the array $A(i{:}m, 1{:}j)$.

We can prove that P is a loop invariant for line 3 of procedure *ARRAY* using induction.

Induction Base: The first time line 3 is reached, $i = 1$ and $j = n$. So, P is trivially true at this time.

Induction Hypothesis: Let a be an arbitrary positive integer. Assume that line 3 is reached an ath time and that P is true this time.

Induction Step: We show that if line 3 is reached an $a+1$st time, then also P is true. If line 3 is reached an $a+1$st time, then it must have also been reached an ath time. From the induction hypothesis, it follows that at this time P is true. Since line 3 is going to be reached an $i+1$st time, the **while** loop is entered for the ath time and one of the cases of lines 6 and 7 is executed. We see that in the case of line 6, X cannot be one of the elements in row i; and in the case of line 7, X cannot be one of the elements of column j. So, updating i or j in the manner of lines 6 and 7 does not affect the truth of P. So, P is true when line 3 is reached for the $a+1$st time.

To complete the correctness proof of *ARRAY*, we observe that if this procedure terminates from line 5, then the correct answer is given. If the procedure does not terminate from line 5, then it must exit the **while** loop after at most $m+n-1$ iterations as line 6 increments i by one while line 7 decrements j by one. Neither i nor j are changed elsewhere in this loop. Hence, after at most $m+n$ iterations, either $i>m$ or $j<1$. If the last time line 3 is reached $i>m$, then from P it follows that X is in $A(1{:}m, 1{:}n)$ iff X is in $A(m+1{:}m, 1{:}j)$. Since there are no elements in $A(m+1{:}m, 1{:}j)$, it follows that X is not in $A(1{:}m, 1{:}n)$ and so the tuple $(0,0)$ is to be returned. The same is the case when $j<1$ the last time line 3 is reached. We see that in both cases, the tuple $(0,0)$ is, in fact, returned from line 10. □

```
line  procedure ARRAY(A,m,n,X)
 1      integer m,n,X,i,j,A(1:m,1:n)
 2      (i,j) ← (1,n)  //start at top right corner//
 3      while i≤m and j≥1 do
 4        case
 5          :A(i,j)=X: return((i,j))  //found X//
 6          :A(i,j)<X: i ← i+1
 7          :A(i,j)>X: j ← j-1
 8        end case
 9      end while
10      return((0,0))  //not found//
11    end ARRAY
```

Algorithm 2.8

2.3.4 The Predicate Transformer Method

We shall develop the *predicate transformer* method of proving programs correct. This method was first proposed by E. Dijkstra in 1975.

A *program* is simply a sequence of program statements. Algorithm 2.9 shows a four line program. Line 1 contains three statements joined together by the connective ";". Lines 2,3 and 4 each contain 1 statement. Line 3 itself is a one line one statement program while line 1 is a one line three statement program. Line 4 is a one line one statement program.

```
line
 1    x ← 2; y ← 4; z ← 6;
 2    x ← x * y;
 3    y ← y + z;
 4    z ← y - x;
```

Algorithm 2.9

A *program state* is described by the values of the variables in the program. Let π be a program that contains the variables x, y, and z. The states π can be in are given by all possible values for the triple (x, y, z). Consider the program π of Algorithm 2.9. The program state after the execution of line 1 is $(2, 4, 6)$. After the execution of line 2, it is $(8, 4, 6)$, after the execution of line 3, it is $(8, 10, 6)$, and after the execution of line 4, it is $(8, 10, 2)$.

We shall restrict our discussion to *deterministic* programs. These programs have the property that every execution of the program that begins in the same state proceeds in exactly the same manner. Programs that do not enjoy this property are called *nondeterministic*. As an example, consider the one line program:

$$x \leftarrow [1, 2]$$

which has the interpretation x is assigned the value 1 or 2. The computer executing this program may assign either of these values to x. In particular, one execution of this program might result in $x = 1$ and another in $x = 2$. (Since we shall be dealing with deterministic programs only, it is not very important for you to understand nondeterminism well at this point. This concept is introduced here only because most treatments of the predicate transformer method consider nondeterminism.)

Let I and R be two predicates defined on the program variables. We use the notation:

$$\{I\} \; \pi \; \{R\}$$

to mean the following:

> If the execution of program π begins in a state that satisfies I, then π terminates (i.e., terminates in a finite amount of time) in a state that satisfies R.

I and R are, respectively, called the *precondition* (or *initial assertion*) and *postcondition* (or *result assertion*) of π. Note that we use the term *terminates* as an abbreviation for *terminates in a finite amount of time*.

Example 2.25: Let α, β, γ, and δ, respectively, be the programs that consist solely of line 1, line 2, line 3, and line 4 of Algorithm 2.9. The following are valid statements. (T and F, respectively, denote the predicates that are always true or false.)

(a) $\{T\} \; \alpha \; \{x = 2, y = 4, z = 6\}$

(b) $\{x = 5, y = 6\} \; \beta \; \{x = 30\}$

(c) $\{x = 2, y = 4\} \; \beta \; \{x = 8\}$

(d) $\{x = 5, y = 6\} \; \beta \; \{x = 30, y = 6\}$

(e) $\{y = 10, z = 8\} \; \gamma \; \{y = 18\}$

(f) $\{x = 10, y = 30\} \; \delta \; \{z = 20\}$

(g) $\{x = 10, y = 30\}\ \delta\ \{x = 10, y = 30, z = 20\}$

(h) $\{T\}\ \pi\ \{x = 8, y = 10, z = 2\}$

(i) $\{y > 10, z > 20\}\ \gamma\ \{y > 30, z > 20\}$

The above statements may be restated in English as below:

(a) No matter what the values of x, y, and z before the program consisting solely of line 1 is executed, the execution of this program will terminate and upon termination, the values of x, y, and z will, respectively, be 2, 4, and 6.

(b) If $x = 5$ and $y = 6$ before line 2 is executed, then the exceution of this line terminates and upon termination, $x = 30$.

(c) If $x = 2$ and $y = 4$ before line 2 is executed, then the exceution of this line terminates and upon termination, $x = 8$.

(d) If $x = 5$ and $y = 6$ before line 2 is executed, then the exceution of this line terminates and upon termination, $x = 30$ and $y = 6$.

(e) If $y = 10$ and $z = 8$ before line 3 is executed, then the execution of this line terminates and upon termination, $y = 18$.

(f) If $x = 10$ and $y = 30$ before line 4 is executed, then the execution of this line terminates and upon termination, $z = 20$.

(g) If $x = 10$ and $y = 30$ before line 4 is executed, then the execution of this line terminates and upon termination, $x = 10$, $y = 30$, and $z = 20$.

(h) When the program π consisting of line 1-4 is executed, it terminates with $x = 8$, $y = 10$, and $z = 2$.

(i) If $y > 10$ and $z > 20$ before line 3 is executed, then the execution of this line terminates and upon termination, $y > 30$ and $z > 20$. $\square$

A program π is said to be *totally correct* (or *strongly verifiable*) with respect to the initial and result assertions I and R iff $\{I\}\ \pi\ \{R\}$ is valid. The program π of Algorithm 2.9 is readily seen to be totally correct with respect to the initial assertion $I = T$ and the result assertion $R = (x = 8, y = 10, z = 2)$.

The notation:

$I\ \{\pi\}\ R$

is used to mean the following:

> If program π begins execution in a state for which the initial assertion I is true and if the execution of π terminates, then on termination, the program state satisfies the result assertion R.

A program π is said to be *partially correct* (or *consistent*) with respect to the initial and result assertions I and R iff $I \{\pi\} R$ is valid. The program given in Algorithm 2.10 is partially correct with respect to every initial and result assertion as it never terminates. The program of Algorithm 2.11 is partially correct with respect to the initial assertion $x > 0$ and the result assertion $x = 2$.

while true do	10: **if** $x = 3$ **then** $x \leftarrow 2$
$x \leftarrow 2;$	**else goto** 10
Algorithm 2.10	**Algorithm 2.11**

Partial correctness makes a statement about a program only for those initial states that both satisfy the initial assertion I and lead to termination. It does not guarantee that the program will terminate whenever execution begins in a state that satisfies the initial assertion. Total correctness, on the other hand, states that if a program is begun in a state that satisfies the initial assertion I, then termination is guaranteed. Both notions of correctness assert that upon termination, the program state satisfies the result assertion R.

Note that if a program is totally correct with respect to I and R, then it is also partially correct with respect to I and R. The reverse is not true for all programs. In further discussion, we shall be concerned only with the notion of total correctness. We shall therefore use the term correct as an abbreviation for total correctness.

Let P and Q be two predicates defined on the states of a program π. We say that Q is *weaker* than P iff $P \Rightarrow Q$ is true. I.e., Q is true for every state for which P is true. For example, the predicate $x \geq 0$ is weaker than the predicate $x = 0$. The predicate $x + y > 2$ is weaker than the predicate $x + y > 8$.

Define WP to be a predicate transformer which for every program π and every postcondition Q, yields the weakest precondition P for which $\{P\} \pi \{Q\}$ is true. Hence, if $\{S\} \pi \{R\}$ for some predicate R, then $S \Rightarrow WP(\pi, R)$. WP is called the *weakest precondition predicate transformer*. $WP(\pi, R)$ transforms the postcondition R into the weakest precondition P for which $\{P\} \pi \{R\}$ is true.

Alternatively, we may say that $WP(\pi, R)$ is a predicate P which is true for every program state with the property that if π begins execution from this state then π terminates in a state that satisfies R. P is true for no other states.

Example 2.26: Let α, β, γ, and δ be the one line programs defined in Example 2.25. Let π be the four line program of Algorithm 2.9. The following equalities are readily seen to be true:

(a) $WP(\alpha, (x = 2, y = 4, z = 6)) = T$
       I.e., every initial state leads to a state that satisfies the given result predicate.

(b) $WP(\alpha, (x = 4)) = F$
       I.e., there is no initial state that causes program α to terminate in a state that satisfies the result condition $x = 4$.

(c) $WP(\beta, x = 6) = (xy = 6)$
       The weakest precondition on β in order that $x = 6$ be satisfied on termination is that $xy = 6$ before the execution of β begins.

(d) $WP(\beta, x > 10) = x > 10/y$
       In order for x to be greater than 10 following the execution of β, x must be greater than $10/y$ before execution begins.

(e) $WP(\gamma, 0 \le y \le 10) = -z \le y \le 10 - z$

(f) $WP(\delta, z > 0) = y > x$

(g) $WP(\delta, (x = 2, z \ge 4)) = (x = 2, y \ge 6)$

(h) $WP(\pi, x = 8) = T$

(i) $WP(\pi, (x = 8, y = 10)) = T$ $\square$

In the predicate transformer method to prove that a program π is correct with respect to the initial and result assertions I and R, we show that $I \Rightarrow WP(\pi, R)$. Theorem 2.3 establishes the validity of the predicate transformer method.

Theorem 2.3: Program π is correct with respect to the initial and result assertions I and R iff $I \Rightarrow WP(\pi, R)$.

Proof: From the definition of WP, we see that if the truth of I does not imply the truth of $WP(\pi, R)$, then $\{I\}$ π $\{R\}$ cannot be true. Furthermore, if $I \Rightarrow WP(\pi, R)$, then $\{I\}$ π $\{R\}$ is true. Hence, program π is correct with respect to I and R iff $I \Rightarrow WP(\pi, R)$. $\square$

The predicate transformer permits a clean definition of program equivalence. Two programs α and β are *equivalent* iff for every result assertion R, $WP(\alpha, R) = WP(\beta, R)$.

Before we can use the predicate transformer method to prove programs correct, we need to develop some expertise in obtaining $WP(\pi, R)$ from π and R. In developing this expertise, it is helpful to study some properties of the predicate transformer WP and also study the form of WP for some of the commonly occurring program statement types.

Theorem 2.4: [Properties of WP] In the following, R and S are predicates and π is a program. Since, R, S, and π occur as free variables, universal quantification is implied by default. So, the following statements are true for every program π (recall that we are dealing with deterministic programs only), every R, and every S.

(a) $WP(\pi, F) = F$

(b) $WP(\pi, R) \lor WP(\pi, S) = WP(\pi, R \lor S)$

(c) $WP(\pi, R) \land WP(\pi, S) = WP(\pi, R \land S)$

(d) $[R \Rightarrow S] \Rightarrow [WP(\pi, R) \Rightarrow WP(\pi, S)]$

Proof:

(a) Observe that there can be no initial state such that π begins execution in this state and terminates in a state that satisfies F (recall that F is false for every state). Hence, the weakest precondition for the postcondition F is a predicate that is true for no state. Consequently, $WP(\pi, F) = F$.

(b) [Distributivity of **or**] Let s be any state for which the left hand side of the equality is true. So, s is a state for which either $WP(\pi, R)$ or $WP(\pi, S)$ or both are true. Suppose that $WP(\pi, R)$ is true for s. This means that program π begun in state s always terminates in a state for which R is true. Consequently, program π begun in state s terminates in a state satisfying $R \lor S$. So, s satisfies $WP(\pi, R \lor S)$. Similarly, if $WP(\pi, S)$ is true for s, then $WP(\pi, R \lor S)$ is true for s. Consequently, $WP(\pi, R) \lor WP(\pi, S) \Rightarrow WP(\pi, R \lor S)$ is true.

To complete the proof, we need to show the implication in the other direction. Let s be a state for which $WP(\pi, R \lor S)$ is true. Since the program π is assumed to be deterministic, every execution of π beginning in the state s must result in π terminating in the same state t. For this state t, $R \lor S$ is true. If t satisfies R, then s satisfies $WP(\pi, R)$. If t satisfies S, then s satisfies $WP(\pi, S)$. So, s satisfies $WP(\pi, R) \lor WP(\pi, S)$.

Remark: The proof of the preceding paragraph is not valid when π is nondeterministic. Since the execution of a nondeterministic program need not be the same each time it begins from a given initial state, the result state may vary from one execution to the next. As an example, consider the nondeterministic program:

$x \leftarrow [1, 2]$

which is to be interpreted as x is assigned the value 1 or 2. As mentioned earlier, no rule is specified as to how the selection between 1 and 2 is carried out. In some executions of this program the postcondition $x = 1$ is true while in others $x = 2$ is true. For every execution, however, the postcondition $x = 1 \lor x = 2$ is true. Therefore, $WP(\pi, x = 1) =$ $WP(\pi, x = 2) = F$ and $WP(\pi, x = 1 \lor x = 2) = T$. So, the implication $WP(\pi, R \lor S) \Longrightarrow WP(\pi, R) \lor WP(\pi, S)$ is not true for this nondeterministic program.

We leave the proofs of (c) and (d) as exercises. $\square$

Theorem 2.5: Let I be an initial assertion for program π. π terminates whenever execution begins in a state satisfying I iff $I \Longrightarrow WP(\pi, T)$.

Proof: When the result assertion is T then the meaning of the predicate transformer becomes: $WP(\pi, T)$ is the weakest precondition such that execution of π begun in a state that satisfies this precondition terminates. The theorem follows from this and the meaning of *weakest*. $\square$

Now that we have studied some of the properties of the predicate transformer WP, we turn our attention to the calculus of WP. Specifically, we consider the program statements: assignment, **if-then**, **if-then-else**, and **while** and show how to obtain the weakest preconditions for these statements given any postcondition. Most other program statements in use can be represented in terms of these. We also consider the composition of several statements. In what follows, we shall often use "," in place of the connective $\wedge$. This will provide us with a slightly clearer notation.

Assignment Statements

Consider the assignment statement:

$x \leftarrow exp$

where exp is an expression. Let R be any postcondition for this assignment statement and let R^x_{exp} be the predicate obtained by substituting the expression exp for every occurrence of x in R and then simplifying the resulting predicate.

Example 2.27: Consider the assignment statement:

$x \leftarrow x + 2$

and the postcondition $R = (x = 10)$. Substituting $x + 2$ for x in R, we get $x + 2 = 10$. Simplifying, we get $R^x_{x+2} = (x = 8)$.

For the statement:

$x \leftarrow x + y$

and the postcondition $R = x>y$, we get $R^x_{x+y} = x+y>y = x>0$.

Finally, consider the statement:

$$z \leftarrow y-x$$

and the postcondition $R = (x = 10, y = 30, z = 20)$. Substituting $y-x$ for z, we get $(x = 10, y = 30, y-x = 20)$. Simplifying, we get $R^z_{y-x} = (x = 10, y = 30)$. Note that R^z_{y-x} is true iff $(x = 10, y = 30, y-x = 20)$ is true. $\square$

From the definitions provided above, it follows that:

(2.2) $WP(x \leftarrow exp, R) = R^x_{exp}$

If-then

In the program statement:

if B then α;

B is a Boolean expression (i.e., it has value **true** or **false**), and α is a program. Let R be any postcondition for this **if** statement. When B is true, the program α is executed. So, for the postcondition R to be true following the execution of this statement, $WP(\alpha, R)$ must be true preceding the execution. When B is false, the postcondition R must be true before execution inorder for it to be true following execution. So, we obtain:

(2.3) $WP(\textbf{if } B \textbf{ then } \alpha, R) = [B \wedge WP(\alpha, R)] \vee [\bar{B} \wedge R]$

Example 2.28: Let π be the program:

if $x>0$ then $z \leftarrow 3$;

If R is the predicate $z = 6$, $R^z_3 = (3=6) = F$. So, $WP(\pi, R) = [x>0 \wedge R^z_3] \vee [x\leq0 \wedge R] = [x>0 \wedge F] \vee [x\leq0 \wedge z=6] = [x\leq0 \wedge z=6]$.

Next, consider the program δ:

if $x = 2$ then $y = z+w$;

Let R be the predicate $x+y = 4$. We see that $R^y_{z+w} = (x+z+w = 4)$. So, $WP(\delta, R) = [x=2 \wedge x+z+w=4] \vee [x\neq2 \wedge x+y=4] = [x=2 \wedge z+w=2] \vee [x\neq2 \wedge x+y=4]$. $\square$

If-then-else

Consider the program statement:

if B **then** α
    else β;

where B is a Boolean expression and α and β are programs. Let R be any postcondition for this statement. When B is true, α is executed and for R to be true following execution, it must be the case that $WP(\alpha, R)$ is true before execution. When B is false, $WP(\beta, R)$ must be true before execution in order for R to be true after the execution of this statement. So, we get:

(2.4) $WP(\text{if } B \text{ then } \alpha \text{ else } \beta, R)$
    $= [B \wedge WP(\alpha, R)] \vee [\bar{B} \wedge WP(\beta, R)]$

Example 2.29: Let π be the program:

if $x>0$ **then** $y \leftarrow 3$
    else $z \leftarrow 2$;

For the post condition $R = (y=10, z=6)$, we have $WP(y \leftarrow 3, R) = (3=10, z=6) = F$, and $WP(z \leftarrow 2, R) = (y=10, 2=6) = F$. So, $WP(\pi, R) = [x>0 \wedge F] \vee [x\leq0 \wedge F] = F$.

For the post condition $R = (y=3, z=5)$, $WP(y \leftarrow 3, R) = (3=3, z=5) = (z=5)$ and $WP(z \leftarrow 2, R) = (y=3, 2=5) = F$. Hence, $WP(\pi, R) = [x>0 \wedge z=5] \vee [x\leq0 \wedge F] = [x>0 \wedge z=5]$.

Finally, consider the post condition $R = (y>0, z>1)$. $WP(y \leftarrow 3, R) = (3>0, z>1) = (T, z>1) = (z>1)$ and $WP(z \leftarrow 2, R) = (y>0, 2>1) = (y>0, T) = (y>0)$. Therefore, $WP(\pi, R) = [x>0 \wedge z>1] \vee [x\leq0 \wedge y>0]$.

Now, consider the program π:

if $x > y$ **then** $z \leftarrow x$
    else $z \leftarrow y$;

Let R be the postcondition $z = \max(x, y)$. From equalities (2.4) and (2.2), we get:

$$WP(\pi, R) = [x>y \wedge WP(z \leftarrow x, R)] \vee [x\leq y \wedge WP(z \leftarrow x, R)]$$

$$= [x>y \wedge x=\max(x, y)] \vee [x\leq y \wedge y=\max(x, y)]$$

$$= \textbf{true} \quad \square$$

While

In the program π:

while B **do**
   α;
end while

B is a Boolean expression and α is a program. Because of the specific instruction set we are considering, it is not possible for one to exit from within this loop. The program α is repeatedly executed so long as B remains true. Let R be any postcondition for this statement. Define $DO(\alpha, R, i)$ as below:

$$DO(\alpha, R, i) = \begin{cases} \overline{B} \wedge R & i=0 \\ B \wedge WP(\alpha, DO(\alpha, R, i-1)) & i>0 \end{cases}$$

Intuitively, $DO(\alpha, R, i)$ is the weakest precondition on π so that π begun in a state that satisfies this predicate will enter its **while** loop exactly i times and terminate in a state that satisfies R.

Theorem 2.6: The following equality is correct:

(2.5) $WP(\pi, R) = [\exists i, i \geq 0 \; DO(\alpha, R, i)]$

Proof: We explicitly establish the implication:

(2.6) $WP(\pi, R) \Longrightarrow [\exists i, i \geq 0 \; DO(\alpha, R, i)]$

The proof that:

(2.7) $[\exists i, i \geq 0 \; DO(\alpha, R, i)] \Longrightarrow WP(\pi, R)$

is left as an exercise.

The proof is by induction on the number of iterations of the **while** loop of π. Let s be any initial state that satisfies $WP(\pi, R)$. Consider the execution of π beginning in this state.

If the **while** loop is executed zero times, then B is false and $WP(\pi, R) = R$. So, $\overline{B} \wedge R$ is true. Hence, $DO(\alpha, R, 0)$ is true and the implication follows.

For the induction hypothesis, let j be an arbitrary natural number. Assume that the implication is correct whenever j iterations of the **while** loop are made.

We shall now show that the implication is correct whenever $j+1$ iterations of the **while** loop are made. Observe that in this case the execution of π is equivalent to the execution of the program δ given in Algorithm 2.12.

α;
while B **do**
  α;
end while

Algorithm 2.12

By assumption on s, s satisfies $WP(\pi, R)$. Further assume that π begun in state s makes exactly $j+1$ iterations of the **while** loop. It is clear that s also satisfies B, and $WP(\delta, R)$, and that program δ begun in the state s will have exactly j iterations of its **while** loop. Let t be the state of program δ after the execution of line 1 (i.e., after the execution of α). Note that the execution of this line must terminate as π begun in the state s terminates. Let γ represent the **while** loop of Algorithm 2.12. Clearly, t satisfies $WP(\gamma, R)$. From the induction hypothesis, it follows that t satisfies $DO(\alpha, R, j)$. Consequently, s satisfies $B \wedge WP(\alpha, DO(\alpha, R, j))$ which is equal to $DO(\alpha, R, j+1)$. So, the implication is valid when $j+1$ iterations are made. □

Example 2.30: Consider the program π:

while $x \leq 0$ **do**
  $x \leftarrow x + 1$;
end while
Let R be the postcondition $x > 0$. From Theorem 2.6, we obtain:

$$WP(\pi, R) = [\exists i, i \geq 0, DO(x \leftarrow x + 1, R, i)]$$

It is not too difficult to see that $DO(x \leftarrow x + 1, R, j)$ is true for $j = 0$ when $x > 0$ in the initial state and $j = \lceil -x \rceil$ when $x \leq 0$ in the initial state. So, $WP(\pi, R) = $ **true**. □

Statement Composition

Algorithm 2.9, is the composition of the statements in lines 1 - 4. The program of Algorithm 2.12 is the composition of the program α and the succeeding **while** loop. In general, a program π may be the composition α; β; γ; ...; δ of several programs.

The equality:

(2.8) $WP(\alpha; \beta, R) = WP(\alpha, WP(\beta, R))$

is easily proved correct.

Example 2.31: Let π be the composition of the programs α and β given below:

α: $x \leftarrow x + y$;
β: $y \leftarrow 2 * x$;

Let R be the predicate $(x > 0, y > 0)$.
$WP(\pi, R) = WP(\alpha, WP(\beta, R))$ (equality 2.8)

$\qquad = WP(\alpha, R^y_{2x})$ (equality 2.2)

$\qquad = WP(\alpha, (x>0, 2x>0))$

$\qquad = WP(\alpha, x>0)$

$\qquad = (x>0)^x_{x+y}$ (equality 2.2)

$\qquad = (x+y > 0)$ $\square$

Example 2.32: Consider the program π:

α: **if** $x > 3$ **then** $x \leftarrow 3$;
β: $y \leftarrow 6 - x$;
γ: $z \leftarrow x + y$;

Let R be the postcondition $(x\leq3, y\geq0, z=6)$. For program π and R, we get:

$WP(\pi, R) = WP(\alpha, WP(\beta, WP(\gamma, R)))$

$\qquad = WP(\alpha, WP(\beta, R^z_{x+y}))$

$\qquad = WP(\alpha, WP(\beta, (x\leq3, y\geq0, x+y=6)))$

$\qquad = WP(\alpha, (x\leq3, y\geq0, x+y=6)^y_{6-x})$

$\qquad = WP(\alpha, (x\leq3, 6-x\geq0, x+6-x=6))$

$\qquad = WP(\alpha, (x\leq3, x\leq6, 6=6))$

$$= WP(\alpha, (x \le 3))$$

$$= [(x>3 \wedge WP(x \leftarrow 3, x \le 3)) \vee (x \le 3 \wedge x \le 3)]$$

$$= [(x>3 \wedge x \le 3) \vee (x \le 3)]$$

$$= [x>3 \vee x \le 3]$$

$$= \textbf{true} \quad \square$$

The next few examples provide program correctness proofs. From Theorem 2.3, we see that to prove that a program π is correct with respect to the initial and result assertions I and R, we need merely prove that $I \implies WP(\pi, R)$. In our examples, we first compute $WP(\pi, R)$ and then show that $I \implies WP(\pi, R)$.

Example 2.33:   We wish to establish the correctness of the following program π. This program finds the largest and smallest of three numbers x, y, and z.

α: $b \leftarrow x$;
β: $s \leftarrow x$;
γ: **if** $y>b$ **then** $b \leftarrow y$
         else if $y<s$ **then** $s \leftarrow y$
δ: **if** $z>b$ **then** $b \leftarrow z$
         else if $z<s$ **then** $s \leftarrow z$

The result assertion R is the conjunction $V \wedge W$ of the two clauses $V = [b = \max\{x, y, z\}]$ and $W = [s = \min\{x, y, z\}]$. Since the program is supposed to work properly for all initial values of x, y, and z, the initial assertion is $I = \textbf{true}$. From Theorem 2.4 part (c), we see that it is sufficient to prove $I \implies WP(\pi, V) \wedge WP(\pi, W)$. So, let us proceed to determine $WP(\pi, V)$ and $WP(\pi, W)$.

Let γ' and δ', respectively, denote the **else** clauses of the statements labeled γ and δ.

$WP(\pi, V)$ will be obtained using statement composition. First, we compute $WP(\delta, V)$ in two steps as below.

$$WP(\delta', V) = [z<s \wedge V_z^s] \vee [z \ge s \wedge V]$$

$$= [z<s \wedge V] \vee [z \ge s \wedge V]$$

$$= V$$

$$WP(\delta, V) = [z>b \wedge V_z^b] \vee [z \le b \wedge WP(\delta', V)]$$

$$= [z > b \wedge z = \max\{x, y, z\}] \vee [z \leq b \wedge (b = \max\{x, y, z\}]$$

Next, we obtain $WP(\gamma; \delta, V)$.

$$WP(\gamma', WP(\delta, V)) = [y < s \wedge WP(\delta, V)_y^s] \vee [y \geq s \wedge WP(\delta, V)]$$

$$= WP(\delta, V)$$

$WP(\gamma; \delta, V)$

$$= WP(\gamma, WP(\delta, V))$$

$$= [y > b \wedge WP(\delta, V)_y^b] \vee [y \leq b \wedge WP(\gamma', WP(\delta, V))]$$

$$= [y > b \wedge z > y \wedge z = \max\{x, y, z\}] \vee [y > b \wedge z \leq y \wedge y = \max\{x, y, z\}]$$
$$\vee [y \leq b \wedge z > b \wedge z = \max\{x, y, z\}] \vee [y \leq b \wedge z \leq b \wedge b = \max\{x, y, z\}]$$

Now, $WP(\beta; \gamma; \delta, V)$ is easily obtained.

$$WP(\beta; \gamma; \delta, V) = WP(\gamma; \delta, V)_x^s$$

$$= WP(\gamma; \delta, V)$$

Finally, we obtain:

$WP(\pi, V)$

$$= WP(\beta; \gamma; \delta, V)_x^b$$

$$= [y > x \wedge z > y \wedge z = \max\{x, y, z\}] \vee [y > x \wedge z \leq y \wedge y = \max\{x, y, z\}]$$
$$\vee [y \leq x \wedge z > x \wedge z = \max\{x, y, z\}] \vee [y \leq x \wedge z \leq x \wedge x = \max\{x, y, z\}]$$

$$= [y > x \wedge z > y] \vee [y > x \wedge z \leq y] \vee [y \leq x \wedge z > x] \vee [y \leq x \wedge z \leq x]$$

$$= [y > x \wedge (z > y \vee z \leq y)] \vee [y \leq x \wedge (z > x \vee z \leq x)]$$

$$= [y > x] \vee [y \leq x]$$

$$= \textbf{true}$$

We leave the proof that $WP(\pi, W) = \textbf{true}$ as an exercise. Once this has been established, we obtain $WP(\pi, R) = \textbf{true}$. Consequently, $I \implies WP(\pi, R)$ and the program is correct. $\square$

Example 2.34: In this example, we consider a simple program π that contains a

while loop. This program is:

α: $x \leftarrow 0$;
β: **while** $x \leq y$ **do**
      $x \leftarrow x + 2$
   end while

This program computes the smallest even number (i.e., natural number divisible by zero) that is larger than y. In this example, we shall only prove that on termination, $R = EVEN(x) \wedge x \geq y$. The initial assertion I is **true**.

Let γ denote the body of the **while** loop. We may obtain $DO(\gamma, R, i)$ for the first few values of i as below.

$$DO(\gamma, R, 0) = x > y \wedge EVEN(x) \wedge x \geq y$$

$$= x > y \wedge EVEN(x)$$

$$DO(\gamma, R, 1) = x \leq y \wedge WP(\gamma, DO(\gamma, R, 0))$$

$$= x \leq y \wedge DO(\gamma, R, 0)_{x+2}^{x}$$

$$= x \leq y \wedge x + 2 > y \wedge EVEN(x+2)$$

$$DO(\gamma, R, 2) = x \leq y \wedge WP(\gamma, DO(\gamma, R, 1))$$

$$= x \leq y \wedge DO(\gamma, R, 1)_{x+2}^{x}$$

$$= x \leq y \wedge x + 2 \leq y \wedge x + 4 > y \wedge EVEN(x+4)$$

$$= x + 2 \leq y \wedge x + 4 > y \wedge EVEN(x+4)$$

At this time, we might suspect that:

$$DO(\gamma, R, i) = \begin{cases} x > y \wedge EVEN(x) & i = 0 \\ x + 2(i-1) \leq y \wedge x + 2i > y \wedge EVEN(x+2i) & i > 0 \end{cases}$$

We can verify the correctness of this equality by induction on i. Once this has been done, we can obtain the equality:

$$WP(\beta, R) = [x > y \wedge EVEN(x)] \\ \vee \exists i, i \geq 1 \ [x + 2(i-1) \leq y \wedge x + 2i > y \wedge EVEN(x+2i)]$$

For the complete program π, we obtain:

$$WP(\pi, R) = WP(\alpha;\beta, R)$$

$$= WP(\alpha, WP(\beta, R))$$

$$= WP(\beta, R)_0^x$$

$$= [0{>}y \wedge EVEN(0)] \vee \exists i,\, i{\geq}1\, [2(i{-}1){\leq}y \wedge 2i{>}y \wedge EVEN(2i)]$$

$$= [0{>}y] \vee \exists i,\, i{\geq}1\, [2(i{-}1){\leq}y \wedge 2i{>}y]$$

$$= \textbf{true}$$

The implication $I \Longrightarrow WP(\pi, R)$ is therefore valid. Consequently, the program is correct with respect to the given initial and result assertions. □

Example 2.35: Consider the recursive procedure *SEARCH* of Algorithm 2.1. Let i denote the result computed by this procedure. The desired result assertion R is $[0{\leq}i{\leq}n] \wedge [i{=}0 \Longrightarrow A(j) \neq X,\, 1{\leq}j{\leq}n] \wedge [i \neq 0 \Longrightarrow A(i){=}x]$. While R does not explicitly require that i be an integer, this is implied as we assume that all array indices are integer. Hence, the condition $0{\leq}i{\leq}n$ really means that i is an integer in the range $[0,n]$.

The initial assertion is "n is a natural number" as procedure *SEARCH* is supposed to terminate in a state satisfying R for all initial states in which n is a natural number. Note that when $n{<}0$ initially, R cannot be satisfied on termination as there is no i such that $0{\leq}i{\leq}n$. The correctness of *SEARCH* will be established by first determining $WP(SEARCH, R)$ and then showing $I \Longrightarrow WP(SEARCH, R)$.

Let U, V, and W, respectively denote the predicates: $[0{\leq}i{\leq}n]$, $[i{=}0 \Longrightarrow A(j) \neq X,\, 1{\leq}j{\leq}n]$, and $[i \neq 0 \Longrightarrow A(i){=}x]$. It is easy to see that $R = U \wedge V \wedge W$. In order to simplify matters, we shall use property (c) of Theorem 2.4 (i.e., distributivity of $\wedge$) and obtain $WP(SEARCH, U)$, $WP(SEARCH, V)$, and $WP(SEARCH, W)$. The conjunction of these three predicates is $WP(SEARCH, R)$.

Procedure *SEARCH* consists solely of a **case** statement. This statement is just a convenient way to write a nested **if** statement. The **case** statement of *SEARCH* is equivalent to the following **if** statement:

if $n{\leq}0$ **then return**(0)
       else if $A(n){=}X$ **then return**(n)
             else return($SEARCH(A,\, n{-}1,\, X)$)

Let β denote the second **if** statement above. I.e.,

if $A(n)=X$ **then return**(n)
    else return$(SEARCH(A, n-1, X))$

Since, the **return** statements of *SEARCH* just return a value for i, they are equivalent to assignment statements (i.e., **return**(0) is equivalent to $i \leftarrow 0;$). With these preliminaries taken care of, we are ready to obtain the weakest preconditions for $U, V,$ and W.

$WP(SEARCH(A, n, X), U)$

$$= [n \leq 0 \wedge WP(i \leftarrow 0, U)] \vee [n > 0 \wedge WP(\beta, U)]$$

$$= [n \leq 0 \wedge 0 \leq 0 \leq n]$$
$$\vee [n > 0 \wedge [(A(n)=X \wedge WP(i \leftarrow n, U))$$
$$\vee (A(n) \neq X \wedge WP(SEARCH(A, n-1, X), U))]]$$

$$= [n=0] \vee [n > 0 \wedge [(A(n)=X \wedge 0 \leq n \leq n)$$
$$\vee (A(n) \neq X \wedge WP(SEARCH(A, n-1, X), U))]]$$

$$= [n=0] \vee [n > 0 \wedge [(A(n)=X \wedge n \geq 0)$$
$$\vee (A(n) \neq X \wedge WP(SEARCH(A, n-1, X), U))]]$$

$$= [n=0] \vee [n > 0 \wedge A(n)=X]$$
$$\vee [n > 0 \wedge A(n) \neq X \wedge WP(SEARCH(A, n-1, X), U)] \quad (2.9)$$

Equations such as (2.9) which define a function in terms of itself are called *recurrence equations*. Methods to solve these equations are studied in Chapter 7. The simplest of these methods, *substitution*, will be used here to solve (2.9). While the following discussion is self contained, the wary student might wish to look at Chapter 7 for further examples on the use of this technique.

  The occurrence of *SEARCH* on the right hand side of (2.9) may be eliminated by noting that (2.9) is valid for all n. So by substituting $n-1$ for n in (2.9) we obtain $WP(SEARCH(A, n-1, X), U)$ in terms of $WP(SEARCH(A, n-2, X), U)$. Now, by substituting $n-2$ for n in (2.9), we obtain $WP(SEARCH(A, n-2, X), U)$ in terms of $WP(SEARCH(A, n-3, X), U)$. By carrying out this substitution n times as below, the value of $WP(SEARCH, U)$ is obtained.

$WP(SEARCH(A, n, X), U)$

$$= [n=0]$$
$$\vee [n > 0 \wedge A(n)=X]$$
$$\vee [n > 0 \wedge A(n) \neq X \wedge [[n-1=0] \vee [n-1 > 0 \wedge A(n-1)=X]$$
$$\vee [n-1 > 0 \wedge A(n-1) \neq X \wedge WP(SEARCH(A, n-2, X), U)]]$$

$$= [n=0]$$
$$\bigvee [n>0 \wedge A(n)=X]$$
$$\bigvee [n=1 \wedge A(n)\neq X]$$
$$\bigvee [n>1 \wedge A(n)\neq X \wedge A(n-1)=X]$$
$$\bigvee [n>1 \wedge A(n)\neq X \wedge A(n-1)\neq X \wedge WP(SEARCH(A, n-2, X), U)]$$

$$= [n=0]$$
$$\bigvee [n>0 \wedge A(n)=X]$$
$$\bigvee [n=1 \wedge A(n)\neq X]$$
$$\bigvee [n>1 \wedge A(n)\neq X \wedge A(n-1)=X]$$
$$\bigvee [n>1 \wedge A(n)\neq X \wedge A(n-1)\neq X \wedge$$
$$[[n-2=0] \vee [n-2>0 \wedge A(n-2)=X]$$
$$\vee [n-2>0 \wedge A(n-2)\neq X \wedge WP(SEARCH(A, n-3, X), U)]]]]$$

$$= [n=0]$$
$$\bigvee [n>0 \wedge A(n)=X]$$
$$\bigvee [n=1 \wedge A(n)\neq X]$$
$$\bigvee [n>1 \wedge A(n)\neq X \wedge A(n-1)=X]$$
$$\bigvee [n=2 \wedge A(n)\neq X \wedge A(n-1)\neq X]$$
$$\bigvee [n>2 \wedge A(n)\neq X \wedge A(n-1)\neq X \wedge A(n-2)=X]$$
$$\bigvee [n>2 \wedge A(n)\neq X \wedge A(n-1)\neq X \wedge A(n-2)\neq X$$
$$\wedge WP(SEARCH(A, n-3, X), U)]$$

.
.
.

$$= [n=0]$$
$$\bigvee [n>0 \wedge A(n)=X]$$
$$\bigvee [n=1 \wedge A(n)\neq X]$$
$$\bigvee [n>1 \wedge A(n)\neq X \wedge A(n-1)=X]$$
$$\bigvee [n=2 \wedge A(n)\neq X \wedge A(n-1)\neq X]$$
$$\bigvee [n>2 \wedge A(n)\neq X \wedge A(n-1)\neq X \wedge A(n-2)=X]$$
$$\bigvee [n=3 \wedge A(n)\neq X \wedge A(n-1)\neq X \wedge A(n-2)\neq X]$$
$$\bigvee [n>3 \wedge A(n)\neq X \wedge A(n-1)\neq X \wedge A(n-2)\neq X \wedge A(n-3)=X]$$
$$\bigvee [n=4 \wedge A(n)\neq X \wedge A(n-1)\neq X \wedge A(n-2)\neq X \wedge A(n-3)\neq X]$$
$$\bigvee [n>4 \wedge A(n)\neq X \wedge A(n-1)\neq X \wedge A(n-2)\neq X \wedge A(n-3)\neq X$$
$$\wedge A(n-4)=X]$$
$$\bigvee \dots$$

$$= \bigvee_{j=0}^{\infty} (n=j)$$

Hence, procedure *SEARCH* terminates in a state with $0\leq i \leq n$ (i.e., returns a value in this range) whenever it is started in a state in which n is a natural number.

Next, let us determine $WP(SEARCH, V)$. We use the symbol Q to denote the predicate $[A(j){\neq}X,\ 1{\leq}j{\leq}n]$.

$WP(SEARCH(A,\ n,\ X),\ V)$

$= [n{\leq}0 \wedge WP(i \leftarrow 0,\ V)] \vee [n{>}0 \wedge WP(\beta,\ V)]$

$= [n{\leq}0 \wedge 0{=}0 \Longrightarrow Q]$
$\quad \vee [n{>}0 \wedge [(A(n){=}X \wedge WP(i \leftarrow n,\ V))$
$\quad\qquad\qquad \vee (A(n){\neq}X \wedge WP(SEARCH(A,\ n{-}1,\ X),\ V))]]$

$= [n{\leq}0 \wedge Q]$
$\quad \vee [n{>}0 \wedge [(A(n){=}X \wedge (n{=}0 \Longrightarrow Q))$
$\quad\qquad\qquad \vee (A(n){\neq}X \wedge WP(SEARCH(A,\ n{-}1,\ X),\ V))]]$

$= [n{\leq}0]$
$\quad \vee [n{>}0 \wedge A(n){=}X]$
$\quad \vee [n{>}0 \wedge A(n){\neq}X \wedge WP(SEARCH(A,\ n{-}1,\ X),\ V)]$

$= [n{\leq}0]$
$\quad \vee [n{>}0 \wedge A(n){=}X]$
$\quad \vee [n{>}0 \wedge A(n){\neq}X \wedge [[n{-}1{\leq}0] \vee [n{-}1{>}0 \wedge A(n{-}1){=}X]$
$\quad\qquad\qquad \vee [n{-}1{>}0 \wedge A(n{-}1){\neq}X \wedge WP(SEARCH(A,\ n{-}2,\ X),\ V)]]$

$= [n{\leq}0]$
$\quad \vee [n{>}0 \wedge A(n){=}X]$
$\quad \vee [n{=}1 \wedge A(n){\neq}X]$
$\quad \vee [n{>}1 \wedge A(n){\neq}X \wedge A(n{-}1){=}X]$
$\quad \vee [n{>}1 \wedge A(n){\neq}X \wedge A(n{-}1){\neq}X \wedge WP(SEARCH(A,\ n{-}2,\ X),\ V)]$

$= [n{\leq}0]$
$\quad \vee [n{>}0 \wedge A(n){=}X]$
$\quad \vee [n{=}1 \wedge A(n){\neq}X]$
$\quad \vee [n{>}1 \wedge A(n){\neq}X \wedge A(n{-}1){=}X]$
$\quad \vee [n{>}1 \wedge A(n){\neq}X \wedge A(n{-}1){\neq}X \wedge$
$\quad\qquad [[n{-}2{\leq}0] \vee [n{-}2{>}0 \wedge A(n{-}2){=}X]$
$\quad\qquad\quad \vee [n{-}2{>}0 \wedge A(n{-}2){\neq}X \wedge WP(SEARCH(A,\ n{-}3,\ X),\ V)]]$

$= [n{\leq}0]$
$\quad \vee [n{>}0 \wedge A(n){=}X]$
$\quad \vee [n{=}1 \wedge A(n){\neq}X]$
$\quad \vee [n{>}1 \wedge A(n){\neq}X \wedge A(n{-}1){=}X]$
$\quad \vee [n{=}2 \wedge A(n){\neq}X \wedge A(n{-}1){\neq}X]$
$\quad \vee [n{>}2 \wedge A(n){\neq}X \wedge A(n{-}1){\neq}X \wedge A(n{-}2){=}X]$
$\quad \vee [n{>}2 \wedge A(n){\neq}X \wedge A(n{-}1){\neq}X \wedge A(n{-}2){\neq}X$

$$\wedge \, WP(SEARCH(A, n-3, X), V)]$$

.
.
.

$$= [n \leq 0]$$
$$\vee \, [n>0 \wedge A(n)=X]$$
$$\vee \, [n=1 \wedge A(n) \neq X]$$
$$\vee \, [n>1 \wedge A(n) \neq X \wedge A(n-1)=X]$$
$$\vee \, [n=2 \wedge A(n) \neq X \wedge A(n-1) \neq X]$$
$$\vee \, [n>2 \wedge A(n) \neq X \wedge A(n-1) \neq X \wedge A(n-2)=X]$$
$$\vee \, [n=3 \wedge A(n) \neq X \wedge A(n-1) \neq X \wedge A(n-2) \neq X]$$
$$\vee \, [n>3 \wedge A(n) \neq X \wedge A(n-1) \neq X \wedge A(n-2) \neq X \wedge A(n-3)=X]$$
$$\vee \, [n=4 \wedge A(n) \neq X \wedge A(n-1) \neq X \wedge A(n-2) \neq X \wedge A(n-3) \neq X]$$
$$\vee \, [n>4 \wedge A(n) \neq X \wedge A(n-1) \neq X \wedge A(n-2) \neq X \wedge A(n-3) \neq X$$
$$\wedge A(n-4)=X]$$
$$\vee \, ...$$

$$= \bigvee_{j=-\infty}^{\infty} (n=j)$$

Hence, $WP(SEARCH, V) = $ "n is an integer". The proof that $WP(SEARCH, W)$ $= $ "n is an integer" is left as an exercise. It follows that $WP(SEARCH, R) = $ "n is a natural number". Since, this is also the initial assertion, it follows that procedure $SEARCH$ is correct with respect to the given initial and result assertions. □

Example 2.36: In this example, we use the predicate transformer method to prove that procedure $SEQSRCH$ given in Algorithm 2.4 is correct with respect to the initial and result assertions used in Example 2.35. Let U, V, and W also be as in Example 2.35.

For the **while** loop γ:

while $A(i) \neq X$ **do**
    $i \leftarrow i-1;$

we obtain:

$$DO(i \leftarrow i-1, U, 0) = [A(i)=X] \wedge [0 \leq i \leq n]$$

$$DO(i \leftarrow i-1, U, 1)$$

$$= [A(i) \neq X] \wedge WP(i \leftarrow i-1, DO(i \leftarrow i-1, U, 0))$$

$$= [A(i) \neq X] \wedge WP(i \leftarrow i-1, [A(i)=X] \wedge [0 \leq i \leq n])$$

$$= [A(i){\neq}X] \bigwedge [A(i-1){=}X] \bigwedge [0{\leq}i-1{\leq}n]$$

$DO(i \leftarrow i-1, U, 2)$

$$= [A(i){\neq}X] \bigwedge WP(i \leftarrow i-1, DO(i \leftarrow i-1, U, 1))$$

$$= [A(i){\neq}X] \bigwedge WP(i \leftarrow i-1, [A(i){\neq}X] \bigwedge [A(i-1){=}X] \bigwedge [0{\leq}i-1{\leq}n])$$

$$= [A(i){\neq}X] \bigwedge [A(i-1){\neq}X] \bigwedge [A(i-2){=}X] \bigwedge [0{\leq}i-2{\leq}n]$$

From the form of the above equations one may guess that:

$$DO(i \leftarrow i-1, U, j) = [\bigwedge_{k=i-j+1}^{i} A(k){\neq}X] \bigwedge [A(i-j){=}X] \bigwedge [0{\leq}i-j{\leq}n]$$

One may verify the correctness of this equality for all natural numbers j by induction on j. Having determined the value of $DO(i \leftarrow i-1, U, j)$, we can proceed to determine $WP(\gamma, U)$. Using equality (2.5), we get:

$WP(\gamma, U)$

$$= \exists j, j{\geq}0 \ DO(i \leftarrow i-1, U,j)$$

$$= \exists j, j{\geq}0 \ ([\bigwedge_{k=i-j+1}^{i} A(k){\neq}X] \bigwedge [A(i-j){=}X] \bigwedge [0{\leq}i-j{\leq}n])$$

$$= \exists j, 0{\leq}j{<}i \ ([\bigwedge_{k=i-j+1}^{i} A(k){\neq}X] \bigwedge [A(i-j){=}X] \bigwedge [0{\leq}i-j{\leq}n])$$
$$\bigvee ([\bigwedge_{k=1}^{i} A(k){\neq}X] \bigwedge [A(0){=}X] \bigwedge [0{\leq}n])$$
$$\bigvee [\exists j, 0{\leq}i{<}j \ ([\bigwedge_{k=i-j+1}^{i} A(k){\neq}X] \bigwedge [A(i-j){=}X] \bigwedge [0{\leq}i-j{\leq}n])$$

$$= \exists j, 0{\leq}j{<}i \ ([\bigwedge_{k=i-j+1}^{i} A(k){\neq}X] \bigwedge [A(i-j){=}X] \bigwedge [0{\leq}i-j{\leq}n])$$
$$\bigvee ([\bigwedge_{k=1}^{i} A(k){\neq}X] \bigwedge [A(0){=}X] \bigwedge [0{\leq}n])$$

From this, we obtain:

$WP(SEQSRCH, U)$

$$= WP(i \leftarrow n, WP(A(0) \leftarrow X, WP(\gamma, U)))$$

$$= WP(i \leftarrow n,$$
$$\exists j, 0{\leq}j{<}i \ ([\bigwedge_{k=i-j+1}^{i} A(k){\neq}X] \bigwedge [A(i-j){=}X] \bigwedge [0{\leq}i-j{\leq}n])$$

$$\bigvee ([\bigwedge_{k=1}^{i} A(k)\neq X] \wedge [X=X] \wedge [0\leq n])]$$

$$= \exists j,\, 0\leq j<n\ ([\bigwedge_{k=n-j+1}^{n} A(k)\neq X] \wedge [A(n-j)=X] \wedge [0\leq n-j\leq n])$$
$$\bigvee ([\bigwedge_{k=1}^{n} A(k)\neq X] \wedge [0\leq n])$$

$$= \exists j,\, 0\leq j<n\ ([\bigwedge_{k=n-j+1}^{n} A(k)\neq X] \wedge [A(n-j)=X])$$
$$\bigvee ([\bigwedge_{k=1}^{n} A(k)\neq X] \wedge [0\leq n])$$

$$= \text{"}n \text{ is a natural number"}$$

The last equality uses the fact that n is given to be an integer.

In obtaining $WP(SEQSRCH, V)$, we use Q to denote the predicate $[A(j)\neq X,\ 1\leq j\leq n]$. The following development follows that for U.

$DO(i \leftarrow i-1, V, 0) = [A(i)=X] \wedge [i=0 \Rightarrow Q]$

$DO(i \leftarrow i-1, V, 1)$

$\quad = [A(i)\neq X] \wedge WP(i \leftarrow i-1, DO(i \leftarrow i-1, V, 0))$

$\quad = [A(i)\neq X] \wedge WP(i \leftarrow i-1, [A(i)=X] \wedge [i=0 \Rightarrow Q])$

$\quad = [A(i)\neq X] \wedge [A(i-1)=X] \wedge [i-1=0 \Rightarrow Q]$

$DO(i \leftarrow i-1, V, 2)$

$\quad = [A(i)\neq X] \wedge WP(i \leftarrow i-1, DO(i \leftarrow i-1, V, 1))$

$\quad = [A(i)\neq X] \wedge WP(i \leftarrow i-1, [A(i)\neq X] \wedge [A(i-1)=X] \wedge [i-1=0 \Rightarrow Q])$

$\quad = [A(i)\neq X] \wedge [A(i-1)\neq X] \wedge [A(i-2)=X] \wedge [i-2=0 \Rightarrow Q]$

From the form of the above equations one may guess that:

$$DO(i \leftarrow i-1, V, p) = [\bigwedge_{k=i-p+1}^{i} A(k)\neq X] \wedge [A(i-p)=X] \wedge [i-p=0 \Rightarrow Q]$$

One may verify the correctness of this equality for all natural numbers p by induction on p. Using equality (2.5), we get:

$WP(\gamma, V)$

$= \exists p, p \geq 0 \ DO(i \leftarrow i-1, V, p)$

$= \exists p, p \geq 0 \ ([\bigwedge_{k=i-p+1}^{i} A(k) \neq X] \wedge [A(i-p)=X] \wedge [i-p=0 \Rightarrow Q])$

$= \exists p, 0 \leq p < i \ ([\bigwedge_{k=i-p+1}^{i} A(k) \neq X] \wedge [A(i-p)=X] \wedge [i-p=0 \Rightarrow Q])$
$\quad \vee \ ([\bigwedge_{k=1}^{i} A(k) \neq X] \wedge [A(0)=X] \wedge [i=i \Rightarrow Q])$
$\quad \vee \ [\exists p, 0 \leq i < p \ ([\bigwedge_{k=i-p+1}^{i} A(k) \neq X] \wedge [A(i-p)=X] \wedge [i-p=0 \Rightarrow Q])$

$= \exists p, 0 \leq p < i \ ([\bigwedge_{k=i-p+1}^{i} A(k) \neq X] \wedge [A(i-p)=X])$
$\quad \vee \ ([\bigwedge_{k=1}^{i} A(k) \neq X] \wedge [A(0)=X] \wedge Q)$
$\quad \vee \ [\exists p, 0 \leq i < p \ ([\bigwedge_{k=i-p+1}^{i} A(k) \neq X] \wedge [A(i-p)=X])$

From this, we obtain:

$WP(SEQSRCH, V)$

$= WP(i \leftarrow n, \ WP(A(0) \leftarrow X, \ WP(\gamma, V)))$

$= WP(i \leftarrow n,$
$\quad [\exists p, 0 \leq p < i \ ([\bigwedge_{k=i-p+1}^{i} A(k) \neq X] \wedge [A(i-p)=X])$
$\quad \quad \vee \ ([\bigwedge_{k=1}^{i} A(k) \neq X] \wedge [X=X] \wedge Q)$
$\quad \quad \vee \ \exists p, 0 \leq i < p \ ([\bigwedge_{k=1}^{i} A(k) \neq X] \wedge [X \neq X] \ \bigwedge_{k=i-p+1}^{-1} [A(k) \neq X]$
$\quad \quad \wedge [A(i-p)=X])$

$= \exists p, 0 \leq p < n \ ([\bigwedge_{k=n-p+1}^{n} A(k) \neq X] \wedge [A(n-p)=X])$
$\quad \vee \ ([\bigwedge_{k=1}^{n} A(k) \neq X] \wedge Q)$

$= \exists p, 0 \leq p < n \ ([\bigwedge_{k=n-p+1}^{n} A(k) \neq X] \wedge [A(n-p)=X])$
$\quad \vee \ [\bigwedge_{k=1}^{n} A(k) \neq X]$

$\quad = \textbf{true}$

The proof that $WP(SEQSRCH, W) = \textbf{true}$ is left as an exercise. Using this, we get $WP(SEQSRCH, R) = $ "n is a natural number". So, $SEQSRCH$ is correct with respect to the given initial and result assertions. $\square$

Example 2.37: In this example, we establish the correctness of procedure $INSERT$ (Algorithm 2.5) with respect to the initial assertion $I = $ "n is a natural number and A is sorted in nondecreasing order" and the result assertion $R = U \bigwedge W$ where $U = [A(1) \leq A(2) \leq \cdots \leq A(n+1)]$ and $V = $ "$A(1:n+1)$ is a permutation of $A(1:n)$ and X". Observe that simply having $R = U$ is not sufficient to assert that $INSERT$ is a correct insertion program. We use the notation $i-j$ to denote the program comprised solely of lines i through j of procedure $INSERT$. So, $2-7$ denotes the entire body of this procedure and $3-6$ denotes the **while** loop only.

To simplify the representation of the weakest precondition predicates, we define predicates $R(i, j)$, for all integers i and all natural numbers j. Intuitively, $R(i, j)$ represents the requirement on the $A(i)$s and X so that if the **while** loop begins with this value of i and is entered exactly j times, then following the execution of line 7, U is satisfied. So, $R(i, 0)$ is the weakest precondition for lines 3-7 to terminate in a state that satisfies U if the **while** loop is not entered. In other words, $R(i,0)$ is $WP(7-7, U)$. One may verify that the following equality is correct:

$$R(i,0) = \begin{cases} A(1)\leq\ldots\leq A(n)\leq X & i=n \\ A(1)\leq\ldots\leq A(n+1) & i<0 \bigvee i>n \\ X\leq A(1)\leq\ldots\leq A(n+1) & i=0 \\ A(1)\leq\ldots\leq A(i)\leq X\leq A(i+2)\leq\ldots\leq A(n+1) & 0<i<n \end{cases}$$

Let $W = WP(7-7, U)$. From our previous remarks,

$$W = WP(A(i+1) \leftarrow X, U) = R(i,0)$$

For $R(i,i)$, $i>0$, we see that:

$$R(i,i) = \begin{cases} X\leq A(1)\leq\ldots\leq A(n) & i\geq n \\ X\leq A(1)\leq\ldots\leq A(i)\leq A(i+2)\leq\ldots\leq A(n+1) & 0<i<n \end{cases}$$

Since the number of iterations of the **while** loop cannot be fewer than 0, we need not consider the case $i<0$ in obtaining the value of $R(i,i)$. The case $i=0$ is covered by the values for $R(i,0)$.

For j in the range $0<j<i\leq n$, we have:

$$R(i,j) = \begin{cases} A(1)\leq\ldots\leq A(n-j)\leq X\leq A(n-j+1)\leq\ldots\leq A(n) & i=n \\ A(1)\leq\ldots\leq A(i-j)\leq X\leq A(i-j+1)\leq\ldots & 0<i<n \\ \qquad\leq A(i)\leq A(i+2)\leq\ldots\leq A(n+1) & \end{cases}$$

The cases $i<0$, $i>n$, and $j>i$ are left as an exercise. For $DO(4-5, W, j)$, we get:

$$DO(4-5, W,0) = [A(i)\leq X] \wedge R(i,0)$$

$DO(4-5, W,1)$

$= [A(i)>X] \wedge WP(4-5, DO(4-5, W, 0))$

$= [A(i)>X] \wedge WP(4-5, [A(i)\leq X] \wedge R(i,0))$

$= [A(i)>X] \wedge WP(4-4, [A(i-1)\leq X \wedge R(i-1,0))$

$= [A(i)>X] \wedge [A(i-1)\leq X] \wedge R(i,1)$

$DO(4-5, W, 2)$

$= [A(i)>X] \wedge WP(4-5, DO(4-5, W, 1))$

$= [A(i)>X] \wedge WP(4-5, [A(i)>X] \wedge [A(i-1)\leq X] \wedge R(i,1))$

$= [A(i)>X] \wedge WP(4-4, [A(i-1)>X] \wedge [A(i-2)\leq X] \wedge R(i-1,1))$

$= [A(i)>X] \wedge [A(i-1)>X] \wedge [A(i-2)\leq X] \wedge R(i,2))$

And, in general we have:

$$DO(4-5, W, j) = [\bigwedge_{k=0}^{j-1} A(i-k)>X] \wedge [A(i-j)\leq X] \wedge R(i,j)$$

for every natural number j. So,

$WP(3-7, U)$

$= \exists j, j\geq 0 \; DO(4-5, W, j)$

$= \exists j, j\geq 0 \; ([\bigwedge_{k=0}^{j-1} A(i-k)>X] \wedge [A(i-j)\leq X] \wedge R(i,j))$

Hence,

$WP(INSERT, U)$

$$= WP(i \quad n, \exists j, 0 \leq j < i \ ([\bigwedge_{k=0}^{j-1} A(i-k) > X] \land [A(i-j) \leq X] \land R(i,j)))$$
$$\lor ([\bigwedge_{k=1}^{i} A(k) > X] \land [X \leq X] \land R(i,i)))$$

$$= \exists j, 0 \leq j < n \ ([\bigwedge_{k=0}^{j-1} A(n-k) > X] \land [A(n-j) \leq X] \land R(n,j)))$$
$$\lor ([\bigwedge_{k=1}^{n} A(k) > X] \land R(n,n))$$

It is easy to see that $I \implies WP(INSERT, U)$. The derivation of $WP(INSERT, V)$ is left as an exercise. The correctness of *INSERT* with respect to I and R will then follow from a proof of $I \implies WP(INSERT, V)$ and Theorem 2.4. □

Remarks The predicate transformer method gets quite unmanageable when one is dealing with complex programs. In particular, computing the weakest precondition for programs with nested loops is quite laborious. So, there are severe limitations to the use of this method in practice. These limitations are greatly reduced if one places assertions at the start and end of blocks of the program and uses the predicate transformer method to show that if the given precondition is true, then the specified postcondition is also true. An alternative is to construct the correctness proof while one is building the program. The text *The Science of Programming*, by David Gries provides an in depth account of how programs and their proofs can be built hand in hand.

REFERENCES AND SELECTED READINGS

The definition of Ackermann's function used in Example 2.11 is due to R. Tarjan and appears in:

> Efficiency of a good but not linear set union algorithm, by R. Tarjan, *Journal of the ACM*, Vol. 22, No. 2, April 1975, pp. 215-225.

Some advanced readings on the subject of proving programs correct are:

> A *discipline of programming*, by E.W. Dijkstra, Prentice-Hall Inc., Englewood Cliffs, New Jersey, 1976.

> *The Science of Programming*, by D. Gries, Springer Verlag, 1981.

and

> A *programming logic*, by R.L. Constable and M.J. O'Donnell, Winthrop Publishers, Massachusetts, 1978.

EXERCISES

1. Give a constructive proof of the following statements:

    (a) There exists an integer that is a perfect square.
    (b) For all n, $n \geq 7$, n cents of postage can be made using 2 cent and 7 cent stamps only.
    (c) Every integer greater than 2 can be represented as the product of prime numbers.

2. A chessboard is *defective* iff it contains exactly one "shaded in" square. Figure 2.2(a) shows a defective 4×4 chessboard. Figure 2.2(b) shows a *triomino* and Figure 2.2(c) shows a tiling of a defective 4×4 chessboard by triominoes. Angled arrows indicate triominoes. Note that in a tiling all non-defective squares are covered by exactly one triomino. The defective square has no triomino over it.

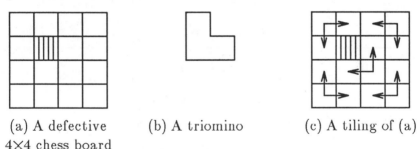

(a) A defective
4×4 chess board

(b) A triomino

(c) A tiling of (a)

Figure 2.2

Provide constructive proofs showing that the 8×8 defective chessboards of Figures 2.3(a) and (b) can be tiled by triominoes.

3. Prove inference rules MI2 and MI3.

4. Prove the following using the inference rule MI1. In each case, n is a natural number.

    (a) $\displaystyle\sum_{i=0}^{n} i^2 = n(n+1)(2n+1)/6$

    (b) $\displaystyle\sum_{i=0}^{n} i^3 = n^2(n+1)^2/4$

    (c) $\displaystyle\sum_{i=0}^{n} (2i-1)^2 = n(2n-1)(2n+1)/3+1$

    (d) $\displaystyle\sum_{i=0}^{n} 2^i = 2^{n+1}-1$

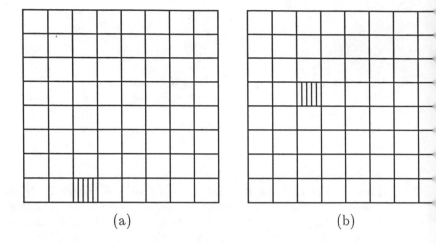

(a)   (b)

Figure 2.3

(e) $\displaystyle\sum_{i=0}^{n} (x+i*a) = (n+1)x + an(n+1)/2$

(f) $\displaystyle\sum_{i=1}^{n} \frac{1}{i(i+1)} = n/(n+1)$

(g) $\displaystyle\sum_{i=0}^{n} \frac{1}{2^i} = 2 - \frac{1}{2^n}$

(h) $\displaystyle\sum_{i=0}^{n} \frac{i}{2^i} = 2 - \frac{n}{2^n} - \frac{2}{2^n}$

5.   [Harmonic Numbers] Let $H_n = \displaystyle\sum_{i=1}^{n} \frac{1}{i}$, $n\geq 1$. H_n is the nth *Harmonic* number. Show that:

(a) $H_{2^m} \geq 1 + m/2$, $m\geq 0$.

(b) $(m+1)H_m - m = \displaystyle\sum_{i=1}^{m} H_i$, $m\geq 1$.

(c) $\displaystyle\sum_{i=2}^{m} \frac{1}{i(i-1)} H_i = 2 - \frac{H_{m+1}}{m} - \frac{1}{m+1}$, $m\geq 2$.

(d) $\displaystyle\sum_{i=0}^{m} \frac{1}{2i+1} = H_{2m+1} - \frac{1}{2}H_m$, $m\geq 0$.

(e) $\displaystyle\sum_{i=1}^{m} H_i^2 = (m+1)H_m^2 - (2m+1)H_m + 2m$, $m\geq 1$.

6.   Let $A(i,j)$ be the Ackermann's function defined in Example 2.11. Use mathematical induction to show the following:

(a) $A(i,2) = 4, i \geq 1$

(b) $A(1,j) = 2^j, j \geq 1$

(c) $A(2,j) = \left. 2^{2^{2^{\cdot^{\cdot^{\cdot^{2}}}}}} \right\} j$ twos, $j \geq 1$

(d) $A(i,j) \geq j, i \geq 0, j \geq 0$

(e) $A(i+1,j) \geq A(i,j), i \geq 0, j \geq 0$.

7. Prove the following using strong mathematical induction. In each case, n is an integer.

(a) If $t(0) = 2$ and $t(n) = 3 + t(n-1), n > 0$, then $t(n) = 3n+2, n \geq 0$.

(b) If $t(1) = c_1$ and $t(n) = c_2 n + t(n-1), n > 1$, then $t(n) = c_2 n(n+1)/2 + c_1 - c_2, n > 1$.

(c) If $t(0) = 0, t(1) = 4\sqrt{5}$, and $t(n) = 6t(n-1) - 4t(n-2), n \geq 2$, then $t(n) = 2(3+\sqrt{5})^n - 2(3 - \sqrt{5})^n, n \geq 0$.

(d) If $f(0) = 2.5, f(1) = 4.5, f(n) = 5f(n-1) - 6f(n-2) + 3n^2, n \geq 2$, then $f(n) = 22.5 + 10.5n + 1.5n^2 - 30*2^n + 10*3^n, n \geq 0$.

(e) If $f(0) = f(1) = f(2) = f(3) = 1$ and $f(n) = 10f(n-1) - 37f(n-2) + 60f(n-3) - 36f(n-4) + 4, n \geq 4$, then $f(n) = 1, n \geq 0$.

(f) If $f(0) = f(1) = f(2) = 1, f(3) = 4$, and $f(n) = 10f(n-1) - 37f(n-2) + 60f(n-3) - 36f(n-4) + 4, n \geq 4$, then $f(n) = (6+1.5n)2^n + (n-6)3^n + 1, n \geq 0$.

(g) If $f(0) = 1, f(1) = 2, f(2) = 20$, and $f(n) = 6f(n-1) - 12f(n-2) + 8f(n-3), n \geq 3$, then $f(n) = (2n^2 - 2n + 1)2^n, n \geq 0$.

(h) If $f(0) = 0$ and $f(n) = 2f(n-1) + 7n, n \geq 1$, then $f(n) = 14*2^n - 7n - 14, n \geq 0$.

(i) If $f(0) = 0$ and $f^2(n) - 2f^2(n-1) = 1$, then $f(n) = \sqrt{2^n - 1}, n \geq 0$.

8. Let $m_i, 1 \leq i \leq n$ be n positive integers. Let $p = \sum_{i=1}^{n} m_i$. Show that:

$$\sum_{i=1}^{n} (m_i / \sum_{j=i}^{n} m_j) \leq \sum_{i=1}^{p} 1/i$$

Use induction on n.

9. Use mathematical induction to show that all $2^n \times 2^n$ defective chessboards can be tiled by triominoes (see Exercise 2).

10. What is wrong with the following proof (reproduced from Opus 1961):

Theorem: All horses are the same color.

Proof: It is obvious that one horse is the same color. Let us assume the predicate $P(k)$ that k horses are the same color and use this to imply that k + 1 horses are the same color. Given the set of $k + 1$ horses, we remove one horse; then the remaining k horses are the same color by hypothesis. We remove another horse and replace the first; the k horses, by hypothesis, are again the same color. We repeat this until by exhaustion the $k + 1$ sets of k horses have each been shown to be the same color. It follows then that since every horse is the same color as every other horse, $P(k)$ entails $P(k + 1)$. But since we have shown $P(1)$ to be true, P is true for all succeedings values of k, that is, all horses are the same color. $\square$

11. Show that:
    (a) $\displaystyle\bigoplus_{1 \leqslant i \leqslant k} X_i = \overline{X}_j \oplus X_1 \oplus ... \oplus X_{j-1} \oplus X_{j+1} \cdots \oplus X_k$

    for all j, $1 \leqslant j \leqslant k$.

    (b) $\displaystyle\bigoplus_{1 \leqslant i \leqslant k} \overline{X}_i = \begin{cases} \displaystyle\bigoplus_{1 \leqslant i \leqslant k} X_i & \text{if } k \text{ is even} \\[2mm] \displaystyle\overline{\bigoplus_{1 \leqslant i \leqslant k} X_i} & \text{if } k \text{ is odd} \end{cases}$

    $\oplus$ denotes the exclusive or operator: $P \oplus Q = (P \lor Q) \land (\overline{P} \lor \overline{Q})$

12. Prove that inference rules MD1 and MD2 are correct.

13. Prove the following using either MD1 or MD2.
    (a) $\displaystyle\sum_{i=-n}^{-1} i = -\frac{n(n+1)}{2}$, n a natural number.
    (b) $[F_{1000} = 1000] \land [F_i = F_{i+1} - 1, i < 1000] \Rightarrow F_i = i$, i an integer and i less than or equal to 1000.
    (c) $[F_{1000} = 1000 \land F_{999} = 1998 \land F_i = F_{i+2} - 2, i < 1000] \Rightarrow [F_i = i$, if $i \leq 1000$ and even and $F_i = 2i$ if $i \leq 1000$ and odd]

14. Prove that the following inference rules are correct ($D = \{a, a+1, ...\}$ and $E = \{b, b+1, ...\}$).
    (a) $[P(a,b) \land \forall x, x+1 \in D(P(x,b) \Rightarrow P(x+1,b) \land \forall x \in D \; \forall y, y+1 \in E(P(x,y) \Rightarrow P(x,y+1))] \Rightarrow \forall x \in D \; \forall y \in E \; P(x,y)$
    (b) $[P(a,b) \land \forall y, y+1 \in E(P(a,y) \Rightarrow P(a,y+1)) \land \forall x \in D, \forall y \in E (P(x,y) \Rightarrow P(x+1,y))] \Rightarrow \forall x \in D, \forall y \in E \; P(x,y)$

15. Prove the validity of inference rule WM2.

16. Prove the following (n is a natural number and c is a positive constant):

(a) If $T(n) = \begin{cases} T(\lfloor n/5 \rfloor) + T(\lfloor 3n/4 \rfloor) + cn & n>0 \\ 0 & n=0 \end{cases}$

then $T(n) \le 20cn$, $n \ge 0$.

(b) If $T(n) = \begin{cases} T(\lfloor n/9 \rfloor) + T(\lfloor 63n/72 \rfloor) + cn & n>0 \\ 0 & n=0 \end{cases}$

then $T(n) \le 72cn$, $n \ge 0$.

(c) If $T(n) = \begin{cases} T(\lfloor n/7 \rfloor) + T(\lfloor 27n/35 \rfloor) + cn & n>0 \\ 0 & n=0 \end{cases}$

then $T(n) \le 35cn$, $n \ge 0$.

(d) If $T(n) = \begin{cases} T(\lfloor 3n/14 \rfloor) + T(\lfloor 2n/3 \rfloor) + cn & n>0 \\ 0 & n=0 \end{cases}$

then $T(n) \le 42cn$, $n \ge 0$.

(e) If $f(n) = \begin{cases} 2f(n/2) + 3 & n>2 \\ 2 & n=2 \end{cases}$

then $f(n) = 5n\,/\,2 - 3$ for $n = 2^k$, $k \ge 1$ and integer.

(f) If $R(1) = c$, $R(2) = 2.5c$, $R(n) = cn + \dfrac{2}{n}\displaystyle\sum_{i=n/2}^{n-1} R(i)$ when n is even and greater than 2, and $R(n) = cn + \dfrac{2}{n}\displaystyle\sum_{i=(n+1)/2}^{n-1} R(i)$ when n is odd and greater than 2, then $R(n) \le 4cn$, $n \ge 1$.

17. Prove or disprove the following inference rule:

$\{P(a),\ \forall x \in D,\ x>a\ [\forall y, a \le y < x\ P(y) \Rightarrow P(x)]\} \models \forall x \in D\ P(x)$

where D consists of all real numbers greater than or equal to a.

18. Use mathematical induction to show that every natural number can be written as the product of prime numbers.

19. Prove that $\gcd(m'+1, b) = \gcd(b, \text{rem}(m'+1, b))$ in the proof of Example 2.21.

20. Use MI1 to prove that predicate P of Example 2.23 is true each time line 3 of procedure *EUCLID* is reached.

21. Write a recursive algorithm corresponding to the iterative version of Euclid's gcd algorithm given in Algorithm 2.7. Prove the correctness of your algorithm using the method of Section 2.3.1.

22. Write a recursive version of Algorithm 2.8. Prove its correctness using the method of Section 2.3.1.

23. Prove that Algorithm 2.8 is correct using the method of Section 2.3.2.

24. Use the method of loop invariants to show that Algorithm 2.4 is correct.

25. Use the method of loop invariants to show that Algorithm 2.5 is correct.

26. Let $A(1:m)$ and $B(1:n)$ be two nondecreasing sequences of numbers, $m \geq 0$ and $n \geq 0$. Let X be a third number. We wish to find a pair (i,j), $1 \leq i \leq m$ and $1 \leq j \leq n$ such that $A(i)+B(j) = X$. Show that procedure SUM (Algorithm 2.13) returns $(0,0)$ iff there is no pair (i,j) such that $A(i)+B(j) = X$; otherwise SUM returns (i,j) such that $A(i)+B(j) = X$. Do this using a loop invariant.

```
line  procedure SUM(A,B,m,n,X)
        //A and B are nondecreasing sequences//
 1      integer m,n,X,A(m),B(n),i,j
 2      (i,j) ← (1,n)
 3      while i≤m and j≥1 do
 4        case
 5          :A(i)+B(j) = X: return((i,j))
 6          :A(i)+B(j) < X: i ← i+1
 7          :else: j ← j−1
 8        end case
 9      end while
10      return((0,0))
11    end SUM
```

Algorithm 2.13

27. Let $P(x)$ be a polynomial of degree n as in Example 2.3. Let $a(i)$ be the coefficient of x^i. Procedure $EVAL$ computes the value of the polynomial at any point X. I.e., it computes $P(x)$.
    (a) Establish the correctness of procedure $EVAL$ using the method of Section 2.3.2.
    (b) Prove that $EVAL$ is correct using the loop invariant method.
    (c) Obtain a recursive version of $EVAL$ and prove its correctness using the method of Section 2.3.1.

28. Procedure $PERM$ (Algorithm 2.15) generates all permutations of the characters in the string A. n is the length of A. The initial call is $PERM(A, 1, n)$. $INTERCHANGE(A, k, i)$ swaps the kth and ith

```
line   procedure EVAL(a, x, n)
 1         decalre a(0:n), n, x, value
 2         value ← a(n);
 3         for i ← n downto 0 do
 4             value ← value*x + a(i−1)
 5         end for
 6         return(value)
 7     end EVAL
```

Algorithm 2.14 Evaluate a polynomial

```
line   procedure PERM(A, k, n)
           //Generate all permutations of characters k through n of string A//
 1         declare A, B, k, n
 2         if k=n then [print(A); return]
 3         B ← A
 4         for i ← k to n do
 5             INTERCHANGE(A, k, i)
 6             PERM(A, k+1, n)
 7             A ← B
 8         end for
 9     end PERM
```

Algorithm 2.15 Recursive permutation generator

characters of A. Prove that this procedure is correct for all natural numbers n. Use induction on n.

29.   Let α, β, γ, and δ be the following program statements:

α: $x \leftarrow u + v$;
β: $y \leftarrow x + u$;
γ: **if** $y<0$ **then** $x \leftarrow y - v$;
δ: **if** $u \geq 5$ **then** $y \leftarrow x - 6$
           else $y \leftarrow x + 4$;

Obtain the following weakest preconditions:

(a) $WP(\alpha, x>0)$
(b) $WP(\alpha, x>3 \wedge u<5)$
(c) $WP(\beta, 5 \leq y \leq 10 \wedge x=3)$
(d) $WP(\gamma, x>3)$
(e) $WP(\gamma, x=y-v \wedge y>0)$

    (f) $WP(\delta, y = x - 6 \lor y = x + 4)$

    (g) $WP(\gamma, 4 \leq y \leq 25)$

    (h) $WP(\alpha;\beta, x = 4 \land y = 6)$

    (i) $WP(\beta;\gamma, x > 0 \land y > 0)$

    (j) $WP(\gamma;\delta, x > 0 \land y > 0)$

    (k) $WP(\alpha;\beta;\gamma;\delta, u = 3 \land x > 2 \land y > 3)$

    (l) $WP(\alpha;\beta;\gamma;\delta, u < 0 \land x > 0 \land y < 0)$

    (m) $WP(\alpha;\beta;\gamma;\delta, v = 6 \land x < 0 \land y < 5)$

30.   Prove parts (c) and (d) of Theorem 2.4.

31.   Prove the implication (2.7).

32.   Show that $WP(\pi, W) = $ **true** where π, and W are as defined in Example 2.33.

33.   Use the predicate transformer method to show that the program of Example 2.34 does, in fact, find the smallest even number that is larger than y.

34.   Consider the following program:

α: $x \leftarrow 1$;
β: **while** $x \leq y$ **do**
      $x \leftarrow x * 2$
    end while

Show that on termination, x is the smallest power of two that is larger than y. Use the predicate transformer method.

35.   Let π be the following program:

$i \leftarrow 2; b \leftarrow A(1)$
while $i \leq n$ **do**
    if $b < A(i)$ **then** $b \leftarrow A(i)$
end while

max Let $R = [b = \max_{1 \leq i \leq n} \{A(i)\}]$. Obtain $WP(\pi, R)$.

36.   Let π be as below:

$i \leftarrow 2; b \leftarrow A(1); s \leftarrow A(1)$
while $i \leq n$ **do**
   if $b < A(i)$ **then** $b \leftarrow A(i)$
         else if $s > A(i)$ **then** $s \leftarrow A(i)$
end while

Let $U = [b = \max_{1 \leq j \leq n} \{A(j)\}]$, $V = [s = \max_{1 \leq j \leq n} \{A(j)\}]$, and $R = U \bigwedge V$.
Obtain $WP(\pi, R)$.

37.   Complete Example 2.35 by determining $WP(SEARCH, W)$.

38.   Complete Example 2.35 by determining $WP(SEQSRCH, W)$.

39.   Obtain $R(i,j)$ for the remaining cases when $0 < j < i$ and also for the case $j > i$. See Example 2.37.

40.   Complete Example 2.37 by obtaining $WP(INSERT, V)$.

41.   Use the predicate transformer method to prove that Algorithm 2.2 is correct with respect to the initial and result assertions given in Example 2.37.

42.   Use the predicate transformer method to prove that procedure $EUCLID$ (Algorithm 2.7) is correct with respect to the initial asertion $m \geq n > 0$ and the result assertion $v = \gcd(m,n)$. v denotes the value returned by the procedure.

43.   Use the predicate transformer method to prove that Algorithm 2.8 is correct. Formulate adequate initial and result assertions.

44.   (a)   Obtain a formula similar to (2.5) for the weakest precondition of the **for** statement:

       for $i \leftarrow n$ **downto** 0 **do**
          α
       end for

    (b)   Use the predicate transformer method to prove that Algorithm 2.14 is correct. Formulate adequate initial and result assertions.

45.   (a)   Obtain a formula similar to (2.5) for the weakest precondition of the **until** statement:

       repeat
          α
       until B

(b)   Consider the program π:

repeat
   $x \leftarrow x + 1;$
until $x > 0$

Let R be the postcondition $x > 0$. Use the predicate transformer method to show the $WP(\pi, R) =$ **true**.

CHAPTER 3

SETS

3.1 SETS, MULTISETS, AND SUBSETS

The notion of a set is fundamental to many branches of study. In fact, we have already used this term in earlier chapters of this book. The following phrases are perhaps familiar to many of us:
  (i)   The set of reserved words in a programming language.
  (ii)   The set of integers.
  (iii) The set of natural numbers.
  (iv)   The domain of a predicate parameter is the set of values that can be assigned to it.
  (v)   The set of students who failed the discrete structures final.

A *set* is formally defined to be a collection of distinct elements. The set of suits in a card deck is {spades, hearts, diamonds, clubs}. The set of U.S. coin denominations is {1¢, 5¢, 10¢, 25¢, 50¢, $1}. {red, blue, green} is a set of primary colors and {violet, indigo, blue, green, yellow, orange, red} is the set of colors in a rainbow.

The term *multiset* denotes a collection of elements. The only difference between the notion of a set and that of a multiset is that a multiset is permitted to have multiple occurrences of the same element whereas no two elements in a set can be the same. Thus the collection {1¢, 1¢, 1¢, 5¢, 5¢}, of coins in my pocket is a multiset but not a set. An obvious consequence of the definitions just given is that all sets are multisets while some multisets are not sets.

Even though the elements in each of the examples considered so far share some common characteristics, this is not required by the definition of a set or multiset. It is perfectly legitimate to refer to collections of unrelated elements as sets or multisets. So, {Jimmy, teeth, peanut, smart, adept}, {red, $1, monday, banana}, and {monkey, sand} are all sets. It is also possible for the elements in a set to themselves be sets. So, {{1,2}, {3,4}, {-1,1/2,3}} is a set containing the three elements: {1,2}, {3,4}, and {-1,1/2,3}. {{{1},{2,3}}} is a one element set. Its only member is {{1}, {2,3}}, which in turn is a two element set.

129

The *size* or *cardinality* of a set, S, denoted $|S|$, is the number of elements in S. Thus, $|\{a,b,d,z\}| = 4$, and $|\{1,2\},\{3,4\},\{-1,1/2,6\}| = 3$. For any set S, the notation $x \in S$ will mean x is an element (or member) of S. $x \notin S$ means that x is not a member of S. A *finite set* is a set containing a finite number of elements. Likewise, an *infinite set* is any set with an infinite size. The *empty set* is a set containing no elements and is denoted as $\varnothing$. Note that the set $\{\varnothing\}$ is not empty as it contains one element. This element is the empty set. If my pocket had a large hole in it, then $S = \varnothing$ would define the set of objects in my pocket.

A set S is a *subset* of the set T if and only if (abbreviated iff) every element in S is an element of T. For example, $\{2,4,6\}$ is a subset of $\{1,2,3,4,5,6\}$. The notation $S \subseteq T$ will mean S is a subset of T. If S is a subset of T and T contains at least one element that is not in S, then S is a *proper subset* of T. This will be denoted as $S \subset T$. So, $\{1,2,3\} \subseteq \{1,2,3\}$, $\{1,2,3\} \subseteq \{1,2,3,4\}$, and $\{1,2,3\} \subset \{1,2,3,4\}$. Note that if $S \subset T$ then $S \subseteq T$ but the reverse may not be true. If $S \subseteq T$ and $T \subseteq S$ then the two sets S and T are said to be *equal* (or identical). In this case, we write $S = T$. Thus, $\{1,2,3\} = \{3,2,1\} = \{3,1,2\}$. Note that the order in which elements are listed within the braces ($\{\ \}$) is not important. As far as a set is concerned, changing the order of elements does not change the set.

In some applications, the order in which set elements are listed is important. Hence, the need to speak of *ordered sets*. Two ordered sets are equal iff they are equal as sets and if the elements appear in the same order. Ordered sets will be represented using parentheses rather than braces. Let $S = (1,2,3)$ and $T = (3,1,2)$ be two ordered sets. $S \neq T$.

It is often desirable to focus attention on some set U of elements. All the sets to be considered are subsets of this set. In this case, we shall refer to U as the *universal set*.

3.2 SET SPECIFICATION

3.2.1 Explicit

We have already used this method to specify a set. Here the elements of a set are listed and the list enclosed in a pair of braces ($\{\ \}$). Some examples are:

(i) WEEKDAYS = {monday, tuesday, wednesday, thursday, friday}
(ii) COURSES = {discrete structures, data structures, compilers, operating systems}

   (iii) $X = \{1,2,4,9,3\}$
   (iv) DECIMAL-DIGITS $= \{0,1,2,3,4,5,6,7,8,9\}$

Clearly, this specification method can be used only for finite sets and will become quite cumbersome for large finite sets.

3.2.2 By Properties

In this specification method the properties of the elements in the set are provided either explicity or implicity. An element is a member of the set iff it satisfies the given properties for that set. When the properties are stated explicity, the notation:

$S = \{ x | \text{property list} \}$

or

$S = \{ x : \text{property list} \}$

is used.

This notation means that S consists of exactly those elements x that satisfy the given properties. Some examples are:

   (i) $N = \{x | x \text{ is an integer and } x \geqslant 0\}$. This defines N to be the set of nonnegative integers. I.e., N is the set of natural numbers.
   (ii) $E = \{x | x \text{ is an even integer and } 0 < x \leqslant 6\}$. This is equivalent to $E = \{2,4,6\}$.
   (iii) $P = \{x : x \text{ is a prime number or } x \text{ is even}\}$.
   (iv) $Y = \{x : x \notin Y\}$

The last set defined, Y, is interesting. Y is defined to be the set of all elements not in Y. Is 2 (or any other element), an element of Y? From (iv), it is clear that 2 cannot be an element of Y and at the same time it cannot be not an element of Y. Thus, it is possible to define paradoxical sets (i.e., ones for which neither $x \in S$ nor $x \notin S$ is true for some element x). In this text we shall not dwell upon how to eliminate such paradoxical sets. However, we shall be concerned solely with sets for which exactly one of $x \in S$ and $x \notin S$ is true for all elements x chosen from the universe of elements.

When the properties of the set elements are provided implicitly, the set is specified by listing enough elements of the set and using an ellipsis (...) to denote the remaining elements. Some examples of this method of specifying a set are:

(i)   EVEN = {0,2,4,6,...} denotes the set of even numbers.
(ii) X = {0,2,4,6,...,100} denotes the set of even numbers up to 100.
(iii)   ALPHABET = {a,b,c,...,z}
(iv)   MONTHS = {January,February,...,December}

3.2.3 Grammars

While all sets can be specified using the methods discussed so far, the resulting specification is unsatisfying in several cases. For example, consider the following specification of the set of valid English sentences:

English Sentences = {x|x is an English sentence}

This specification is quite adequate for some one who either knows all the sentences in English or knows how to tell whether or not a given sequence of English words forms a grammatically correct sentence. For the remaining people it is desirable to provide a set of rules that will enable one to determine if a given sequence of words does in fact form a sentence of English.

A *grammar* is a set of rules that enables one to generate all the elements of a given set. The grammars we shall be dealing with are called *formal grammars*. Henceforth, by a grammar, we shall mean a formal grammar. Every formal grammar, G, is specified by providing four quantities: terminal symbols, nonterminal symbols, grammar rules, and a start symbol.

(1)   *Terminal Symbols (T)*
   T is a finite set of symbols called *terminals*. All elements in the set specified by the grammar G must be composed of terminal symbols only. For example, if G is to define the set of nonnegative numbers in binary then T = {0,1}. If negative numbers are also to be in the set then T = {-,0,1}. Other examples are T = {0,1, 2,3,4,5,6,7,8,9}; T = {+,-,/,*}; T = {a,b,c}; T = {**do, if, then, else, end, for, while, loop, stop, until, case**}.

(2)   *Nonterminal Symbols (N)*
   N is a finite set of symbols different from those in T. The symbols in N are referred to as *nonterminals* and will be distinguished from those in T by the use of angle brackets (<>). Some examples for N are: {<A>, <B>, <C>}; {<exp>, <factor>, <term>, <digit>, <assignment>}; and {<noun>, <verb>, <noun phrase>, <adjective>, <adverb>}.

(3) *Grammar Rules (R)*

R is a finite set of rules of the form $\alpha \to \beta$ where α and β are finite sequences of symbols from T and N. β can be the null sequence ϵ, but α must be made up of at least one symbol. We shall use Greek letters to denote sequences of symbols. A sequence $\alpha\beta\gamma$ is *directly derivable in G* from the sequence $\alpha\delta\gamma$ iff $\delta \to \beta$ is a rule in R. We shall write $\alpha\delta\gamma \to \alpha\beta\gamma$ to mean $\alpha\beta\gamma$ is directly derivable from $\alpha\delta\gamma$. If there exist strings α_1, α_2, ..., α_n such that $\alpha_1 \to \alpha_2$, $\alpha_2 \to \alpha_3$, ..., $\alpha_{n-1} \to \alpha_n$, $n \geqslant 2$, then we shall say that α_n is *derivable* from α_1 through the use of a nonempty sequence of rules in G. The sequence of rules $\alpha \to \beta_1$, $\alpha \to \beta_2$, ..., $\alpha \to \beta_k$ may be abbreviated $\alpha \to \beta_1|\beta_2|..|\beta_k$.

(4) *Start Symbol (I)*

I is an element of N.

A sequence, σ, of terminal symbols is an element of the set, $S(G)$, *generated by the grammar G* iff it is derivable from I using the rules of the grammar. $S(G)$ contains no other elements.

A grammar G can therefore be specified as a four tuple $G = (T, N, R, I)$. The following examples illustrate the use of grammars in specifying sets.

Example 3.1: Consider the grammar $G = (\{0, 1, 2, 3, 4, 5, 6, 7, 8, 9\}, \{<I>\}, \{<I> \to 0|1|2|3|4|5|6|7|8|9\}, <I>)$. The only elements derivable from the start symbol $<I>$ are the decimal digits 0, 1, 2, 3, 4, 5, 6, 7, 8, and 9. Hence, $S(G) = \{0, 1, 2, 3, 4, 5, 6, 7, 8, 9\}$. □

Example 3.2: Let $G = (\{0,1\}, \{<U>\}, \{<U> \to 0 \mid 1 \mid 0<U> \mid 1<U>\}, <U>)$. 0 and 1 are readily seen to be derivable from the start symbol $<U>$. Hence, $0 \in S(G)$ and $1 \in S(G)$. The sequence 01001 is in $S(G)$. To see this consider the derivation: $<U> \to 0<U> \to 01<U> \to 010<U> \to 0100<U> \to 01001$. It should be easy to see that $S(G)$ is exactly the set of all binary sequences. If these sequences are being interpreted as the binary representations of the nonnegative integers then the sequences 0, 00, 000, 0000, etc. all represent the same number. Similarly, 1, 01, 001, etc. also represent the same number.

The set of binary representations with leading zeroes removed (except the last one in the case of 0) is given by the grammar $G1 = (\{0,1\}, \{<V>,<U>\}, \{<V> \to 0 \mid 1 \mid 1<U>,<U> \to 0 \mid 1 \mid 0<U> \mid 1<U>\}, <V>)$. The rules involving $<V>$ generate either 0, 1 or $1<U>$. From the preceding grammar G, we know that the rules for $<U>$ generate all binary sequences. However, to use the rules for $<U>$, we must first obtain $1<U>$ from $<V>$. So, all sequences of length two or more must have a 1 as the first digit. Consequently, $S(G1)$ is the desired set.

If we wish to generate the set of binary representations of all integers, then we can use the grammar $G2 = (\{0,1,-\}, \{<V>,<U>\}, \{<V> \rightarrow 0|1 |-1|1<U>|-1<U>, <U> \rightarrow 0|1|0<U>|1<U>\}, <V>)$.

The set of all even numbers in binary is generated by the grammar $G3 = (\{0,1\}, \{<V>,<U>\}, \{<V> \rightarrow 0 | 1<U>, <U> \rightarrow 0 | 1<U> | 0<U>\}, <V>)$. $\square$

Example 3.3: Let $S = \{ab, aabb, aaabbb, \dots\} = \{a^n b^n \mid n \geqslant 1\}$ where a^n is a sequence of n a's. This set is generated by the grammar $G = (\{a,b\}, \{<I>\}, \{<I> \rightarrow ab|a<I>b\}, <I>)$.

The set $S = \{x : x$ is an intermixed sequence of a's, b's and c's$\}$ is generated by the grammar $G = (\{a,b,c\}, \{<I>\}, \{<I> \rightarrow a|b|c|a<I>|b<I>| c<I>\}, <I>)$.

Let w be an intermixed sequence of a's, b's, and c's. Let w^r denote the reverse of w (i.e., if $w = abccbaa$ then $w^r = aabccba$). The grammar $G = (\{a,b,c\}, \{<I>\}, \{<I> \rightarrow aa|bb|cc|a<I>a|b<I>b|c<I>c\}, <I>)$ generates the set $\{x : x = ww^r$ and w is an intermixed sequence of a's, b's, and c's$\}$. $\square$

Example 3.4: Let us find a grammar that generates the set $S = \{a^n b^n c^n \mid n \geqslant 1\}$. The set of terminals is clearly $T = \{a,b,c\}$. The set of nonterminals can be determined only after all the rules have been written. We shall use $<I>$ as the start symbol. If we use the rules $<I> \rightarrow abc|a<I>bc$, then the grammar will generate the set $\{a^n(bc)^n \mid n \geqslant 1\}$. While this has the right number of a's, b's and c's in each element, the b's and c's are not in the right order.

So, let us try $<I> \rightarrow abc \mid a<I><B><C>$ and rearrange the $<B>$s and $<C>$s into the desired order before converting them into terminals. The rule $<I> \rightarrow abc \mid a<I><B><C>$ allows us to generate $a^n(<B><C>)^n$ for any n, $n \geqslant 1$. To bring all the $<B>$'s to the left of the $<C>$'s, we introduce the rule $<C><B> \rightarrow <B><C>$. This allows us to transform $a^n(<B><C>)^n$ into $a^n<B>^n<C>^n$. At this point we may be tempted to complete the set of rules by adding in the rules $<B> \rightarrow b$ and $<C> \rightarrow c$. While the set of rules $\{<I> \rightarrow abc \mid a<I><B><C>, <C><B> \rightarrow <B><C>, <B> \rightarrow b, <C> \rightarrow c\}$ does generate all elements of the form $a^n b^n c^n$, $n \geqslant 1$, it also generates other elements. For example $aabcbc$ can be derived as follows: $<I> \rightarrow$

$a<I><B><C> \rightarrow aabc<B><C> \rightarrow aabcb<C> \rightarrow aabcbc.$

The difficulty, of course, stems from the fact that the rules may be used in any order. To enforce the requirement that all b's appear to the left of all c's, we need to replace the rules $<B> \rightarrow b$ and $<C> \rightarrow c$ by others. Since each b is to be preceded either by an a or another b, we could introduce the rules $a<B> \rightarrow ab$ and $b<B> \rightarrow bb$. As a result of these two rules, no $<B>$ preceded by a $<C>$ or a $<B>$ can be converted into a terminal symbol unless the preceding $<C>$ is moved to its right or the preceding $<B>$ converted to a b (note that a $<B>$ cannot be converted into an a or a c). If a $<B>$ is preceded by a c, then it cannot be converted to a terminal. Finally, a rule to convert $<C>$s is needed. The rule $<C> \rightarrow c$ is sufficient. If the rule is ever used to convert a $<C>$ which has some $<B>$s to its right then the first of these $<B>$s can never be converted to a terminal b and so a set element will not result. Hence, the grammar $G = (\{a,b,c\},$ $\{<I>,<B>,<C>\}, \{<I> \rightarrow abc \mid a<I><B><C>, <C><B>$ $\rightarrow<B><C>, a<B> \rightarrow ab, b<B> \rightarrow bb, <C> \rightarrow c\}$ is such that $S(G) = \{a^nb^nc^n \mid n \geq 1\}$. $\square$

Grammars are often characterized by the kinds of rules they contain. If all the rules are of the form $\alpha \rightarrow \beta$ where α and β are sequences of terminals and nonterminals, then the grammar is said to be of *type 0*. If in addition, all rules are of the form $|\alpha| = 1$ ($|\alpha|$ denotes the number of symbols in α, i.e., its length), $\alpha \in N$ and β is either a terminal symbol or a terminal symbol followed by a nonterminal symbol (e.g. $<I> \rightarrow 0$ or $<I> \rightarrow 0<B>$), then the grammar is *regular* (e.g., the grammar of Example 3.1 and grammars G and $G1$ of Example 3.2 are regular). If $|\alpha| = 1$, $\alpha \in N$ and $|\beta| \geq 1$ for all rules in the grammar, then the grammar is *context free* (also known as Backus Naur Form (BNF) grammar). The grammars of Example 3.3 are context free or BNF grammars. If $|\alpha| \leq |\beta|$ for all rules, then we have a *context sensitive* grammar (see Example 3.4).

Context free or BNF grammars are particularly useful in specifying the set of syntactically correct statements in a programming language. For example, the set of legal variable names (identifiers) is given by the rules:

(R1)	$<\text{digit}> \rightarrow$	$0\mid1\mid2\mid3\mid4\mid5\mid6\mid7\mid8\mid9$
(R2)	$<\text{letter}> \rightarrow$	$A\mid B\mid C\mid...\mid Z$
(R3)	$<\text{alphanumeric}> \rightarrow$	$<\text{letter}>\mid<\text{digit}>$

(R4)   <alphastring> →   <alphanumeric>
                                         |<alphanumeric><alphastring>
(R5)   <identifier> →   <letter>|<letter><alphastring>

The start symbol for the grammar is <identifier>. All symbols enclosed in angle brackets are nonterminals and the remaining symbols are terminals.

The set of all syntactically valid arithmetic expressions involving only identifiers, binary plus (i.e. + with two operands as in a + b), binary minus, multiply, divide, and left and right parenthesis is generated by the following rule (together with those just given for an identifier):

<exp> → <exp> + <exp> | <exp> - <exp> | <exp> ∗ <exp>
       | <exp> / <exp> | (<exp>) | <identifier>

The start symbol is <exp>. The expression $(A+B)/C$ can be derived as follows:

<exp> → <exp> / <exp>
    → (<exp>) / <exp>
    → (<exp> + <exp>) / <exp>
    .
    .
    .
    → (<identifier> + <identifier>) / <identifier>
    .
    .
    .
    → $(A+B)/C$

This derivation can be represented diagramatically as in Figure 3.1.

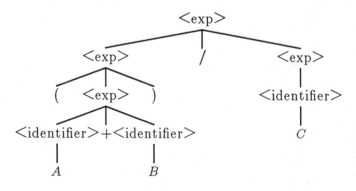

Figure 3.1 Derivation tree for $(A+B)/C$.

A diagramatic representation of a derivation (as in Figure 3.1) in a context free grammar will be referred to as a *derivation tree*. The start symbol is placed at the top of the tree (called the root). Beneath each nonterminal is placed the sequence of symbols it is replaced by via an application of exactly one rule.

Figure 3.2 gives some additional derivation trees. Figures 3.2(c) and (d) give two different derivation trees for the same arithmetic expression $A*B+C$. This means that this expression can be derived in at least two different ways using the grammar given earlier. A context free grammar, G, is said to be *ambiguous* iff there exists at least one x, $x \in S(G)$ such that x has two or more distinct derivation trees using G. The set $S(G)$ is *inherently ambiguous* iff there exists no unambiguous grammar that generates it.

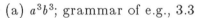

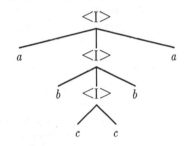

(a) a^3b^3; grammar of e.g., 3.3   (b) $abccba$; grammar of e.g. 3.3

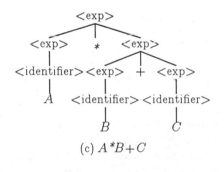

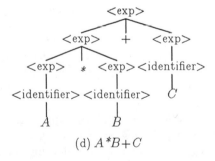

(c) $A*B+C$ (d) $A*B+C$

Figure 3.2 Derivation trees.

Figures 3.2(c) and (d) show that our grammar for arithmetic expressions is ambiguous. This ambiguity is not of much consequence if the grammar is being used solely to describe syntactically correct arithmetic expressions. If the grammar is also to be used to determine the meaning of these expressions, then of course, we have a problem. Let $A = 2$, $B = 3$, and $C = 4$. The derivation tree of Figure 3.2(c) implies that $A*B+C$ may be computed as $A*(B+C) = 14$ while the tree of Figure 3.2(d) allows one to compute $A*B+C$ as $(A*B)+C = 10$. Thus the ambiguity in the grammar for arithmetic expressions permits one to compute certain expressions in one of many different ways. These different evaluations of the same expression may produce different values. This is, of course, undesirable in a programming language where the programmer must know exactly how each expression will be evaluated. So, we desire an unambiguous grammar for expressions.

In our quest for an unambiguous grammar for arithmetic expressions, we shall limit ourselves to expressions containing only the following quantities: identifiers, (,), binary $+$, binary $-$, $*$, and $/$. Before attempting to write some grammar rules, we need to clarify the semantics of an expression. The rules for evaluation become unambiguous when the following statements are included.

(1) $*$ and $/$ have priority 2
(2) $+$ and $-$ have priority 1
(3) Let $\odot$ and $\oplus$ denote two operators (e.g., $+$, -, $/$, or $*$). If $a \odot b \oplus c$ is a sub-expression, then b is the right operand of $\odot$ iff the priority of $\odot$ is greater than or equal to that of $\oplus$.
(4) b is the left operand of $\oplus$ iff it is not the right operand of $\odot$.

With these additions to our knowledge of the meaning of the left and right parentheses, it is possible to unambiguously determine how each expression is to be evaluated. For any grammar obeying these additional rules, there can be only one derivation tree for each expression. The derivation tree of Figure 3.2(c) now becomes invalid. The tree of Figure 3.2(d) satisfies the rules.

We can arrive at an unambiguous grammar for arithmetic expressions using the following line of reasoning. The right operand of a multiply or divide must be either an identifier or an expression enclosed in parenthesis. To see this note that a right operand of the form $[a\odot...]$ is ruled out by rule 4 above. Since $\odot$ cannot have a priority more than 2, a cannot be the left operand of $\odot$. So, $[a\odot...]$ cannot together form the right operand of a multiply or divide. Using $<$ROP $*,/>$ to denote the right operand of a multiply or a divide, we get the rule:

(R6) $<$ROP$*,/> \rightarrow (<$exp$>)|<$identifier$>$

From statements (3) and (4), we see that if σ is a legal right operand for an add or subtract, then so also are σ * <ROP *,/> and σ /<ROP *,/>. Also, <ROP *,/> can be a right operand for a + and a −. This leads to the rule

(R7) <ROP −,+> → <ROP *,/> |<ROP −,+> * <ROP *,/>
$$|<ROP −,+> / <ROP *,/>$$

Finally, we observe that <ROP −,+> and <ROP *,/> are both expressions and that any expression can be the left operand of a + or a −. This gives the rule:

<exp> → <ROP −,+> |<ROP *,/> |<exp> + <ROP −,+>
$$|<exp> − <ROP −,+>$$

This can be simplified to:

(R8) <exp> → <ROP −,+> |<exp> + <ROP −,+>
$$|<exp> − <ROP −,+>$$

as any right operand of a * or / can also be the right operand for a + and −.

The new grammar for arithmetic expressions has $T = \{A, ..., Z, 0, ..., 9, (,), −, +, *, /\}$; N = {<exp>, <ROP −,+> , <ROP *,/>, <identifier>, <digit>, <letter>, <alphanumeric>, <alphastring>}; and <I> = <exp>. The set of rules is {R1, ..., R8}.

3.3 SET OPERATIONS

3.3.1 Definitions

Now that we all know what sets are and how they may be specified, we can study the more common operations performed on sets. These operations can be defined either using set notation (as below) or in terms of Venn diagrams (Figure 3.3). In a Venn diagram the universe, U, from which the set elements are chosen is represented as a square. The individual sets are represented by circles. The shaded areas in the diagrams of Figure 3.3 give the elements in the set resulting from the corresponding operation.

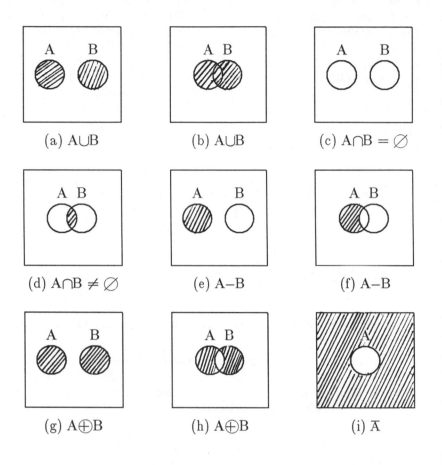

Figure 3.3 Venn diagrams for set operations.

Definition 3.1: [Set Operations] Let *A* and *B* be two arbitrary sets whose elements are members of the universal set *U*.

(a) *Set Union* (Figures 1.3(a) and (b)): The *union* of *A* and *B* (denoted *A* ∪ *B*) is the set:

$$A \cup B = \{ x \mid x \in A \text{ or } x \in B \}$$

(b) *Set Intersection* (Figures 3.3(c) and (d)): The *intersection* of *A* and *B* is denoted *A* ∩ *B* and is defined as:

$A \cap B = \{x \mid x \in A \text{ and } x \in B\}$

(c) *Set Difference*: $A - B$ denotes set difference. It is defined as follows:

$A - B = \{x \mid x \in A \text{ and } x \notin B\}$

(d) *Exclusive Or*: $A \oplus B$ denotes the *exclusive or* of the sets A and B. It is defined as:

$A \oplus B = \{x \mid x \in (A \cup B) - (A \cap B)\}$

(e) *Set Complement*: The *complement* of A is denoted $\overline{A}$. $\overline{A}$ consists of all elements that are in $U - A$. So,

$\overline{A} = U - A$

(f) *Cartesian Product*: The *cartesian product* $A \times B$ of two sets A and B is given by:

$A \times B = \{(x,y) \mid x \in A \text{ and } y \in B\}$

Note that $|A \times B| = |A| * |B|$. $A \times B = \emptyset$ iff at least one of A and B is empty. If $A_1, A_2, ..., A_n$ are sets then:

$A_1 \times A_2 \times \cdots \times A_n = \{(a_1, a_2, ..., a_n) \mid a_i \in A_i, 1 \leqslant i \leqslant n\}$

The cartesian product $A \times A \times ... \times A$ (containing i A's) is often abbreviated A^i. To avoid confusion with the notation for the next operation we shall use A_x^i to denote this cartesian product.

(g) *Set Power*: A^i denotes the *i*th power of the set A. Let ϵ denote the *null string*. By definition, the null string ϵ has the property that $\epsilon x = x \epsilon = x$ for all x. Note that ϵ does not denote a set. It denotes a possible set element. So, $\epsilon \neq \emptyset \neq \{\epsilon\}$ and $1\{\epsilon\}1 = 1$. The *i*th *power* of a set is defined as below:

$A^0 = \{\epsilon\}$

$A^i = \{x_1 x_2 \cdots x_i \mid x_j \in A, 1 \leqslant j \leqslant i\}, i > 0$

(h) *Closure*: A^+ denotes the *closure* of set A. It is given by:

$A^+ = A \cup A^2 \cup A^3 \cup .... = \bigcup_{i \geq 1} A^i$

(i)   *Reflexive Closure*: The *reflexive closure* of A is denoted A^*. It is given by:

$$A^* = A^0 \cup A^+ = \bigcup_{i \geq 0} A^i$$

(j)   *Power Set*: The *power set* of A is the set of all subsets of A. It is denoted 2^A.

Example 3.5: If $A = \{1,2,6,9\}$ and $B = \{2,9,11,13,16\}$, then $A \cup B = \{1,2,6,9,11,13,16\}$, $A \cap B = \{2,9\}$, $A - B = \{1,6\}$, $A \oplus B = \{1,6,11,13,16\}$, $A \times B = \{(1,2),(1,9), (1,11), (1,13), (1,16), (2,2), (2,9), (2,11), (2,13), (2,16), (6,2), (6,9), (6,11), (6,13), (6,16), (9,2), (9,9), (9,11), (9,13), (9,16)\}$. If $U = \{1, 2, ..., 16\}$, then $\bar{A} = \{3, 4, 5, 7, 8, 10, 11, 12, 13, 14, 15, 16\}$ and $\bar{B} = \{1, 3, 4, 5, 6, 7, 8, 10, 12, 14, 15\}$. If $C = \{0,1\}$ then $C^2 = \{00, 01, 10, 11\}$, $C^3 = \{000, 001, 010, 011, 100, 101, 110, 111\}$, $C^+ = \{0, 1, 00, 01, 10, 11, 000, 001, ...\}$, $C^* = \{\varnothing, 0, 1, 00, 01, 10, 11, ...\}$, and $2^C = \{\varnothing, \{0\}, \{1\}, \{0, 1\}\}$. Let $D = \{a, b, c\}$. $2^D = \{ \varnothing, \{a\}, \{b\}, \{c\}, \{a,b\}, \{a,c\}, \{b,c\}, \{a,b,c\}\}$. $\square$

3.3.2 Properties

Two sets A and B are *disjoint* iff $A \cap B = \varnothing$. A collection of sets A_i, $1 \leq i \leq n$ is *pairwise disjoint* iff $A_i \cap A_j = \varnothing$ for all i, j, $i \neq j$.

Definition 3.2 [Property definitions] Let a, b, and c be elements from some domain D (i.e., $a \in D$, $b \in D$, and $c \in D$). Let $\odot$ and $\oplus$ be any binary operators (not necessarily set operators). We define the following properties with respect to $\odot$, $\oplus$, and D:

(a)   *commutativity*: $\odot$ is *commutative* over D iff $a \odot b = b \odot a$ for all $a \in D$ and for all $b \in D$.

(b)   *associativity*: $\odot$ is *associative* over D iff $a \odot b \odot c = a \odot (b \odot c) = (a \odot b) \odot c$ for all a, b, and c in D.

(c)   *distributive property*: $\odot$ is *left distributive* with respect to $\oplus$ over the domain D iff for every a,b, and c in D, $a \odot (b \oplus c) = (a \odot b) \oplus (a \odot c)$. $\odot$ is *right distributive* with respect to $\oplus$ over the domain D iff for every a, b, and c in D, $(a \oplus b) \odot c = (a \odot c) \oplus (b \odot c)$.

Example 3.6: [arithmetic operators] $a + b = b + a$ for all integers a and b. Hence $+$ is commutative over the integers. $6 - 8 \neq 8 - 6$. So $-$ is not commutative over the integers. $a * b = b * a$ for all real a and b. Hence $*$ is commutative over the reals. However, $a * b$ is not necessarily equal to $b * a$ when a and b are matrices (in fact, one or both of $a * b$ and $b * a$ may not even be defined). $*$ is not commutative over the domain of matrices. $*$ is associative over the domain of integers as $a * b * c = (a * b) * c = a * (b * c)$ for all a, b, and c that are integer. $*$ is also associative over the domain of matrices. On the other hand, $/$ is not associative over the integers or reals. For example, $(2/3)/2 = 0$ while $2/(3/2) = 2$ in integer arithmetic. In real arithmetic $(2./3.)/2. = 0.333...$ while $2./(3./2.) = 1.333...$ For integer values of a, b, and c, $a * (b+c) = a * b + a * c$ and $(a+b) * c = a * c + b * c$. Therefore, $*$ is both left and right distributive with respect to $+$ over the integers. $(a+b)/c = a/c + b/c$ for all real assignments to a, b and c. So, $/$ is right distributive with respect to $+$ over the reals. However, $/$ is not right distributive with respect to $+$ over the integers. This is so because $(1+1)/2 = 1$ while $1/2 + 1/2 = 0$ in integer arithmetic. $/$ is not left distributive with respect to $+$ over the reals as $10/(1+1)$ $\neq 10/1 + 10/1$. $\square$

Example 3.7: [logical operators] Let U be the set of all well formed formulas. From the equivalences E9 and E10 (see Figure 1.6), we see that both $\wedge$ and $\vee$ are commutative over the domain U. Left distributivity of $\wedge$ with respect to $\vee$ and of $\vee$ with respect to $\wedge$ is apparent from the equivalences E11 and E12. Right distributivity is a consequence of commutativity and left distributivity. $\square$

Of the set operators defined in the preceding section, only $\cup$, $\cap$, $-$, $\oplus$, and $\times$ are binary (set complement is equivalent to $-$ as by definition, $\bar{A}$ $= U - A$. However, we will choose to regard it as unary as the left operand of the $-$ is fixed to be the universal set U.). Theorem 3.1 establishes some of the properties of these operators. Other properties are considered in the exercises.

Theorem 3.1: The following are true over the domain of sets:
(i) $\cup$, $\cap$, and $\oplus$ are commutative and associative
(ii) $-$ is neither commutative nor associative
(iii) $\times$ is neither commutative nor associative
(iv) $\cup$ is both left and right distributive with respect to $\cap$
(v) $\cap$ is both left and right distributive with respect to $\cup$
(vi) $\cup$ is not left distributive with respect to x.

Proof: (i) Let A and B be arbitrary sets. $A \cup B = \{x \mid x \in A \text{ or } x \in B\} = \{x \mid x \in B \text{ or } x \in A\} = B \cup A$. $A \cap B = \{x \mid x \in A \text{ and } x \in B\} = \{x \mid x \in B \text{ and } x \in A\} = B \cap A$. $A \oplus B = \{x \mid x \in (A \cup B) - (A \cap B)\} = \{x \mid x \in (B \cup A) - (B \cap A)\}$ (note: this follows from the just established commutativity of $\cup$ and $\cap$) $= B \oplus A$.

(ii) Let $A = \{1,2,6\}$ and $B = \{1\}$. $A - B = \{2,6\}$ while $B - A = \varnothing$. So, $-$ is not commutative.

(iii) Let A and B be as in (ii). $A \times B = \{(1,1), (2,1), (6,1)\}$ while $B \times A = \{(1,1), (1,2), (1,6)\}$. $(A \times B) \times B = \{((1,1),1), ((2,1),1), ((6,1),1)\}$, $A \times (B \times B) = \{(1,(1,1)), (2,(1,1)), (6,(1,1))\}$, and $A \times B \times B = \{(1,1,1), (2,1,1), (6,1,1)\}$. Hence, $\times$ is neither commutative nor associative.

(iv and v) We shall only show that $\cup$ is right distributive with respect to $\cap$. The remaining three parts are left as an exerise. To establish that $\cup$ is right distributive with respect to $\cap$ we need to show that $(A \cap B) \cup C = (A \cup C) \cap (B \cup C)$ for all sets A, B, and C. Let A, B, and C be three arbitrary sets. From the definitions of $\cap$ and $\cup$, it follows that:

$$
\begin{aligned}
(A \cap B) \cup C &= \{x \mid x \in A \text{ and } x \in B\} \cup \{x \mid x \in C\} \\
&= \{x \mid (x \in A \text{ and } x \in B) \text{ or } x \in C\} \\
&= \{x \mid (x \in A \text{ or } x \in C) \text{ and } (x \in B \text{ or } x \in C)\} \\
&= \{x \mid x \in A \text{ or } x \in C\} \cap \{x \mid x \in B \text{ or } x \in C\} \\
&= (A \cup C) \cap (B \cup C)
\end{aligned}
$$

(vi) Let $A = \{1\}$, $B = \{2\}$, and $C = \{3\}$. $B \times C = \{(2,3)\}$, $A \cup (B \times C) = \{1,(2,3)\}$, $A \cup B = \{1,2\}$, $A \cup C = \{1,3\}$, and $(A \cup B) \times (A \cup C) = \{(1,1), (1,3), (2,1), (2,3)\}$. Clearly, $A \cup (B \times C) \neq (A \cup B) \times (A \cup C)$.
□

Theorem 3.2: [Boolean algebra] The triple $(S, \cap, \cup)$ where S is the set of all subsets of the universal set U is a Boolean algebra.

Proof: We need to show that $(S, \cap, \cup)$ satisfies Huntington's postulates. With respect to these postulates, S, $\cap$, and $\cup$, respectively, correspond to K, ., and $+$. It is readily verified that S is closed under $\cap$ and $\cup$. Theorem 3.1 established the commutativity and distributivity of $\cap$ and $\cup$. The identity element is $I = U$ and the zero element is $Z = \varnothing$. Both of these are members of S. It is easy to see that $P \cap I = P$ and $P \cup Z = P$ for all P in S. For any P in S, $\bar{P} = U - P$, $P \cap \bar{P} = Z$, and $P \cup \bar{P} = I$. So, postulate 5 is also satisfied. Hence, $(S, \cap, \cup)$ is a Boolean algebra. □

3.3.3 Set Expressions

In Chapter 1, we saw that several English sentences could be represented in symbolic form using propositional letters, predicates, logical operators, and quantifiers. We shall now see some examples of English statements that may be represented using set variables and operators. As a first example, consider a standard card deck with 52 cards. This set of 52 cards is the universal set, U, for this example. Let S, H, D, and C respectively denote the subsets of spades, hearts, diamonds, and clubs in U. Let HON be the set of honor cards (i.e, aces, kings, queens and jacks). The following is a list of some equivalent English sentences and set expressions:

(i)   The set of non honor cards $= \overline{HON} = U - HON$
(ii) The set of spade honors $= S \cap HON$
(iii) The set of hearts that are not honors $= H \cap \overline{HON}$
(iv) The set of spade and diamond honors $= (S \cup D) \cap HON$

As another example, suppose that a questionnaire was sent out to 1000 married individuals and that this questionnaire contained exactly two questions:

(a) How long have you been married?
(b) Should the marriage institution be abolished?

The responses to this questionaire were tabulated and the results are shown in Table 3.1. Let U be the set of 1000 individuals to whom the questionaire was sent. Let A, B, C, and D and E be as defined below:

(a)	(b)			
YEARS	YES	NO	Don't Know	TOTAL
≤ 1	10	180	10	200
>1 & ≤ 2	100	100	100	300
>2 & ≤ 5	80	5	15	100
>5	300	55	45	400
TOTAL	490	340	170	1000

Table 3.1

A = set of individuals who answered yes to question (b)
B = set of individuals who answered no to question (b)
C = set of individuals married up to 1 year
D = set of individuals married no more than 2 years but more than 1 year.
E = set of individuals married no more than 5 years but more than 2 years.

$B \cap (C \cup D)$ denotes the set of individuals who answered "No" to question (b) and who have been married no more than two years. $\overline{(A \cup B)}$ denotes the set of individuals who answered "Don't know" to question (b). The set of individuals who answered "Yes" and who have been married more than five years is given by $A \cap \overline{(C \cup D \cup E)}$. $|B \cap (C \cup D)| = 280$, $|\overline{(A \cup B)}| = 170$, and $|A \cap \overline{(C \cup D \cup E)}| = 300$. How about $|A \cup C|$? We know that $|A| = 490$ and $|C| = 200$. Since there are some individuals who are members of both A and C, $|A \cup C| \neq |A| + |C|$. Rather, it is easy to see that $|A \cup C| = |A| + |C| - |A \cap C| = 680$. Since A and B are disjoint, $|A \cup B| = |A| + |B| = 490 + 340 = 830$. However, $A \cup B$ and C are not disjoint and $|A \cup B \cup C| \neq |A \cup B| + |C|$. Instead, we have $|A \cup B \cup C| = |A \cup B| + |C| - |(A \cup B) \cap C| = 830 + 200 - (10 + 180) = 840$.

The following theorem obtains an imporant equality for the size of the union of sets.

Theorem 3.3: [Inclusion-Exclusion Theorem] Let A_i, $1 \leqslant i \leqslant n$ be a collection of n, $n \geqslant 2$, arbitrary sets. $\left| \bigcup_{i=1}^{n} A_i \right| = \sum_{i=1}^{n} |A_i| - \sum_{1 \leqslant i < j \leqslant n} |A_i \cap A_j| + \sum_{1 \leqslant i < j < k \leqslant n} |A_i \cap A_j \cap A_k| - \sum_{1 \leqslant i < j < k < m \leqslant n} |A_i \cap A_j \cap A_k \cap A_m| + \dots + (-1)^{n-1} |A_1 \cap A_2 \cap \dots \cap A_n|$.

Proof: The proof is by induction on n.

Induction Base: ($n = 2$) When $n = 2$, there are only two sets. $A_1 \cup A_2$ contains all the elements in A_1 and A_2. However, A_1 and A_2 may have some common elements (i.e., all elements in the set $A_1 \cap A_2$). Only one copy of each common element is retained in $A_1 \cup A_2$. So, $|A_1 \cup A_2| = |A_1| + |A_2| - |A_1 \cap A_2|$. This agrees with the formula in the theorem when n is set to 2.

Induction Hypothesis: Let m be an arbitrary natural number greater than 2. Assume that the theorem is true when $n = m - 1$.

Induction Step: We shall prove that the theorem is true when $n = m$. Let $S = \bigcup_{i=1}^{m-1} A_i$. From the induction base, it follows that $|S \cup A_m| = |S| + |A_m| - $

$|S \cap A_m|$. From the induction hypothesis, we know that $|S| = \sum_{i=1}^{m-1} |A_i| - \sum_{1 \leqslant i < j < m} |A_i \cap A_j| + ... + (-1)^{m-2}|A_1 \cap A_2 \cap ... \cap A_{m-1}|$.

Since $\cap$ is right distributive with respect to $\cup$, $S \cap A_m = \bigcup_{i=1}^{m-1} (A_i \cap A_m) = \bigcup_{i=1}^{m-1} G_i$, where $G_i = A_i \cap A_m$, $1 \leqslant i < m$. Using the induction hypothesis once again, we obtain $|S \cap A_m| = \sum_{i=1}^{m-1} |G_i| - \sum_{1 \leqslant i < j < m} |G_i \cap G_j| + ... + (-1)^{m-2} |G_1 \cap G_2 \cap ... \cap G_{m-1}| = \sum_{i=1}^{m-1} |A_i \cap A_m| - \sum_{1 \leqslant i < j < m} |A_i \cap A_j \cap A_m| + ..., + (-1)^{m-2}|A_1 \cap A_2 \cap ... \cap A_{m-1} \cap A_m|$. The last equality follows from the fact that $A_m \cap A_m = A_m$. Combining the equalities derived, we obtain:

$$\left|\bigcup_{i=1}^{m} A_i\right| = |S| + |A_m| - |S \cap A_m|$$
$$= \sum_{i=1}^{m-1} |A_i| - \sum_{1 \leqslant i < j < m} |A_i \cap A_j| + ...$$
$$+ (-1)^{m-2}|A_1 \cap A_2 \cap ... \cap A_{m-1}|$$
$$+ |A_m|$$
$$- \left(\sum_{i=1}^{m-1} |A_i \cap A_m| - \sum_{1 \leqslant i < j < m} |A_i \cap A_j \cap A_m| + ...\right.$$
$$\left. + (-1)^{m-2}|A_1 \cap A_2 \cap ... \cap A_m|\right)$$
$$= \sum_{i=1}^{m} |A_i| - \sum_{1 \leqslant i < j \leqslant m} |A_i \cap A_j| + \sum_{1 \leqslant i < j < k \leqslant m} |A_i \cap A_j \cap A_k| + ... + (-1)^{m-1}|A_1 \cap A_2 \cap ... \cap A_m|. \qquad \square$$

For clarity, we explicity state the form of $\left|\bigcup_{i=1}^{n} A_i\right|$, for $n = 2$ and 3.

(1) $|A_1 \cup A_2| = |A_1| + |A_2| - |A_1 \cap A_2|$

(2) $|A_1 \cup A_2 \cup A_3| = |A_1| + |A_2| + |A_3| - |A_1 \cap A_2| - |A_1 \cap A_3| - |A_2 \cap A_3| + |A_1 \cap A_2 \cap A_3|$

As an application of these formulas, consider our earlier determination of $|A \cup B \cup C|$ (cf. questionnaire example). From (2) above, we obtain $|A \cup B \cup C| = |A| + |B| + |C| - |A \cap B| - |A \cap C| - |B \cap C| + |A \cap B \cap C|$. Since A and B are disjoint, $|A \cap B| = |A \cap B \cap C| = 0$. So, $|A \cup B \cup C| = |A| + |B| + |C| - |A \cap C| - |B \cap C| = 840$.

3.4 CORRESPONDENCE, COUNTABILITY, AND DIAGONALIZATION

In Section 3.1, we made the distinction between a finite and an infinite set. The class of infinite sets may itself be divided into two disjoint subsets. One of these contains those sets that are countable or enumerable and the other contains those that are not countable. To define the notion of a countable set, we need to first introduce two other terms. Let A and B be two sets. An *association* (or relation), C, between A and B is a set of pairs (a,b) such that $a \in A$, and $b \in B$. If $(a,b) \in C$, then we say that a is associated with b and b is associated with a. If $(a,b) \in C$ and $(a,c) \in C$ then a is associated with both b and c. A *one-to-one correspondence* between A and B is an association C between A and B such that:

(a) $\forall a \in A \ (|\{b : (a,b) \in C\}| = 1)$

and

(b) $\forall b \in B \ (|\{a : (a,b) \in C\}| = 1)$

I.e., in a one-to-one correspondence, every element of A is associated with exactly one element of B and every element of B is associated with exactly one element of A.

Two sets A and B *can be put into* one-to-one correspondence iff there exists a one-to-one correspondence C between A and B. Observe that A and B can be put into one-to-one correspondence iff B and A can. A one-to-one correspondence associates each element of A with a distinct element of B and vice-versa.

Example 3.8: The sets $A = \{1,2,3\}$ and $B = \{C,D,F\}$ can be put into one-to-one correspondence. To see this, just consider the association $S = \{(1,C), (2,D), (3,F)\}$.   □

A set S is *countably infinite* iff there exists a one-to-one correspondence between S and the set N of natural numbers. If S is infinite but not countably infinite, then it is *uncountable*. S is a *discrete set* iff it is either finite or countably infinite.

Several sets are easily seen to be countably infinite. The set $E = \{0,2,4,...\}$ of even numbers is countably infinite because the association $\{(i, 2i)| i \in N\}$ associates each element of N with a distinct element of E and vice versa. Likewise, the association $\{(i, 2i+1) | i \in N\}$ implies that the set of odd numbers is countably infinite. The set, Z, of integers can be shown to be countably infinite by using the association $\{(i,j) \mid i \in N, j \in Z, j=\lfloor i/2 \rfloor$ if i is even and $j = -\lfloor i/2 \rfloor$ if i is odd$\}$.

It isn't always quite as easy to show that a given set is countably infinite. As the next example, we consider the set, Q^+, of non-negative rational forms which is defined as:

$$Q^+ = \{p/q \mid p \in N \text{ and } q \in N - \{0\}\}$$

Note that the set Q^+ contains several elements that equal the same rational number (e.g., 2/1, 4/2, 6/3, ...). The set Q of non-negative rational numbers is given by:

$$Q - \{0\} \cup \{p/q \mid p/q \in Q^+, p \neq 0, \text{ and } \gcd(p,q) = 1\}$$

Q^+ is readily seen to be an infinite set. But, is it countable or uncountable? Q^+ is countable iff there is a one-to-one correspondence C between Q^+ and N. This one-to-one correspondence may be established by enumerating the elements in Q^+ in an order such that the first element listed is associated with 0, the second with 1, the third with 2 and so on. This was implicitly done for the examples considered above. Figure 3.4 explicitly displays the enumeration and resulting pairings for the set of even numbers and the set of integers. These enumerations have the following properties:

N	E	N	Z
0	0	0	0
1	2	1	-1
2	4	2	1
3	6	3	-2
4	8	4	2
5	10	5	-3
⋮	⋮	⋮	⋮

Figure 3.4 Enumeration of E and Z

(a) Each element of the set being enumerated is listed exactly once.
(b) The first element listed is paired with 0, the second with 1, the third with 2, etc.
(c) For each element $x \in N$, we can effectively determine which element it is paired with.

(d)   For each element in the set being enumerated, we can effectively determine the element in N with which it is paired.

If for any set S we can generate an enumeration method such that the resulting list of elements satisfies properties (a) to (d), then S is countably infinite. Likewise, if S is countably infinite, such a list can be produced by simply listing for each $i \in N$, $i = 0,1,2,...$, the element in S with which it is paired.

We shall prove that Q^+ is countably infinite by describing how to list the elements of Q^+ so as to satisfy properties (a) to (d). One may be tempted to list the elements of Q^+ as outlined below:

(1)   list 0/1, 1/1, 2/1, 3/1, 4/1, ...
(2)   list 0/2, 1/2, 2/2, 3/2, 4/2, ...
(3)   list 0/3, 1/3, 2/3, 3/3, 4/3, ...

The difficulty with this enumeration scheme is that we will never finish listing the rational forms in (1). Hence, no pairing is generated for rational forms p/q for $q > 1$. This listing of Q^+ does not satisfy properties (a) and (d) but it does satisfy properties (b) and (c). Every x, $x \in N$ is paired with the rational form $x/1$.

Another approach is to list the elements of Q^+ as shown in Figure 3.5. In this figure the elements of Q^+ have been listed in a two dimensional array. It is evident that each element of Q^+ appears exactly once in this two dimensional array. If the rows are numbered 1, 2, 3, ... top to bottom and the columns 0, 1, 2, ... left to right, then element p/q appears in row q and column p of the array. The elements of Q^+ are to be enumerated in the order given by the arrows. The resulting list is: 0/1, 0/2, 1/1, 0/3, 1/2, 2/1, 0/4, 1/3, 2/2, 3/1, 0/5, 1/4, 2/3, 3/2, 4/1, 0/6, ... Note that the lines of Figure 3.5 join together all p/q having the same sum $p+q$. The position of any p/q in our enumeration is not too difficult to determine. The number of elements with $p+q = 1$ is 1. The number with $p+q = 2$ is 2; with $p+q = 3$ is 3; and in general there are p+q elements with sum $p+q$. Thus, preceding any p/q in the list there are $\sum_{i=1}^{p+q-1} i = (p+q-1)(p+q)/2$ elements with sum less than $p+q$. Also, p/q is the $(p+1)$st element listed with sum $p+q$. So, p/q is the $((p+q-1)(p+q)/2 + p + 1)$th element in the list of Q^+ elements. Hence, p/q is paired with $(p+q-1)(p+q)/2+p$ which is an element of N. Note that this enumeration of Q^+ satisfies properties (a) to (d) and Q^+ is therefore countably infinite.

The proof just provided can be extended to show that the set Q is also countably infinite. In this extension, the pairing of elements of Q with those in N is done by going in the direction of the arrows of Figure 3.5 and omitting elements in $Q^+ - Q$.

p →							
0/1	1/1	2/1	3/1	4/1	5/1	6/1	...
0/2	1/2	2/2	3/2	4/2	5/2	6/2	
0/3	1/3	2/3	3/3	4/3	5/3	6/3	
0/4	1/4	2/4	3/4	4/4	5/4	6/4	
0/5	1/5	2/5	3/5	4/5	5/5	6/5	
0/6	1/6	2/6	3/6	4/6	5/6	6/6	

Figure 3.5 Enumerating Q^+.

As a final example of a countably infinite set, we consider the set, $P(x)$, of polynomials in one variable (i.e., the variable x) and having only integer coefficients. This set is defined as:

$$P(x) = \{\sum_{i=0}^{n} a_i x^i \mid n \in N, a_i \in Z, a_n \neq 0\} \cup \{0\}$$

The *degree* of the polynomial $\sum_{i=0}^{n} a_i x^i$ is n. The a_is are called the *coefficients*. We shall use $d(p)$ to denote the degree of polynomial p and $|a_i|$ to denote the absolute value of a_i (i.e., if $a_i < 0$ then $|a_i| = -a_i$, otherwise $|a_i| = a_i$). Let $c(p) = \max_{1 \leq i \leq n} \{|a_i|\}$.

We may first try the following strategy to enumerate $P(x)$:

(1) list all polynomials of degree 0
(2) list all polynomials of degree 1
(3) "   "   "   "   2
    .
    .
    .

This attempt at enumerating all the polynomials fails as the number of polynomials to be listed in (1) is infinite. So, polynomials of degree more than 0 never get listed. To establish countability of $P(x)$, it suffices to list the polynomials in each of the following sets. The polynomials in the first set are listed first, then those in the second set are listed, and so on.

(1) $S_1 = \{p \mid p \in P(x)$ and $d(p) \leqslant 10$ and $c(p) \leqslant 10\}$
(2) $S_2 = \{p \mid p \in P(x) \text{-} S_1$ and $d(p) \leqslant 100$ and $c(p) \leqslant 100\}$
(3) $S_3 = \{p \mid p \in P(x) \text{-} S_1 \text{-} S_2$ and $d(p) \leqslant 1000$ and $c(p) \leqslant 1000\}$
(4) $S_4 = \{p \mid p \in P(x) \text{-} S_1 \text{-} S_2 \text{-} S_3$ and $d(p) \leqslant 10000$ and $c(p) \leqslant 10000\}$

.
.
.

We can make the following observations about the sets S_1, S_2,
...

(a)   The sets S_1, S_2, ... are pairwise disjoint and each $p \in P(x)$ is a member of exactly one of these sets. The set containing any polynomial p is easily determined from $d(p)$ and $c(p)$. Let $r(p) = \max\{d(p), c(p)\}$. Let k be such that $10^{k-1} < r(p) \leqslant 10^k$. Clearly, such a k must exist and is unique. One easily observes that $p \in S_k$.
(b)   Each S_i is finite. In fact, $|S_i| \leqslant (2*10^i+1)^{(10^i+1)}$.
(c)   Since each S_i is finite, each set can be enumerated in a finite amount of time. Hence, by enumerating the members of S_1, S_2, ..., we enumerate $P(x)$. Therefore, we can obtain an association between $P(x)$ and N. This association has all the properties of a one-to-one correspondence. Hence, $P(x)$ is countably infinite.

Having seen a few countably infinite sets, we are now ready to study some important theorems concerning countably infinite sets.

Theorem 3.4: Every infinite set contains a countably infinite subset.

Proof: Let S be an arbitrary infinite set. There must exist an x_1 such that $x_1 \in S$ (as $S \neq \varnothing$). Since S is infinite, $S\text{-}\{x_1\}$ is also infinite. Now, there exists an x_2 such that $x_2 \in S\text{-}\{x_1\}$. Also, $S\text{-}\{x_1, x_2\}$ is infinite. It follows that S contains an infinite subset $T = \{x_1, x_2, x_3, ...\}$. This subset can be put into one-to-one correspondence with N (x_i associates with i, $i \in N$). $\square$

Example 3.9: The set R of real numbers is infinite. It contains several countably infinite subsets. Some of these are Z (the set of integers), E (the set of even numbers), Q, and N. However, as we shall soon see, R is itself not countably infinite. $\square$

Theorem 3.5: Every infinite subset of a countably infinite set is countable infinite.

Proof: Let S be any countably infinite set. Let T be any infinite subset of S. Since S is countably infinite, S can be put into one-to-one correspondence with N. Using this correspondence, a one-to-one correspondence between T and N may be established. Consider the one-to-one correspondence between S and N. This is shown in Figure 3.6. Elements in S marked with an asterisk (*) denote members of T. A one-to-one correspondence between T and N is obtained by associating the first asterisked element with 0, the second with one, the third with 2, etc. Since T is infinite, we will not run out of asterisked elements to associate with any i, $i \in N$. Hence, T is countably infinite. □

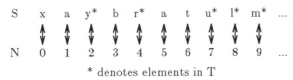

* denotes elements in T

Figure 3.6 One-to-one correspondence between S and N.

Example 3.10: Since N is countably infinite and $E \subset N$ is an infinite set, from Theorem 3.5 it follows that E is countably infinite. The set, O, of odd numbers is a subset of N. Since O is infinite and N is countably infinite, O is also countably infinite. □

Theorem 3.6: The union of a finite number of countably infinite sets is countably infinite.

Proof: Let $A_0, A_1, A_2, ..., A_{k-1}$ be countably infinite sets and let k be finite. Since the A_is are countably infinite, there exists a one-to-one correspondence, C_i, between A_i and N, $0 \leqslant i < k$. Let $f(i,j)$ denote the element of A_j that is associated with i, $i \in N$ in the correspondence C_j. Let $s = \lfloor i/k \rfloor$ and $t = i - s*k$.

When the A_js are mutually disjoint, the association $(i, f(s,t))$ defines a one-to-one correspondence between $\bigcup_{i=0}^{k-1} A_i$ and N.

When the A_js are not mutually disjoint the association $(i, f(s,t))$ associates at least one element of $\bigcup_{i=0}^{k-1} A_i$ with more than one element of N. In this case, a one-to-one correspondence can be established by defining $g(0)$

$= f(0,0)$. For any n, $n > 0$, $g(n)$ is defined in the following way. Let p be the least natural number for which $f(s,t)$ ($s = \lfloor p/k \rfloor$ and $t = p - s * k$) is different from each of $g(0)$, $g(1)$, ..., $g(n-1)$. Since the A_is are infinite sets, such a p must exist. Set $g(n) = f(s,t)$. It is obvious that $(i, g(i))$, $i \in N$, defines a one-to-one correspondence between $\bigcup_{i=0}^{k-1} A_i$ and N. This completes the proof of the theorem. $\square$

Example 3.11: Let $Z_1 = \{0,1,2,...\}$ and $Z_2 = \{-1, -2, ...\}$. Z_1 and Z_2 are readily seen to be countably infinite. From Theorem 3.6, it follows that $Z = Z_1 \cup Z_2$ is countably infinite. $\square$

Theorem 3.7: The union of a countably infitite set of countably infinite sets is countably infinite.

Proof: Left as an exercise. $\square$

Diagonalization

At this point we may be wondering if all infinite sets are countably infinite. In 1874, Cantor showed that the set of real numbers is not countably infinite. The technique used by him to prove this result is known as *diagonalization*. Before showing that the set of real numbers is uncountable, we shall prove a similar result for the set, T, of decimal forms. This set is defined as:

$$T = \{0.x_1 x_2 ... \mid x_j \in \{0,1,2,...,9\}\}$$

If T is countably infinite, then there is a one-to-one correspondence between T and N. Let x^i be the unique element of T associated with i in this one-to-one correspondence. We may enumerate T to obtain the list given below:

$$0.x_0^0 x_1^0 x_2^0...$$

$$0.x_0^1 x_1^1 x_2^1 ...$$

$$0.x_0^2 x_1^2 x_2^2 ...$$

$$0.x_0^3 x_1^3 x_2^3 ...$$

.

.

This list must, by definition of a one-to-one correspondence, contain each element of T exactly once. We shall now show that this cannot be the case, i.e. there is at least one decimal form in T that is not on the above list. Consider the decimal form $y = 0.y_0y_1y_2 \cdots$ defined below:

$$y_i = \begin{cases} 0 & \text{if } x_i^i \neq 0 \\ 1 & \text{if } x_i^i = 0 \end{cases}$$

Clearly, $y \in T$. But, y is not on the above list. The proof of this is by contradiction. Assume that $y = x^j$ for some j. Then, $y_i = x_i^j$, $i \geq 0$. In particular, $y_j = x_j^j$. But, by definition of y, $y_j \neq x_j^j$. So, there can be no j for which $y = x^j$. Hence, y is not on the list and there is no one-to-one correspondence between T and N.

The preceding proof should be studied carefully. The use of 0 and 1 in the construction of y is not important. We could just as well have used

$$y_i = \begin{cases} 3 & \text{if } x_i^i \neq 3 \\ 5 & \text{if } x_i^i = 3 \end{cases}$$

or any other pair of numbers. *The important thing is that y is constructed so that it differs from each of the decimal forms on the list in at least one decimal position.* In the particular construction of y used, the ith digit y_i of y is different from the ith digit x_i^i of x_i. The digits x_i^i, $i \geq 0$ give the diagonal digits of the purported list of elements in T. Hence the term diagonalization. Once again, it is not important to make the ith digit of y different from the ith digit of x^i. We could have constructed a y such that $y_0 \neq x_0^3$, $y_1 \neq x_1^6$, $y_2 \neq x_2^0$, $y_3 \neq x_3^1$, $y_4 \neq x_4^5$, and so on. The important point is that y must differ from each of the x's on the list in at least one digit.

Now, let us return to the problem of showing that the set of real numbers is uncountable. First, consider the set, S, defined as:

$$S = \{x : x \text{ is a real number and } 0 \leq x < 1\}$$

If S is a countable set, then there exists a one-to-one correspondence C between N and S. Let $r_i \in S$ be associated with i, $i \in N$ in this one-to-one correspondence. We shall diagonalize over the list $r_0, r_1, \ldots$ and generate a real number in S that is not equal to any of the r_is, $i \in N$. This will imply that C cannot be a one-to-one correspondence between N and S, as assumed. So, S is not countable.

In order to carry out the diagonalization, we need to use a representation for the numbers in S. Let us use a decimal representation. Clearly, every number in S can be represented in the infinite decimal form $0.x_0x_1x_2$ ... (for numbers in S having a finite decimal representation $0.x_1...x_k$ an equivalent infinite form is $0.x_1...x_k000....$).

We need to be a little more specific about the infinite decimal representation of a real number. Some real numbers have two such decimal representations. For example, consider the representations $0.15999...$ and $0.16000...$. Let $w = 0.15999...$. $100w = 15.999...$ and $1000w = 159.999...$. So, $900w = 1000w-100w = 144.000...$ Hence, $w = .16000...$. For definiteness, we shall always use the decimal representation with an infinite sequence of zeroes at the end rather than the one with an infinite sequence of nines at the end (i.e., for numbers that can be represented in both ways). One may verify that if x and y are two different infinite decimal representations then x and y can represent the same real number only if one has an infinite sequence of zeroes at the end and the other has an infinite sequence of nines at the end.

From the discussion of the preceding paragraph, we see that the uncountability of S is not an immediate consequence of the uncountability of the set T of decimal forms considered earlier. We can however use an identical diagonalization scheme to diagonalize over $r_0, r_1, ...$ Let $x^i = 0.x_0^i x_1^i x_2^i ...$ be the unique infinite decimal representation of the real number r_i. Define the infinite decimal form $y = 0.y_0y_1y_2...$ as below:

$$y_i = \begin{cases} 0 & \text{if } x_i^i \neq 0 \\ 1 & \text{if } x_i^i = 0 \end{cases}$$

From our discussion of the diagonalization of T, it follows that y differs from each of the decimal forms $x^0, x^1, x^2, ...$ in at least one digit. It is also clear that y does not have an infinite sequence of nines at the end. The last two statements together imply that y does not represent the same number as represented by any of the decimal forms $x^0, x^1, x^2, ...$ Since y does indeed represent a number in S, it follows that C does not associate every element of S with an element of N. Hence, C does not define a one-to-one correspondence between S and N.

From Theorem 3.5 and the knowledge that S is an uncountably infinite subset of the set, R, of real numbers, it follows that R is not countably infinite (i.e., R is uncountable).

Before concluding this chapter, we shall look at one more proof by diagonalization. In Chapter 5 we shall see several additional proofs that will use this technique. Consider the set Z of integers. We know that Z is countably infinite. In fact, each of the following orderings defines a one-to-one correspondence between Z and N (the ith element in the ordering is associated with i):

(a) $0, 1, -1, 2, -2, 3, -3, \ldots$
(b) $0, 1, 2, -1, -2, 3, 4, -3, -4, \ldots$
(c) $2, 1, 0, -1, -2, 4, 3, -3, -4, \ldots$

.
.
.

So, we can speak of the set, T, of one-to-one correspondences between Z and N. Is this set countably infinite? The answer is no and the proof is obtained by diagonalizing over the list created by any purported one-to-one correspondence that could possibly be made between T and N. If T is countably infinite then T and N can be put into one-to-one correspondence. This one-to-one correspondence leads to a list $x^0, x^1, x^2, \ldots$ of the elements of T:

$$x^0 = x_0^0 x_1^0 x_2^0 \ldots$$

$$x^1 = x_0^1 x_1^1 x_2^1 \ldots$$

$$x^2 = x_0^2 x_1^2 x_2^2 \ldots$$

$$x^3 = x_0^3 x_1^3 x_2^3 \ldots$$

.
.
.

If there exists a j such that $x_k^k = x_{k+1}^{k+1}$, $k \geq j$, then the sequence x^0, x^1, $x^2, \ldots$ does not contain all the elements of T. To see this, observe that if such a j exists, then there are at most j x's which have x_j^j in a position less than or equal to j (i.e., x's for which $x_a^i = x_j^j$ and $a \leq j$). However, T contains infinitely many elements with x_j^j in a position less than or equal to j.

So, we may assume that there is no j such that $x_k^k = x_{k+1}^{k+1}$, $k \geq j$. A new one-to-one correspondence, y, between Z and N may be defined as below:

$$y_0 = \begin{cases} 0 \ \text{ if } x_0^0 \neq 0 \\ 1 \ \text{ if } x_0^0 = 0 \end{cases}$$

y_i = an integer with least absolute value that is not equal to any of $y_0, y_1, \ldots,$
   y_{i-1}, x_i^i.

Note that there are only one or two candidates for any y_i. For example if x_0^0 = 6 and $x_1^1 = 2$, then y_1 can be either $+1$ or -1. If $x_1^1 = 1$ then y_1 can only be -1. If there are two candidates for y_i we may select either (to be definite, we could always select the non-negative one).

The fact that the y defined in this way is indeed a one-to-one correspondence between Z and N follows from the definition of y and the fact that there is no j such that $x_k^k = x_{k+1}^{k+1}$, $k \geqslant j$.

Finally, we need to show that y is not any of the x's in the list. To see this, suppose $y = x^j$ for some j, $j \geqslant 0$. Then y_i must equal x_i^j for all i. In particular, y_j must equal x_j^j. From the definition of y_j, we know that $y_j \neq x_j^j$. Hence, no j with $y = x^j$ exists. This implies that T is not countably infinite.

EXERCISES

1. Which of the following are sets?
    (a) {(i), (ii), (iii)}
    (b) {house, apartment, duplex, motel, street, parrot}
    (c) {Madrid, Spain, Italy, Paris, Berlin, Spain}
    (d) {Mary, Margaret, Smith, Mary, Jones}

2. Which of the following statements are true?
    (a) $19 \in \{1,5,9,13, ...\}$
    (b) $13 \in \{x \mid x$ is an even number$\}$
    (c) $29 \in \{x \ rx$ is a prime$\}$
    (d) $49 \in \{x^2 \mid 0 \leqslant x < 7\}$
    (e) $101 \in \{1,3,5,9, ...,99\}$

3. Obtain regular grammars for the following sets:
    (a) $\{a^n \mid n$ is even and $n > 0\}$
    (b) $\{x : |x| = 5$ and all symbols in x are English letters$\}$
    (c) $\{a^{3n} \mid n \in N\}$

4. Obtain context free grammars for the following sets:
    (a) $\{x \ \$ \ y \ \$ \ y^r \ \$ \ x^r : x$ and y are intermixed sequences of a's and b's, $|x| \geqslant 1$ and $|y| \geqslant 1\}$
    (b) $\{a^n b^{3n} \mid n \geqslant 1\}$

5. Obtain context sensitive grammars for the sets:
    (a) $\{xyy^rx^r \mid x$ and y are as in Ex 4(a) and $|x| = |y|\}$
    (b) $\{a^n b^n c^n d^n \mid n \geqslant 0\}$
    (c) $\{1^n \$ x \mid x$ is the binary representation of $n, n \geqslant 0\}$

6. Write down the set represented by each of the following:
    (a) {a,b,aa,bcd,acd} $\cup$ {bcd,acd,dac,cad,b,aaa}
    (b) {a,b,aa,bcd,acd} $\cap$ {bcd,acd,dac,cad,b,aaa}
    (c) {a,b,c} $\cup$ $\varnothing$
    (d) {a,b,c} $\cap$ {1,2,4,8,9,11}
    (e) {a,b,aa,bcd,acd} $-$ {bcd,acd,dac,cad,b,aaa}
    (f) {a,b,aa,bcd,acd} $\oplus$ {bcd,acd,dac,cad,b,aaa}
    (g) {a,b,c} $\times$ {1,2}
    (h) {a,b,c} $\times$ {a,b}
    (i) $2^{\{a,b\}}$
    (j) $2^{\varnothing}$
    (k) $\{a,b,1\}^2$
    (l) $\{0,1\}^3$

7. Show that the following statements are true. A, B and C are arbitrary sets from some universe U.
   (a) $A \cup A = A$
   (b) $A - A = \emptyset$
   (c) $A \oplus A = \emptyset$
   (d) If $B = \bar{A}$ then $\bar{B} = A$
   (e) $A \cup (B \cap C) = (A \cup B) \cap (A \cup C)$
   (f) $A \cap (B \cup C) = (A \cap B) \cup (A \cap C)$
   (g) $(A \cap B) \cup C = (A \cup C) \cap (B \cup C)$
   (h) $(A \cup B) \cap C = (A \cap C) \cup (B \cap C)$
   (i) $A \cap B \subseteq A$
   (j) $A - B \subseteq A$
   (k) $(A - B) \cap B = \emptyset$

8. Which of the following statements are true? Prove your answers.
   (a) $+$ is left and right distributive with respect to $-$ over the integers.
   (b) $/$ is left distributive with respect to $*$ over the integers.
   (c) $/$ is left distributive with respect to $*$ over the reals.
   (d) $*$ is left distributive with respect to $/$ over the reals.
   (e) $*$ is right distributive with respect to $/$ over the integers.
   (f) $*$ is left and right distributive with respect to $+$ over the domain of matrices.
   (g) $*$ is associative over the matrices.

9. Show that the following statements are true over the domain of sets:
   (a) $\cup$ is not right distributive with respect to x.
   (b) $\cap$ is neither left nor right distributive with respect to x.
   (c) $\oplus$ is neither left nor right distributive with respect to x.
   (d) $\oplus$ is neither left nor right distributive with respect to $\cup$
   (e) $\oplus$ is neither left nor right distributive with respect to $\cap$

10. (a) Define the following sets:
    A ... set of aristocrats
    B ... set of bad guys
    D ... set of democrats
    R ... set of republicans

    Obtain set expressions that denote the following sets:

    (i) The set of all democrats who are aristocrats but not bad guys.
    (ii) The set containing all the aristocrat republicans and the bad democrats.
    (iii) The set consisting of all aristocrats who are neither democrats nor bad guys.

11. (a) Prove De Morgan's Laws:
     (i) $\overline{(A \cup B)} = \overline{A} \cap \overline{B}$, for all sets A and B.
     (ii) $\overline{(A \cap B)} = \overline{A} \cup \overline{B}$, for all sets A and B.
  (b) Let A_i, $1 \leqslant i \leqslant n$ be n arbitrary sets. Show that:

$$(i) \quad \overline{A_1 \cup A_2 \cup \cdots \cup A_n} = \overline{A}_1 \cap \overline{A}_2 \cap \ldots \cap \overline{A}_n$$

$$(ii) \quad \overline{A_1 \cap A_2 \cap \cdots \cap A_n} = \overline{A}_1 \cup \overline{A}_2 \cup \ldots \cup \overline{A}_n$$

12. Show that if A and B are regular sets, then so also are the sets:
  (a) $A \cup B$
  (b) $A \cap B$
  (c) $\overline{A}$
  (d) $A \oplus B$
  (e) $A \times B$

13. Show that if A is a context free set and B is a regular set then the following sets are context free :
  (a) $A \cup B$
  (b) $A \times B$

14. Show that the following scts are countably infinite:
  (a) $\{i^2 \,|\, i \in N\}$
  (b) $\{p \,|\, p$ is a prime number$\}$
  (c) $\{x \,|\, x$ is an English book$\}$
  (d) $\{3,6,9,12,15, \ldots \}$

15. Show that the set of context free grammars with $T = \{0,1\}$, $N = \{<I>,<J>,<K>,<L>\}$ and start symbol$<I>$ is countably infinite.

16. Show that the set of valid arithmetic expressions containing only the operators $+$, $-$, $*$, and $/$, and the identifiers $A, B, C, \ldots, Z$ is countably infinite.

17. Show that S^+ is countably infinite for all finite sets S.

18. Is the set of two variable polynomials, $P(x,y)$ with integer coefficients countably infinite? Prove your answer.

19. (a) Show that no finite set can be put into one-to-one correspondence with any of its proper subsets.
  (b) Show that every infinite set can be put into one-to-one correspondence with at least one of its proper subsets.

20. Prove Theorem 3.7.

21. Is S^+ countably infinite for all countably infinite sets S? Prove your answer.

22. Is $2^{\{0,1\}^*}$ countably infinite? Prove your answer.

23. x is an algebraic number iff it is the root of at least one polynomial p, $p \in P(x)$ (see Section 3.4 for a definition of $P(x)$). For example, 2 is an algebraic number because it is the root of the polynomial x−2. Is the set of algebraic numbers countably infinite? Prove your answer.

CHAPTER 4

RELATIONS

4.1 INTRODUCTION

Suppose that one of the rear tires of your car develops a flat and that you wish to replace the culprit wheel with the spare one in the trunk. This wheel replacement project involves several operations or tasks. The set S of needed operations is {raise car rear, loosen nuts, remove nuts, remove culprit wheel, get jack from trunk, put jack in place, get spare wheel, put culprit wheel in trunk, tighten nuts, put a block in front of a front wheel, remove block, remove jack, lower car, put jack in trunk, further tighten nuts}.

We shall refer to the fifteen elements of S as A1, A2, ... A15. The elements are numbered in the order in which they have been listed above. It is not too difficult to see that the various operations listed above cannot be performed in any arbitrary order. For example, the car rear cannot be raised until the jack has been put in place; the jack cannot be put in place until it is taken out of the trunk; etc. Hence, there exists a relationship, between pairs of operations. This relationship is one of precedence. Certain operations must be done before others. The elements of S can be paired into pairs of the form (a,b) such that operation a must be done before operation b. The set of all such pairs defines the relationship P given below:

$$P = \{(A1,A2),(A1,A3),(A3,A4),(A2,A3),(A5,A1),(A5,A6),...\}$$

The set P is important in that once all the elements of P are known, we can attempt to order the elements of S to obtain an ordered set $T = (a_1,a_2,...,a_{15})$ (where each a_i denotes a distinct operation). This ordered set has the property that there is no i and j, $i > j$ and $i \leqslant 15$ such that (a_i,a_j) is an element of P. In other words, the operations in S can be carried out in the order specified by T without violating any of the precedences given in P. An ordering with this property is called a *topological order*. We shall see more of topological orders in a later section.

We shall define a *k-tuple*, to be an ordered set $(x_1,x_2,...,x_k)$ of size k. Thus, the pairs of P are 2-tuples. Element x_i of the k-tuple $(x_1,x_2,...,x_k)$ is the *ith element* (or *ith component*) of the tuple. It is in *position i* of the tuple. Position i is also called *co-ordinate i*. A *k-ary relation*, R, is a set of k-tuples.

When $k=2$, the relation R is said to be a *binary relation*; when $k=3$, it is *ternary*; when $k=4$, it is *quarternary*; etc.

The relation P defined above is a binary relation as each element of P is a pair. It is important to note that each pair of P is an ordered pair. The pair (A2,A1) is to be regarded different from (A1,A2). (A2,A1) means that A2 must be done before A1, while (A1,A2) means that A1 must be done before A2.

The set of values that the ith component of a k-tuple may have is called the *domain* of the ith component. Let D_1, D_2, ..., D_k, respectively, be the domains of the components of the k-tuples in a k-ary relation R. R will be said to be a relation on D_1, ..., D_k. If all the D_is are the same, then R is simply a relation on D_1. One readily sees that R is a subset of the cartesian product $D_1 \times D_2 \times D_3 \times ... \times D_k$.

The following examples should convince you that relations are, in fact, common occurrences.

Example 4.1: [Family] The Jones family consists of 12 members: grandfather Joe, grandfather Harry, grandmother Rose, grandmother Mildred, father Bill, mother Terrie, and children Tanya, Tiny Tim, Joe II, Louise, Marcia and Greg (too many children you say? Tell Bill and Terrie.). Joe and Rose are Bill's parents.

We may define the relation PARENT to be the set of 2-tuples (a,b) such that a is a parent of b. The relation CHILD may be defined to be the set of 2-tuples (a,b) such that a is a child of b. The relation, siblings of opposite sex (SOS), is defined to be the set of ordered pairs (a,b) such that a and b are of different sex and either a is a brother of b or a is a sister of b. For the Jones family we get:

PARENT =   {(Joe, Bill), (Harry, Terrie), (Rose, Bill), (Mildred, Terrie), (Bill, Tanya), (Bill, Tiny Tim), (Bill, Joe II), (Bill, Louise) (Bill, Marcia), (Bill, Greg), (Terrie, Tanya), (Terrie, Tiny Tim), (Terrie, Joe II), (Terrie, Louise), (Terrie, Marcia), (Terrie, Greg)}

CHILD = $\{(a, b) | (b, a) \in \text{PARENT}\}$

SOS1 =   {(Tanya, Tiny Tim), (Tanya, Joe II), (Tanya, Greg), (Marcia, Tiny Tim), (Marcia, Joe II), (Marcia, Greg), (Louise, Tiny Tim), (Louise, Joe II), (Louise, Greg)}

SOS = $\{(a,b)|(a,b)\in SOS1$ or $(b,a)\in SOS1\}$

One should note that the set SOS1 does not represent the relation siblings of opposite sex completely. Since the tuples are ordered sets (by definition) the pair (Tanya, Tiny Tim) is not the same as the pair (Tiny Tim, Tanya).

We could further define the relation WIFE to be the set of pairs (a,b) such that a is the wife of b. We get:

WIFE = {(Mildred,Harry), (Terrie,Bill), (Rose,Joe)}. □

Example 4.2: [Numbers and Arithmetic] Several binary relations are commonly defined on numbers. The relation less than $(<)$ consists of all 2-tuples (a,b) such that a is less than b; the relation equal $(=)$ is the set of all 2-tuples a, b such that $a=b$; and the relation greater than $(>)$ is the set of all (a,b) such that $a>b$. The relations, $<$, $=$, $>$, $\leqslant$, $\geqslant$, $\neq$ are defined to be the sets:

$$
\begin{aligned}
< \quad &= \quad \{(1,2),(0,3),(1,6),(1.1,1.3),(-2,2),...\} \\
= \quad &= \quad \{(0,0),(1,1),(1.1,1.1),(-2.3,-2.3),(4.5,4.5),...\} \\
> \quad &= \quad \{(-6,-8),(3/7,1/7),(8,4),...\} \\
&= \quad \{(a, b)|(b, a)\in <\} \\
\leqslant \quad &= \quad \{x|x\in < \text{ or } x\in =\} \\
\geqslant \quad &= \quad \{x|x\in > \text{ or } x\in =\} \\
\neq \quad &= \quad \{x|x\in < \text{ or } x\in >\}
\end{aligned}
$$

Ternary relations are needed to define the binary operations of $+, -,$ $*, /,$ and $\uparrow$ $(x\uparrow y = x^y)$. The tuple (a,b,c) is a member of the relation $+$ iff $a+b = c$; it is a member of the relation $-$ iff $a-b = c$; of $*$ iff $a*b = c$; of $/$ iff $a/b = c$; and of $\uparrow$ iff $a^b = $ c. □

Example 4.3: [Sets] In the chapter on sets we defined the binary relations $\subset$, $\subseteq$, and $=$. (a,b) is a member of the relation $\subset$ iff the set a is properly contained in the set b; it is a member of $\subseteq$ iff a is contained in b; etc. If we restrict the domain of the first component to be the set $\{0,1,2\}$ and its proper subsets (i.e., $D_1 = \{\varnothing,\{0\},\{1\},\{2\},\{0,1\},\{0,2\},\{1,2\},\{0,1,2\}\}$) and if $D_2 = \{\{0,1,2\},\{0,1\}\}$, then we get:

$\subset$ ={$(\varnothing, \{0, 1\})$, $(\{0\}, \{0, 1\})$, $(\{1\}, \{0, 1\})$, $(\varnothing, \{0, 1, 2\})$, $(\{0\}, \{0, 1, 2\})$, $(\{1\}, \{0, 1, 2\})$, $(\{2\}, \{0, 1, 2\})$, $(\{0, 1\}, \{0, 1, 2\})$, $(\{0, 2\}, \{0, 1, 2\})$, $(\{1, 2\}, \{0, 1, 2\})$}

$=$ $= \{(\{0,1\},\{0,1\}), (\{0,1,2\},\{0,1,2\})\}$

$\subseteq$ $= \{x|x\in \subset \text{ or } x\in =\}$ □

Our previous examples have dealt only with binary and ternary relations. k-ary relations for large values of k are often encountered in relational databases.

Example 4.4: In a *relational database*, all information is represented by a set of relations. A student database maintained by a university could consist of the five relations: BIODATA, FEE, DEPT, COURSE, and STAT defined as below:

(i) BIODATA is a set of 6-tuples. Each element of this relation is of the form:

(NAME,ADDR,PH-NUM,BIRTH,SEX,ID-NUM)

where NAME is the student's name; ADDR is the student's address, PH-NUM is the phone number; BIRTH is the date of birth; SEX we are familiar with; and ID-NUM is a unique identification number.

(ii) FEE is a set of 4-tuples of the form:

(NAME,ID-NUM,QTR,$)

The fourth component of each tuple in FEE gives the tuition fee paid for the quarter QTR by the student with name NAME and identification number ID-NUM.

(iii) DEPT is a set of 5-tuples. Each 5-tuple has the format:

(NAME,ID-NUM,MAJOR,YREN,DEG)

DEG gives the degree (B.S.,M.S.,Ph.D, etc.) towards which the student is working. YREN is the year and quarter in which the student joined the department given by MAJOR.

(iv) COURSE is a set of 7-tuples. Each tuple in COURSE describes a course taken by some student. It has the format:

(ID-NUM,COURSE-NUM,DESCP,QTR,GRADE,CR,INSTR)

A possible tuple in COURSE is:

(12594, CSci 3400, DISC.STR., S79, A, 4, Sahni)

This tuple states that the student with ID-NUM 12594 took the Discrete

Structures course (CSci 3400) from Professor Sahni in Spring of 1979. The course carried a 4 credit load and the grade obtained was an A.

(v) The relation STAT is a 3-ary relation. Each tuple in STAT has the format:

(ID-NUM,GPA,#CR)

#CR is the total number of credits so far completed by the student with identification number ID-NUM. GPA is this student's current grade point average.

NAME	ADDR	PH-NUM	BIRTH	SEX	ID-NUM
Ray Pasta	121 Dundee Drive	291-4192	7/28/60	M	196422
Mary Jane	94 Little Puddle Road	442-9916	3/15/58	F	216411
Jo Jo Karate	992 Tough Street	777-7777	5/12/57	F	116118
Clark Kent	1 Superman Avenue	221-6211	9/ 1/35	M	999214

(a) Biodata

NAME	ID-NUM	QTR	$
Ray Pasta	196422	F	800
Ray Pasta	196422	W	900
Mary Jane	216411	S	900
Mary Jane	216411	F	800
Mary Jane	216411	W	700
Clark Kent	999214	F	800
Clark Kent	999214	S	400
Clark Kent	999214	W	800

(b) Fee

NAME	ID-NUM	MAJOR	YREN	DEG
Mary Jane	216411	C.Sci.	1976	MS
Clark Kent	999214	Aero.	1976	BS
Ray Pasta	196422	C.Sci.	1976	BA
Jo Jo Karate	116118	Ph.Ed.	1980	PhD

(c) DEPT

IDNUM	GPA	#CR
116118	3.8	102
999214	4.0	45
196422	2.0	92
216411	3.7	164

(d) STAT

ID-NUM	COURSE-NUM	DESCP	QTR	GRADE	CR	INSTR
196422	3400	Disc.Str.	F	A	4	Sahni
196422	5701	Databases	F	A	4	Hevner
216411	5503	Compilers	W	B	4	Maly
216411	5502	Op.Sys.	W	A	4	Bruell

(e) COURSE

Figure 4.1 Example student database.

The relations of a relational database are normally represented in a table format (as in Figure 4.1). Figure 4.1 gives the sets corresponding to each of the above five relations for an exceptionally small university. Since each relation is a set, the order in which the rows of a table appear is not important. However, the order of the columns is important as each element of a k-ary relation is a k-tuple. When one is dealing with the relations of a database, one often refers to the various components of a tuple by the component name (or attribute). So, we may speak of the ID-NUM field of a tuple in STAT, the QTR field of a tuple in COURSE, the ADDR field (or attribute) of a tuple in BIODATA, etc.

We shall see in a later section that by performing suitable operations on the above relations it is possible to obtain answers to queries such as:

Q1: Which courses did John Adams take in Fall of 1980?
Q2: In which courses did Joe Doe get a C or D grade?
Q3: Who are the students majoring in Computer Science?
etc. □

4.2 BINARY RELATIONS

4.2.1 REPRESENTATIONS

A *binary relation* is a 2-ary relation. Binary relations occur quite frequently. In fact, most of our examples so far have been binary relations. If R is a binary relation and $(a,b) \in R$, then we write aRb (read as "a is R related to b"; or simply "aRb"). For example, if R is the binary relation $<$, then we write $2<3$ to mean 2 is related to 3 by the "less than" relation. This representation of members of a binary relation is called *infix notation* as the relation name (i.e., $R, <$, etc.) appears in between the two related elements a and b. Other examples of infix designation of elements of binary relations are: $4>2$; $5 \leqslant 7$; $\{0,1\} \subset \{0,2,1,6\}$; the jack must be removed from the trunk *before* it can be placed in position; etc. In this last example concerning the jack, the relation is "before" and corresponds to our example about changing wheels. As one observes, the infix representation is a natural representation for the elements of a binary relation.

When the binary relation R is a finite set, it may be specified by a relation matrix or graphically. Let D_1 and D_2 be the domains of the first and second components of the elements of R. Let $D = D_1 \cup D_2$. Note that since R is a finite set, D_1, D_2, and D are all finite sets. The reverse is also true; i.e., if D_1 and D_2 are finite, then R must be a finite set as R is a subset of the cartesian product $D_1 \times D_2$.

Let $n = |D|$. The *relation matrix* for the binary relation R is an $n \times n$ matrix M. Assume some ordering of the elements in D. Let $D = (a_1, a_2, ..., a_n)$. Element $M(i,j)$ (for i and j such that $1 \leqslant i \leqslant n$ and $1 \leqslant j \leqslant n$) of the matrix M is as given below:

$$M(i,j) = m_{ij} = \begin{cases} 1 & \text{if } a_i R a_j \text{ (i.e., } (a_i, a_j)) \in R \\ 0 & \text{otherwise} \end{cases}$$

Example 4.5: Consider the relation $R = \{(1,2), (1,4), (2,2), (3,4), (4,3), (3,1), (3,3), (4,2)\}$ and the ordered set $D = (1,2,3,4)$. The relation matrix corresponding to R is given in Figure 4.2(a). If we use the ordering $(A1,A2,A3,A4,A5)$, then the relation matrix for the relation $S = \{(A1,A2), (A2,A4), (A1,A4), (A2,A5), (A4,A5), (A1,A5), (A3,A4), (A3,A5)\}$ is as given in Figure 4.2(b). □

$$
\begin{array}{c|cccc}
 & 1 & 2 & 3 & 4 \\
\hline
1 & 0 & 1 & 0 & 1 \\
2 & 0 & 1 & 0 & 0 \\
3 & 1 & 0 & 1 & 1 \\
4 & 0 & 1 & 1 & 0 \\
\end{array}
$$

(a) Relation R

$$
\begin{array}{c|ccccc}
 & A1 & A2 & A3 & A4 & A5 \\
\hline
A1 & 0 & 1 & 0 & 1 & 1 \\
A2 & 0 & 0 & 0 & 1 & 1 \\
A3 & 0 & 0 & 0 & 1 & 1 \\
A4 & 0 & 0 & 0 & 0 & 1 \\
A5 & 0 & 0 & 0 & 0 & 0 \\
\end{array}
$$

(b) Relation S

Figure 4.2 Relation matrices.

It should be easy to see that once the ordering of elements in D is fixed, $(D = (a_1, a_2, ..., a_n))$, the relation matrix corresponding to a relation R is unique. The set representation for R is readily obtained from the matrix representation, M; $R = \{(a_i, a_j) | M(i,j) = 1\}$.

The *graphical representation* of a binary relation R consists of vertices and directed edges. A *vertex* is often drawn as a circle with a label in it. Figure 4.3(a) shows vertices that have been labeled 1, 2, and 3. A vertex with label i will be referred to as vertex i. Figure 4.3(b) shows some alternate ways to draw a vertex. A *directed edge* is a line joining two (not necessarily distinct) vertices. This line has an arrowhead on it (as in Figure 4.3(c)). If a directed edge connects vertices i and j and the arrowhead points towards vertex j, then the edge is designated as $<i,j>$. If the arrowhead points towards vertex i, the edge is $<j,i>$. Figure 4.3(c) shows edges $<1,2>$, $<3,4>$, and $<5,5>$.

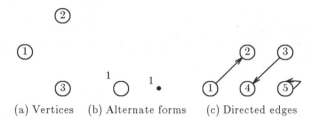

(a) Vertices   (b) Alternate forms   (c) Directed edges

Figure 4.3 Vertices and edges.

In the graphical representation of a binary relation R, there are n vertices ($n = |D| = |D_1 \cup D_2|$) and $e = |R|$ directed edges. Each vertex corresponds to exactly one element of D. The label on a vertex is the element in D to which it corresponds. For each (a,b) in R, the graphical representation contains one directed edge, $<a,b>$. Figures 4.4(a) and (b) show the graphical representation of the relations R and S of Example 4.5.

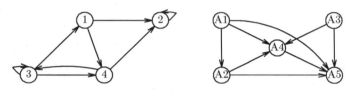

Figure 4.4 Graphical representations.

4.2.2 Properties

The precedence relation, P, obtained in Section 4.1 for the spare wheel replacement problem has several interesting properties. For example, if $(x,y) \in P$ and $(y,z) \in P$, then (x,z) must also be in P. In words, if operation x must be done before operation y, and if operation y must be done before operation z, then x must also be done before z. Another property of P is that P contains no elements of the form (x,x). I.e., there is no operation x which must be done before itself. Furthermore, if $(x,y) \in P$, then $(y,x) \notin P$ as otherwise the wheel changing problem will be infeasible. To see this, note that operation x must be done before y (as $(x,y) \in P$) and operation y must be done before x (as $(y,x) \in P$). Clearly, there is no way to change the wheels so that both these precedences are satisfied.

Let R be a binary relation. Let D_1 and D_2 respectively be the domains of the two components of R. Define $D = D_1 \cup D_2$. Table 4.1 lists the conditions under which R has a certain property Q. Column 1 of this table gives the property name, column 2 gives the adjective and noun forms of the property name used when talking about that property, and column 3 gives the condition that every binary relation R with that property satisfies. This condition is an "iff" condition. Thus, a binary relation has the property of reflexivity iff $(x,x) \in R$ for every element x in D. Another way to say this is: the relation R is reflexive (or R is a reflexive relation) iff $\forall x \in D(xRx)$.

PROPERTY	ADJECTIVE	CONDITION
Reflexivity	Reflexive	$\forall x \in D \ (xRx)$
Irreflexivity	Irreflexive	$\forall x \in D \ (x \not R x)$
Symmetry	Symmetric	$\forall x,y \in D \ (xRy \Longrightarrow yRx)$
Asymmetry	Asymmetric	$\forall x,y \in D \ (xRy \Longrightarrow y \not R x)$
Antisymmetry	Antisymmetric	$\forall x,y \in D \ (xRy \text{ and } yRx \Longrightarrow x=y)$
Transitivity	Transitive	$\forall x,y,z \in D \ (xRy \text{ and } yRz \Longrightarrow xRz)$

Table 4.1 Binary relation properties.

Of the six properties listed in Table 4.1, the precedence relation P (cf. wheel changing example) satisfies the conditions for the properties: irreflexivity, asymmetry, and transitivity. So, P is irreflexive, asymmetric, and transitive. The relation $<$ defined over the domain of real numbers is also irreflexive, asymmetric, and transitive. The relation $\leq$ with domain D being the real numbers is reflexive ($x \leq x$ for all x), antisymmetric, and transitive. The relation $=$ is reflexive, symmetric and transitive.

An examination of the property conditions reveals that a relation which is reflexive cannot also be irreflexive (unless $D = \varnothing$). To carry out a useful discussion of relation properties, we shall assume that $D \neq \varnothing$ in the remainder of this sub-section. A relation which is irreflexive cannot also be reflexive. In addition, it is possible for a relation to be neither reflexive nor irreflexive (e.g., $S = \{(1,1), (2,3)\}$ and $D = \{1,2,3\}$).

A symmetric relation cannot also be asymmetric. Similarly, an asymmetric relation cannot be symmetric. However, all asymmetric relations are also antisymmetric. Note, however, that all antisymmetric relations are not asymmetric (e.g., $\leq$ over the real numbers).

4.2.3 Composition And Closure

Let $R1$ and $R2$ be two binary relations defined on the set S. The composition of $R1$ and $R2$, denoted $R1oR2$, is the set:

$$R1oR2 = \{(x,y)|xR1z \text{ and } zR2y \text{ for some } z \in S\}$$

If $R1 = \{(1,2),(3,4),(2,4),(4,2)\}$ and $R2 = \{(2,4),(2,3),(4,1)\}$, then $R1oR2 = \{(1,4), (1,3), (3,1), (2,1), (4,4), (4,3)\}$ and $R2oR1 = \{(2,2), (2,4), (4,2)\}$. RoR is usually written R^2 and $RoRo...oR$ (with i Rs) is written R^i, $i \geqslant 1$. R^0 denotes the identity relation $\{(x,x) \mid x \in S\}$. One readily observes that xR^iy, $i > 2$, iff there exist $x_1, x_2, ..., x_{i-2}$ (not necessarily distinct) such that $xRx_1, x_1Rx_2,...,x_{i-2}Ry$.

The *transitive closure*, R^+, of a binary relation R is the set $\bigcup\limits_{i=1}^{\infty} R^i$. The *reflexive transitive closure*, R^*, of R is the set $R^0 \cup R^+$. One may show that $R^+ = RoR^* = R^*oR$. Note that if R is a transitive relation, then $R = R^+$. Further, if R is also reflexive then $R = R^*$ (see the exercises). When R is a binary relation on a finite set S of size n, $R^+ = \bigcup\limits_{i=1}^{n} R^i$ (see exercises).

Example 4.6: The transitive closure of R, $R = \{(1,2), (2,3), (3,4), (4,1)\}$ is $R \cup R^2 \cup R^3 \cup R^4$. $R^2 = \{(1,3), (2,4), (3,1), (4,2)\}$; $R^3 = RoR^2 \cup R^2oR = \{(1,4), (2,1), (3,2), (4,3)\}$; and $R^4 = \{(1,1), (2,2), (3,3), (4,4)\}$. So, $R^+ = \{(i,j) \mid 1 \leqslant i \leqslant 4, 1 \leqslant j \leqslant 4\}$. R^+ is shown graphically in Figure 4.5(b). Note that $R^* = R^+$ in this case.

Figure 4.5 gives the graphical representation of one other binary relation $(R1)$ together with the representation of its transitive and reflexive transitive closures. □

The transitive closure of a binary relation R finds application in several situations. One of these is in specifying a transitive relation T. Suppose x_iTx_{i+1}, $1 \leqslant i < n$. Since T is transitive, x_iTx_j for $1 \leqslant i < j \leqslant n$. So, to specify T completely, it is necessary to specify $n(n-1)/2$ pairs. This is rather cumbersome for $n > 4$. A more elegant way to specify T is to say $T = \{(x_i,x_{i+1})|1 \leqslant i < n\}^+$. If x_iTx_i, $1 \leqslant i \leqslant n$, x_iTx_{i+1}, $1 \leqslant i < n$, and T is transitive, then $T = \{(x_i,x_{i+1})|1 \leqslant i < n\}^*$.

An *incomplete specification* of a transitive and reflexive relation T is a set R such that $T = R^*$. If T is transitive but not reflexive, then R is an *incomplete specification* of T iff $T = R^+$. R is a *transitive reduction* of the transitive relation T iff T has no incomplete specification P such that $|P| < |R|$.

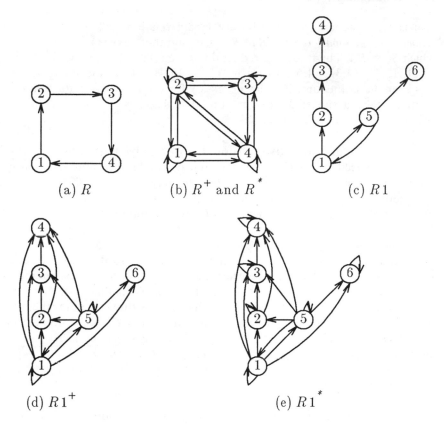

(a) R (b) R^+ and R^* (c) $R1$

(d) $R1^+$ (e) $R1^*$

Figure 4.5 Transitive closures.

Example 4.7: Consider the relation $T = \{(1,1), (2,2), (3,3), (1,2), (2,3), (1,3)\}$. T is readily seen to be reflexive and transitive. The relations $R1 = \{(1,1), (1,2), (2,3), (1,3)\}$, and $R2 = \{(1,2), (2,3)\}$ are incomplete specifications of T as $R1^* = R2^* = T$. $R1$ is not a transitive reduction of T while $R2$ is.

    The relation $T1 = \{(1,2), (2,3), (3,4), (4,5), (1,3), (1,4), (1,5), (2,4), (2,5), (3,5)\}$ is transitive but not reflexive. Some incomplete specifications are: $R3 = \{(1,2), (2,3), (3,4), (4,5), (2,5), (3,5)\}$; $R4 = \{(1,2), (2,3), (3,4), (4,5), (2,4)\}$; and $R5 = \{(1,2), (2,3), (3,4), (4,5)\}$. It can be shown that $R5$ is a transitive reduction of $T1$. $R3$ and $R4$ are not. $\square$

    The least cumbersome way to specify a transitive relation T is by specifying its transitive reduction (by least cumbersome we mean an explicit specification with fewest number of entries). The notion of transitive reduction is easily generalized to arbitrary binary relations. R is the *transitive reduction of the binary relation B* iff $R^* = B^*$ and there exists no S, $|S| < |R|$, for which $S^* = B^*$. The transitive reduction of the relation $\{(1,2), (2,1), (2,3),$ $(3,2), (3,4), (4,3), (4,5), (5,1)\}$ is $\{(1,5), (5,4), (4,3), (3,2), (2,1)\}$. As this example points out, it is not necessary for the transitive reduction of a relation to be a subset of the relation. The same is true for incomplete specifications of transitive relations.

Often, when a transitive relation T is incompletely specified (i.e., an incomplete specification R is provided) it is desirable to compute T from R. As mentioned earlier, this involves obtaining one of R^* and R^+ depending on whether T is reflexive or not. We shall now consider the problem of obtaining R^*. R^+ can be obtained from R^* by computing $R o R^*$.

There is a very elegant algorithm to compute R^* from R. This algorithm is due to Warshall. Assume that R is a binary relation on the set S. Let $S = \{1,2,...,n\}$. As defined earlier, the relation matrix corresponding to R is an $n \times n$ matrix $A(i,j)$, $1 \leqslant i \leqslant n$, $1 \leqslant j \leqslant n$, such that $A(i,j) = 1$ if $(i,j) \in R$ and $A(i,j) = 0$ otherwise. Warshall's algorithm computes the relation matrix C corresponding to R^*. Define the matrix C^k, $k \geqslant 0$, to be such that $C^k(i,j) = 1$ iff either $i = j$, or there exists a sequence $i, x_1, x_2, ..., x_r, j$ such that $x_a \leqslant k$, $1 \leqslant a \leqslant r$ and iRx_1, $x_1 Rx_2$, ..., $x_r Rj$. $C^k(i,j) = 0$ otherwise. Clearly, $C(i,j) = C^n(i,j)$ for all i and j. $C^0(i,j) = 1$ iff either $A(i,j) = 1$ or $i = j$. $C^0(i,j) = 0$ otherwise.

From the definition of C^k one may obtain the following relationship between the elements of C^k and C^{k-1} for $k \geqslant 1$:

(4.1) $C^k(i,j) = C^{k-1}(i,j) \lor (C^{k-1}(i,k) \land C^{k-1}(k,j))$

In the above equation, 1 and 0 are to, repectively, be treated as true and false. Let us verify the correctness of (4.1). By definition, $C^k(i,j) = 1$ iff either $i = j$ or iRx_1, $x_1 Rx_2,...,x_r Rj$ for some $x_1, x_2, ..., x_r$ no greater than k. If such a sequence exists then there also exists a sequence $y_1, y_2, ..., y_p$, of distinct numbers, $y_a \leqslant k$, $1 \leqslant a \leqslant p$ such that iRy_1, $y_1 Ry_2$, ..., $y_p Rj$. This follows from the observation that if $x_b = x_c$ for some b, $b < c$, then i, x_1, $x_2,...,x_a$, $x_{b+1},...,x_r$ is also a sequence for which iRx_1, $x_1 Rx_2,...$, $x_{a-1} Rx_a$, $x_a Rx_{b+1}$, ... $x_r Rj$.

If there is a sequence i, x_1, $x_2,..., x_r$, j of distinct numbers with the desired properties, then either k equals one of the x_is or it does not. In the latter case no x_i is larger than $k-1$ and $C^k(i,j) = C^{k-1}(i,j)$. If one of the x_is equals k, then the sequence has the form i, x_1, ..., k, ..., x_r, j. In this case, $C^k(i,j) = C^{k-1}(i,k) \land C^{k-1}(k,j)$. Combining these two cases, yields (4.1).

$C = C^n$ can be obtained from C^0 by using (4.1) to first compute C^1, then C^2, then C^3,..., and finally C^n. Note that each time a C^k is being computed, we need C^{k-1} (see (4.1)). From the order in which the C^is are being computed, it follows that C^{k-1} will be known when C^k is being computed, $1 \leqslant i \leqslant n$. Procedure CL (Algorithm 4.1) is a direct consequence of (4.1).

procedure $CL(A, C^n, n)$
  //compute $C^n = A^*$//
 Boolean $A(n,n), C^0(n,n), C^1(n,n),..., C^n(n,n)$
  $C^0(i,j) \leftarrow A(i,j)$, $1 \leqslant i \leqslant n$, $1 \leqslant j \leqslant n$
  $C^0(i,i) \leftarrow 1$, $1 \leqslant i \leqslant n$
  for $k \leftarrow 1$ **to** n **do**
   $C^k(i,j) \leftarrow C^{k-1}(i,j) \bigvee (C^{k-1}(i,k) \bigwedge C^{k-1}(k,j))$ $1 \leqslant i \leqslant n$, $1 \leqslant k \leqslant m$
  end
end CL

Algorithm 4.1

Procedure CL may be simplified somewhat by observing that:

(i) $C^k(i,k) = C^{k-1}(i,k)$, $1 \leqslant i \leqslant n$
(ii) $C^k(k,j) = C^{k-1}(k,j)$, $1 \leqslant j \leqslant n$
(iii) $C^{k-1}(i,j)$ for $i \neq k$ and $j \neq k$ is needed only when $C^k(i,j)$ is being computed.

As a result of these three observations, we may eliminate the superscripts from all occurrences of C in procedure CL. The final reflexive transitive closure algorithm is procedure $CLOSURE$ (Algorithm 4.2).

line	procedure $CLOSURE(A, C, n)$
	//compute $C = A^*$//
1	**Boolean** $A(n,n), C(n,n)$
2	$C(i,j) \leftarrow A(i,j)$, $1 \leqslant i \leqslant n$, $1 \leqslant j \leqslant n$
3	$C(i,i) \leftarrow 1$, $1 \leqslant i \leqslant n$
4	**for** $k \leftarrow 1$ **to** n **do**
5	$C(i,j) \leftarrow C(i,j) \bigvee (C(i,k) \bigwedge C(k,j))$, $1 \leqslant i \leqslant n$, $1 \leqslant j \leqslant n$
6	**end**
7	**end** $CLOSURE$

Algorithm 4.2 Reflexive transitive closure.

4.2.4 Equivalence Relations

A binary relation R on the set D (i.e., $D = D_1 \cup D_2$) is an *equivalence relation* iff it is reflexive, symmetric and transitive. If R is an equivalence relation, then aRb is also written $a \equiv b$.

Equivalence relations are not an uncommon occurrence. Consider the modulo (**mod**) function. x **mod** y for integers x and y is defined to be the remainder of x divided by y. 3 **mod** 2 = 1, 4 **mod** 2 = 0, (-5) mod 3 = $(6-5)$**mod**3 = 1, and so on. Two integers i and j are *congruent modulo n* iff i **mod** n = j **mod** n. The congruence modulo n relation is a binary relation that is easily seen to be an equivalence relation.

The = relation is another example of an equivalence relation. In the programming language FORTRAN there is an EQUIVALENCE statement which in its simplest form has the format:

EQUIVALENCE (v_1, v_2, v_3), (v_4, v_5), $(v_6, v_7, v_8, v_9, v_{10})$,.....

where the v_is are variable names. The effect of this statement is that variables v_1, v_2, and v_3 are assigned the same location in memory; variables v_4 and v_5 are assigned the same location; and variables v_6, v_7, ..., v_{10} are assigned the same location.

We may define a binary relation R over the variables appearing in a program. $v_i R v_j$ iff variables v_i and v_j are assigned the same location in memory. Clearly, $v_i R v_i$ for all variables in the program. Also, $v_i R v_j \implies v_j R v_i$ and $v_i R v_k \wedge v_k R v_j \implies v_i R v_j$. Hence, R is an equivalence relation.

The EQUIVALENCE statement of FORTRAN is a means to define an equivalence relation over the set of variables in a program. Suppose that the set V of variables in some FORTRAN program is $A, B, C, D, E, F, G, H, I, J$, and that the only EQUIVALENCE statement in this program is:

EQUIVALENCE (A, D, F), (B, G), (G, A), (C, H)

The relation R is:

$$R = X \cup Y \cup Z$$

where

$$X = \{(x,x) | x \in V\}$$
$$Y = \{(A,D), (A,F), (A,B), (A,G), (D,F), (D,B), (D,G), (F,B), (F,G), (B,G), (C,H)\}$$
$$Z = \{(x,y) | (y,x) \in Y\}$$

A less cumbersome way to describe R is to say that variables A, B, D, F, and G are to be assigned to the same location and so also are variables C

and H. If we let $E_1 = \{A,B,D,F,G\}$, $E_2 = \{C,H\}$, $E_3 = \{E\}$, $E_4 = \{I\}$, and $E_5 = \{J\}$, then R is given by:

$$R = \{(x,y) | x \text{ and } y \in E_i \text{ for some } i, 1 \leqslant i \leqslant 5\}$$

Let R be an equivalence relation over the set T. A subset E of T ($E \subseteq T$) is an *equivalence class* with respect to R and T iff the following conditions hold:

(i)   For all x and y in E, xRy.
(ii) For all x in E and for all y in $T-E$, $x\not{R}y$.

If R is as defined above by our FORTRAN equivalence statement, then each of E_1, E_2, E_3, E_4, and E_5 is an equivalence class. Note that the subset $\{A,D,G\} \subset E_1$ is not an equivalence class with respect to R and T as condition (ii) is not satisified (ARB but $A \in \{A,D,G\}$ and $B \in T-\{A,D,G\}$).

If R is an equivalence relation over the set T, then no element in T can be in two different equivalence classes. This follows from condition (ii) above and the reflexivity of an equivalence relation. Another important consequence of conditions (i) and (ii) is that the equivalence classes resulting from the equivalence relation R are unique. These equivalence classes may be obtained by first starting with each element of T in a distinct set and then using the elements $(x,y) \in R$ for distinct x and y to combine together the sets containing x and y (in case x and y are presently in different sets). For our FORTRAN example, we would begin with the sets:

$$\{A\},\{B\},\{C\},\{D\},\{E\},\{F\},\{G\},\{H\},\{I\},\{J\}$$

Since, $(A,D) \in R$, the sets $\{A\}$ and $\{D\}$ are combined to get the set $\{A,D\}$. Using $(A,F) \in R$, we combine the sets $\{A,D\}$ and $\{F\}$ to get $\{A,D,F\}$. Using $(C,H) \in R$, we get the set $\{C,H\}$. Since (A,B) and (D,G) are in R, the sets $\{A,D,F\}$ and $\{B\}$ can first be combined to get $\{A,D,F,B\}$. Then this set may be combined with $\{G\}$ to get the set $\{A,D,F,B,G\}$. The remaining pairs in R do not result in any further set combinations. So, we are left with the sets $\{A,B,D,F,G\}$, $\{C,H\}$, $\{E\}$, $\{I\}$, and $\{J\}$. Each of these sets defines an equivalence class.

One should note that the process just described to obtain equivalence classes is an algorithm iff R is a finite set. If R is infinite, we will never get done using all the pairs in R. If T is a finite set then R can have at most n^2 ($|T| = n$) pairs and the above equivalence class generation methods is an algorithm.

Let $S_1, S_2, ..., S_k$ be subsets of the set T. The set $S = \{S_1, S_2, ..., S_k\}$ is a *cover* of the set T iff $\bigcup\limits_{i=1}^{k} S_i = T$. The *size* of the cover S is k. A cover S of T is a *partition* of T iff $S_1, ..., S_k$ are pairwise disjoint. If S is a partition, then each S_i, $1 \leqslant i \leqslant k$ is a *block* of the partition.

Our discussion of equivalence relations and classes is summarized by the following theorem.

Theorem 4.1: An equivalence relation R over a set T partitions T into a unique set of equivalence classes. If T is finite, the number of equivalence classes in finite. If T is infinite, the number of equivalence classes may or may not be finite. $\square$

Example 4.8: The modulo function, **mod** n, partitions the set, I, of integers into n equivalence classes. Each of these sets is a countably infinite set. If we label these equivalence classes $E_0, E_1, ..., E_{n-1}$ then E_i for any i, $0 \leqslant i < n$, is given by:

$$E_i = \{i+kn | k \text{ is an integer}\}$$

Consider the set $X = \{p^j | p \text{ is a prime number and } j \text{ is a natural number}\}$. Define R to be a binary relation with the property that aRb iff a and b are in X and are powers of the same prime number p. One may easily verify that R is an equivalence relation. R partitions X into equivalence classes; one equivalence class for each prime number. Since the number of prime numbers is countably infinite, the set of equivalence classes of X under the equivalence relation R is also countably infinite. $\square$

4.2.5 Partial Orders

Any nontrivial project can be broken down into some number of tasks such that the completion of all these tasks results in the completion of the project. We have already seen one such decomposition of a project into tasks, i.e., the wheel replacement problem of Section 4.1. In this problem, the project was decomposed into 15 tasks, A1, A2, ..., A15. The tasks resulting from the decomposition of a project are generally related to one another in terms of a precedence relation. Certain tasks have to be completed before others can start. In the wheel replacement project, for example, task A2 had to be completed before A1 could start; task A1 had to be completed before A3 could start, etc.

As another example of the decomposition of a project into tasks, consider the project of obtaining a bachelors degree in computer science. This project is completed when a certain number of course credits has been successfully completed. Let $S = \{$discrete structures, data structures, design and analysis of algorithms, compilers, operating systems, programming languages, ...$\}$ be the set of courses that must be completed in order to obtain a bachelor's degree. The prerequisites for each course define a precedence relation between pairs of courses. If the discrete structures and data structures courses are both prerequisites to the compilers course, then both of these courses must be completed before (i.e., must precede) the compilers course can be taken.

Let R be a precedence relation on a set $T = \{T_1, T_2, ...\}$ of tasks of a project (i.e., aRb iff task a must be completed before task b can begin). We immediately observe that R is a transitive relation. If R is not asymmetric then there exist a and b in T such that aRb and bRa. The existence of such a pair of elements implies that the project is infeasible as it is not possible to both do a before b and b before a. In addition, the existence of a task a for which aRa implies infeasibility. So, for feasible projects, the precedence relation R is irreflexive, asymmetric and transitive. Note that if R is irreflexive and transitive, then it must also be asymmetric. We shall use "$<$" (read as "precedes" or "less than") to denote relations with these properties.

The preceding discussion of projects and tasks could have just as well been carried out in terms of a "precedes or equals" relation Q. In this case, aQb iff task a either equals task b (i.e., they are the same task) or must be completed before b can start. From our earlier discussion we can easily conclude that for feasible projects, Q is reflexive, antisymmetric, and transitive. Relations that satisfy these three properties will be denoted "$\leqslant$" (read as "precedes or equals", or "is less than or equal to").

For any feasible project with task set T, the binary relations $<$ and $\leqslant$ are related by the following equations:

(1) $< \; = \{(a,b)|(a,b) \in \; \leqslant$ and $a \neq b\}$
(2) $\leqslant \; = \; < \cup \{(a,a)|a \in T\}$

The relations $<$ and $\leqslant$ are *partial ordering relations* and they define a *partial ordering* of the elements of the set T on which they are defined. The set T is *partially ordered* by a partially ordering relation. Partial ordering relations frequently arise in situations other than projects. The binary relations:

$$\leqslant \; = \; \{(a,b)\,|\,a\leqslant b\text{ and } a \text{ and } b \text{ are real numbers}\}$$
$$\subseteq \; = \; \{(a,b)\,|\,a\subseteq b\text{ and } a \text{ and } b \text{ are sets}\}$$
$$D \; = \; \{(a,b)\,|\,a \text{ and } b \text{ are natural numbers greater than 0, and } a \text{ divides}$$
$$b\}$$
$$E \; = \; \{(a,b)\,|\,\text{person } a\text{'s salary is no more than } b\text{'s}\}$$

are all partial ordering relations of type $\leqslant$. The following relations are partial ordering relations of type $<$:

$$< \; = \; \leqslant - \{(a,a)\,|\,a \text{ is a real number}\}$$
$$\subset \; = \; \subseteq - \{(a,a)\,|\,a \text{ is a set}\}$$
$$D1 \; = \; D - \{(a,a)\,|\,a \geqslant 1 \text{ and } a \text{ is a natural number}\}$$
$$E1 \; = \; E - \{(a,a)\,|\,a \text{ is a person}\}$$

R is an *incompletely specified partial ordering relation* of type $<$ (or $\leqslant$) on the set S iff R^{+} (or R^{*} in case of $\leqslant$) is irreflexive, asymmetric, and transitive (or R^{*}, in case of $\leqslant$, is reflexive, antisymmetric, and transitive). In the sequel, we shall often not explicity distinguish between incompletely specified and completely specified partial ordering relations. Whenever an incompletely specified relation is presented, it will be understood that the full relation is its (reflexive) transitive closure. If R is a partial ordering relation on the set S, then (S,R) is a *partial order*.

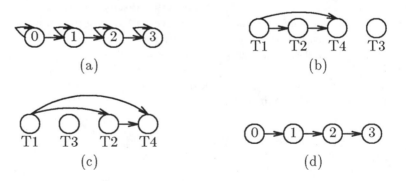

Figure 4.6 Topological graphical representations.

There are two rather special ways to draw the graphical representation of a partially ordered finite set S with partial ordering relation R. In the first of these, the vertices are drawn on a straight line, left to right. Let v_1, v_2, ..., v_n (where $n = |S|$) be the vertex labels left to right. By definition, $S = \{v_1, v_2, ..., v_n\}$. All edges are of the form $<v_i, v_j>$ where $j \geqslant i$. A graphical representation of this type is a *topological graphical representation*.

Figure 4.6(a) gives a topological graphical representation for the partial order $(\{0,1,2,3\}, \leqslant)$. The transitive closure of this representation yields the complete partial order. Figures 4.6(b) and (c) give two different topological representations for the partial order $(\{T1,T2,T3,T4\}, \{(T1,T2), (T2,T4), (T1,T4)\})$. When the partial ordering relation is of the type $\leqslant$, the edges $<x,x>$ are usually not drawn and are implicit. So, the partial order $(\{0,1,2,3\}, \leqslant)$ will usually be drawn as in Figure 4.6(d) rather than as in Figure 4.6(a).

Figure 4.7(a) is the graphical representation of an incompletely specified partial order on the set $\{1,2,3,4,5,6,7,8,9,10\}$. Figures 4.7(b), (c), and (d) give some of the possible topological graphical representations for this partial order. Note that whether the partial order (S,R) is completely specified or not, the number of edges in its topological graphical representation exactly equals the number of pairs in R.

The order v_1, v_2, ..., v_n in which the elements of S appear in a topological graphical representation is called a *topological order*. Observe that an ordering v_1, v_2, ..., v_n of the elements of S, given the partial order (S,R), is a topological order iff there exists no i and j, $1 \leqslant i < j \leqslant n$ for which $v_j R v_i$. Clearly this is necessitated by the requirement that all edges be of the form $<v_i, v_j>$ for $i \geqslant j$. The following theorem establishes the existence of a topological graphical representation for every partial order (S,R).

Theorem 4.2: Let S be a finite set and let (S,R) be a partial order. R may be incompletely specified. The elements of S can be arranged into a topological order.

Proof: The proof is by construction. We shall show how to obtain a topological order of the elements in S. Algorithm 4.3 starts with a finite set S and a partial ordering R and prints out the elements of S in a topological order. In the **for** loop of lines 2-9, the elements of S are printed out one by one. To complete the proof we only need to show that lines 3 and 4 of the algorithm are correct.

Suppose that on some iteration of the loop of lines 2 through 9, S is such that for every $x \in S$ there is a y in S, $y \neq x$ and yRx. At this time, $i < n$

as otherwise $|S| = 1$ and distinct x, y do not exist. Let x_2 and x_1 be such that $x_2 \neq x_1$ and $x_2 R x_1$. For x_2, there exists an $x_3 \in S$, $x_3 \neq x_2$ such that $x_3 R x_2$. If $x_3 \neq x_1$, then for x_3 there exists an $x_4 \in S$, $x_3 \neq x_4$ such that $x_4 R x_3$. If x_4 is not one of x_1 and x_2, then consider $x_5 \in S$ such that $x_4 \neq x_5$ and $x_5 R x_4$. If x_5 is not one of x_3, x_2, and x_1, then consider x_6 and so on.

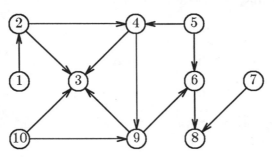

(a) An incompletely specified partial order

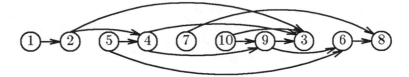

(b) A topological graphical representation

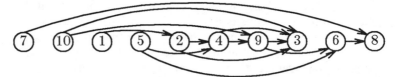

(c) Another topological graphical representation

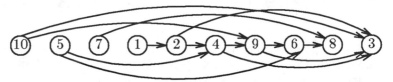

(d) Yet another topological graphical representation

Figure 4.7 An incompletely specified partial order.

```
line   procedure TOPOLOGICAL-ORDER (S,R)
         //print elements of S in topological order//
         //R may be incompletely specified//
1        n ← |S|
2        for i ← 1 to n do
3          if for every x∈ S there is a y∈ S, x ≠ y such that yRx
4            then print ("not a partial order"); stop
5          endif
6          Let x ∈ S be such that there is no y ∈ S, y≠ x for which yRx
7          print(x)
8          S ← S − {x}
9        end for
10     end TOPOLOGICAL-ORDER
```

Algorithm 4.3

This argument can be carried on at most n times before we discover an $x_j \in \{x_1, x_2, ..., x_{j-2}\}$ such that $x_j R x_{j-1}$, $x_{j-1} R x_{j-2}$, ..., $x_2 R x_1$. This is so because $|S| \leq n$ in any iteration of the loop. Now, $x_j = x_r$ for some r, $1 \leq r \leq j-2$. Let T be the completely specified relation corresponding to R (i.e., $T = R^+$ or R^*). From the transitivity of T, we infer $x_j T x_r$ and $x_j R x_{r+1}$. Since $x_j = x_r$, we obtain $x_r T x_r$, $x_r T x_{r+1}$ and $x_r T x_{r+1}$. Hence T is neither irreflexive nor antisymmetric (as $x_r \neq x_{r+1}$). So, T cannot be a partial ordering relation of type $<$ or $\leq$. Hence, if (S,R) is a partial order, the conditional of line 3 will never be true and a topological order will be obtained. $\square$

The second way to draw the graphical representation of a partially ordered set is to draw it such that all edges are directed from bottom to top. Once again, edges of type $< v_i, v_i >$ will be omitted. Figure 4.8 shows some of the possible ways to draw the partially ordered sets of Figures 4.6(a) and (b) so that all edges are directed upwards. Figure 4.9 does the same for the partial order of Figure 4.7(a). One way to obtain such a representation of S is to simply rotate any topological graphical representation of S anticlockwise by 90 degrees (see Figures 4.8(a) and (b). In Figures 4.6, 4.7, 4.8, and 4.9 the direction of edges is fixed (i.e. either left to right or down to up respectively). Hence, the arrowheads on the edges may be omitted. A *Hasse diagram* of a partially ordered set S is a graphical representation in which all edges are oriented upwards. The arrowheads are implicit (see Figures 4.10 and 4.11).

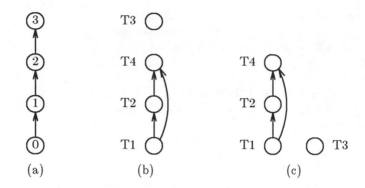

(a)   (b)   (c)

Figure 4.8

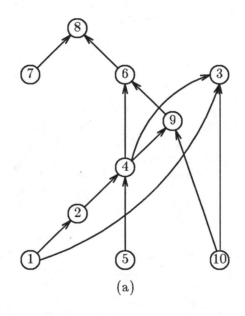

(a)

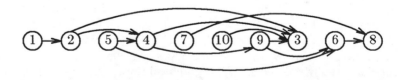

(b)

Figure 4.9

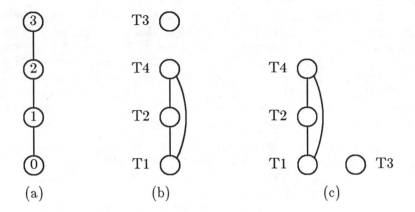

Figure 4.10 Hasse diagrams for Figure 4.8.

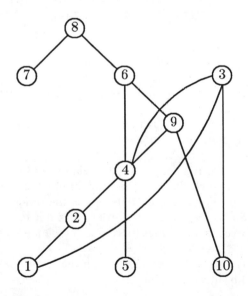

Figure 4.11 Hasse diagram for Figure 4.9.

Let (S,R) be a partial ordering. An element x of S is a *minimal* element iff there exists no y in S such that $x \neq y$ and yRx. x is a *maximal* element iff there is no $y \in S$, $x \neq y$ such that xRy. If $S = \{0,1,2,3\}$ and if R is the relation $<$, then 0 is a minimal element and 3 a maximal element. If $S = \{T1,T2,T3,T4\}$ and R is the precedence relation $\{(T1,T2),\ (T2,T4),\ (T1,T4)\}$, then both T1 and T3 are minimal elements. T3 and T4 are maximal elements. The maximal elements of the partial order of Figure 4.11 are 3 and 8. The minimal elements are 1, 5, 7, and 10.

Let (S,R) be a partial ordering. Let T be the completely specified version of R. Let $\{x_1, \ldots, x_r\}$ be a subset of S. $z \in S$ is an *upper bound* of the elements $x_1, \ldots, x_r$ iff $x_i T z$, $1 \leqslant i \leqslant r$. For the partial order of Figure 4.11, 3 is an upper bound of 1, 2, 4, and 10; 6 is an upper bound of 4, 5, and 9. z is a *least upper bound* (l.u.b.) of x_i, $1 \leqslant i \leqslant r$, iff z is an upper bound and the x_is, $1 \leqslant i \leqslant r$ have no other upper bound w such that wTz. In Figure 4.11, 4 is a l.u.b. of 2 and 5; 8 is a l.u.b. of 7 and 6; 9 is a l.u.b. of 4 and 10; both 6 and 3 are l.u.b.s of 4, 9, and 10.

z is a *lower bound* of x_i, $1 \leqslant i \leqslant r$ iff zTx_i, $1 \leqslant i \leqslant r$. z is a *greatest lower bound* (g.l.b.) of x_i, $1 \leqslant i \leqslant r$, iff z is a lower bound and there is no other lower bound w of x_i, $1 \leqslant i \leqslant r$, such that zTw. In Figure 4.11, 1 is a lower bound of 4, 9, and 3; 10 is a lower bound of 9 and 3; 2 is a g.l.b. of 4, 9, and 3; 10 is a g.l.b. of 9 and 3.

A partial order (S,R) is a *total order* iff the following are true:

(i) S has exactly one minimal and one maximal element (these need not be distinct).

(ii) Every element which is neither maximal nor minimal has a unique g.l.b. and a unique l.u.b.

The partial ordering defined by the relation $<$ (less than) on the set of natural numbers is such that 0 is the only minimal element and ∞ the only maximal element. For every i, $i > 0$, $i-1$ is the unique g.l.b. and $i+1$ the unique l.u.b. So, $(N, <)$ is a total order. It is not too difficult to see that corresponding to each total order there is a unique topological order.

A *lattice* is a partial order (S,R) such that every pair of elements in S has a unique g.l.b. and a unique l.u.b.

4.3 OPERATIONS ON k-ARY RELATIONS

Our discussion of k-ary relations will be carried out in the context of relational databases (see Example 4.4). A *relation scheme* is an ordered set of (distinct) attribute names. In Example 4.4, we had five relation schemes: BIODATA, FEE, DEPT, COURSE, and STAT. These were defined as:

$$
\begin{aligned}
\text{BIODATA} &= \text{(NAME,ADDR,PH-NUM,BIRTH,SEX,ID-NUM)} \\
\text{FEE} &= \text{(NAME,ID-NUM,QTR,\$)} \\
\text{DEPT} &= \text{(NAME,ID-NUM,MAJOR,YREN,DEG)} \\
\text{COURSE} &= \text{(ID-NUM,COURSE-NUM,DESCP.,QTR,GRADE,CR,} \\
&\quad\ \ \text{INSTR)} \\
\text{STAT} &= \text{(ID-NUM,GPA,CR)}
\end{aligned}
$$

Let RS be a relation scheme with the k attributes: $A_1, A_2, ..., A_k$. Let $D_1, D_2, ..., D_k$, respectively, be the domains of these k attributes. A *record* (or tuple) is an ordered set $(v_1, v_2, ..., v_r)$. The record $(v_1, v_2, ..., v_k)$ *corresponds* to the relational scheme RS iff $r = k$ and $v_i \in D_i$, $1 \leqslant i \leqslant k$. A relation R is an *instance* of a relational scheme RS iff it is a set of records (or tuples) that correspond to RS (i.e., it is a subset of $D_1 \times D_2 \times ... \times D_k$). R is also said to be a relation *defined* on the attributes $A_1, A_2, ..., A_k$.

In the context of databases, relations are usually drawn as a table with each row corresponding to a record. Each column is labeled by the corresponding attribute name. The order of the records in a table is not important.

Figure 4.1(a) is an instance of the relation scheme BIODATA. Similarly, the relations depicted in Figures 4.1(b), (c), (d), and (e) are respectively instances of the relation schemes FEE, DEPT, STAT, and COURSE.

Usually, we shall be dealing with only one instance of each relation scheme at any time. So, we shall refer to the relation simply by using the name of the relation scheme (rather than use a new name for the relation). Thus, when we speak of the relation BIODATA, we really mean an instance of the relation scheme BIODATA.

A *relational database scheme* is a set of relation schemes. So, STUDENT-INFO = {BIODATA, FEE, DEPT, COURSE, STAT} is a relational database scheme. An *instance* of a relational database scheme DS = $\{RS_1, RS_2, ..., RS_k\}$ is a set $\{R_1, R_2, ..., R_k\}$ of k relations such that each R_i is an instance of the relation scheme RS_i, $1 \leqslant i \leqslant k$. If S is an instance of the relational database scheme DS, then we shall refer to S simply as the *relational database* DS. The tables of Figure 4.1 define the database STUDENT-INFO (or more accurately, they define an instance of the database scheme STUDENT-INFO).

In order to extract information from a relational database, it is necessary to be able to manipulate k-ary relations. Let R be a relation. By $AT(R)$ we shall mean the attribute set on which R is defined. For any record $r \in R$ and any subset $B \subseteq AT(R)$, $VAL(r,B)$ will denote the ordered multiset of values of the components of r that correspond to the attributes in B. This multiset is ordered according to the ordering of the attributes of B in r.

As an example, consider the relation BIODATA of Figure 4.1(a). If r is the first record in this table and if B is the set {ADDR, PH-NUM}, then $VAL(r,B) = (121$ Dundee Dr., 291-4192). If r is the second record and B = {NAME,SEX}, then $VAL(r,B)$ = (Mary Jane, F). In general, $VAL(r,B)$ will not be a set but will be a multiset as it is possible for many records of a relation to have identical components.

The more frequently used operations on k-ary relations are:

(a) Projection

For any relation R and attribute set B, $B \subseteq AT(R)$ and $|B| = b$, the *projection* of R onto B (denoted $P(R,B)$) is the relation:

$$P(R,B) = \{(v_1,v_2,...,v_b) | VAL(r,B) = (v_1,v_2,...,v_b) \text{ for some } r \in R\}$$

The relative order of attributes in $P(R,B)$ is the same as that in R. The relation $P(R,B)$ can be obtained from R by first deleting all columns in R that correspond to attributes in the set $AT(R)-B$. Next, multiple occurrences of the same row are deleted from the resulting table.

The projection of the relation BIODATA onto the set {NAME, ADDR} is given in Figure 4.12(a). It is obtained by simply deleting columns 3, 4, and 5 of Figure 4.1(a). Since no two rows are identical following this deletion of columns, no rows need to be deleted. Figure 4.12(b) gives the projection of the relation BIODATA onto the set {SEX}.

It is easy to see that for every R, $AT(R)$ and $B \subseteq AT(R)$, the size of the projection. $P(R,B)$ is no more than the size of R. I.e., $|P(R,B)| \leq |R|$.

(b) Selection

Let R, B, and b be as above. Let $v = (v_1, v_2, ..., v_b)$ be a record. $SEL(R,B,v)$ is the relation:

$$SEL(R,B,v) = \{r | r \in R \text{ and } VAL(r,B) = v\}.$$

NAME	ADDR
Ray Pasta	121 Dundee Drive
Mary Jane	94 Little Puddle Road
Jo Jo Karate	992 Tough Street
Clark Kent	1 Superman Avenue

SEX
M
F

(a) P(BIODATA, {NAME,ADDR})   (b) P(BIODATA,{SEX})

NAME	ADDR	PH-NUM	BIRTH	SEX	ID-NUM
Mary Jane	94 Little Puddle Road	442-9916	3/15/58	F	216411
Jo Jo Karate	992 Tough Street	777-7777	5/12/57	F	116118

(c) SEL(BIODATA, {SEX}, (F))

NAME	ID-NUM	MAJOR	YREN	DEG
Mary Jane	216411	C.Sci.	1976	MS
Ray Pasta	196422	C.Sci.	1976	BA

(d) SEL(DEPT, {MAJOR, YREN}, (C.Sci., 1976))

NAME	ADDR	ID-NUM	MAJOR	YREN	DEG
Ray Pasta	121 Dundee Drive	196422	C.Sci.	1976	BA
Mary Jane	94 Little Puddle Road	216411	C.Sci.	1976	MS

(e) Join of Figures (a) and (d)

NAME	ADDR	SEX
Ray Pasta	121 Dundee Drive	M
Ray Pasta	121 Dundee Drive	F
Mary Jane	94 Little Puddle Road	M
Mary Jane	94 Little Puddle Road	F
Jo Jo Karate	992 Tough Street	M
Jo Jo Karate	992 Tough Street	F
Clark Kent	1 Superman Avenue	M
Clark Kent	1 Superman Avenue	F

(f) Join of Figures (a) and (b)

Figure 4.12

Figure 4.12(c) gives the relation resulting from the selection operation
SEL(BIODATA,{SEX},(F)). This gives us the biodata of all female stu-
dents. The relation resulting from the selection SEL(DEPT,
{MAJOR,YREN}, (CSci,1976)) is given in Figure 4.12(d).

(c) Join

The join operation is used to combine together two relations. Let R1 and R2
be two relations and let $S = AT(R1) \cap AT(R2)$. Without loss of generality,
we may assume that the attributes in S appear in the same order in R1 and
R2. For every record $r \in R1$ and every record $t \in R2$, let $r.t$ denote the record
obtained by first deleting from t all components corresponding to attributes
in S and then concatenating the remaining record to the right end of record
r.

When r is the first record in BIODATA and t the first record in DEPT
(see Figures 4.1(a) and (c)), then $r.t$ = (Ray Pasta, 121 Dundee Dr., 291-
4192, 7/28/60, M, 196422, CSci., 1976, M.S.).

The *join* of R1 and R2, denoted $J(R1,R2)$, is defined to be the relation:

$$J(R1.R2) = \{x | x = r.t \text{ for some } r \in R1 \text{ and some } t \in R2 \text{ such that } VAL(r,S) = VAL(t,S)\}$$

Figure 4.12(e) gives the join of the relations of Figures 4.12(a) and
(d). Figure 4.12(f) gives the join of the relations of Figures 4.12(a) and (b).
The join operation as defined here is commonly refered to as the *natural join*.
A more general definition of join can be found in Date's book which is cited
at the end of this chapter.

It is interesting to note that a selection can actually be performed using
the join operation. To select all records of a relation R1 with $VAL(R1,B) = v$, we simply define a new relation $R2 = \{v\}$ with $AT(R2) = B$ and perform
the operation $JOIN(R1,R2)$. Also, note that if R1 and R2 have the same
attribute sets then $JOIN(R1,R2) = R1 \cap R2$. If $R1 \cap R2 = \emptyset$, then
$JOIN(R1,R2)$ is the cross product of R1 and R2.

(d) Semi-Join

The *semi-join* of two relations R1 and R2 with $S = AT(R1) \cap AT(R2)$ is the
relation:

$$SJ(R1,R2) = \{r | r \in R1 \text{ and for some } t \in R2, VAL(r,S) = VAL(t,S)\}$$

The semi-join of the relations of Figures 4.12(c) and (d) is just the relation containing the first record of Figure 4.12(c).

In words, the semi-join of two relations R1 and R2 is a subset of R1. This subset contains only those records of R1 that match with at least one record in R2 on their common attributes S. The semi-join is thus a generalization of selection.

By using the functions just described, it is possible to retrieve very specific information from a relational database. The names of all female students in our hypothetical university example (Figure 4.1) can be obtained by first selecting all female records from the relation BIODATA and then projecting the resulting relation on the attribute NAME. A list of all courses taken by John Adams in fall of 1980 can be obtained as follows:

$$R \leftarrow SEL(\text{BIODATA},\{\text{NAME}\},(\text{John Adams}))$$
$$S \leftarrow P(R,\{\text{ID-NUM}\})$$
$$T \leftarrow J(S,\text{COURSE})$$
$$U \leftarrow P(T,\{\text{COURSE},\text{DESCP}\})$$

4.4 FUNCTIONAL AND MULTIVALUED DEPENDENCIES

An examination of the database of Figure 4.1 reveals that there exist relationships between the values of certain sets of attributes and the values of other attribute sets. The ID-NUM of a student uniquely determines the students name and also the students date of birth, phone number, sex, and address. In fact, for every instance of the database scheme STUDENT-INFO (see Section 4.3) it must be the case that ID-NUM uniquely determines the student name, address, date of birth, phone number, sex, and address. For the database of Figure 4.1, the name uniquely determines the ID-NUM too. This of course is not true for all instances of the database scheme STUDENT-INFO as it is quite possible for a university to have two students having the same name. These students will be given different ID-NUMs. The sex of a student does not uniquely determine the student name in Figure 4.1. Similarly, the year enrolled does not determine the major uniquely. However, the year enrolled and major attributes together uniquely determine the ID-NUM, name and degree in the database instance of Figure 4.1. Once again, there can be instances of the STUDENT-INFO database in which the major and the year enrolled attributes will not uniquely determine any of the attributes ID-NUM, NAME, and DEG.

Let R be a relation. The attribute set $Y \subseteq AT(R)$ is *functionally dependent* on the attribute set $X \subseteq AT(R)$ (written $X \rightarrow Y$) in R iff for every pair of records r, $t \in R$, $[VAL(r,X) = VAL(t,X)] \implies [VAL(r,Y) = VAL(t,Y)]$. I.e., an assignment of values to X uniquely determines the values of the attributes in Y. Note that if $Y = \{Y_1, Y_2, ..., Y_k\}$, then $X \rightarrow Y$ is equivalent to the set of functional dependencies (FDs) $\{X \rightarrow Y_1, X \rightarrow Y_2, ..., X \rightarrow Y_k\}$.

For the relation BIODATA of Figure 4.1, the FD $\{ID\text{-}NUM\} \rightarrow \{NAME,ADDR,PH\text{-}NUM,BIRTH,SEX\}$ holds. This FD is equivalent to the set of FDs: $\{\{ID\text{-}NUM\} \rightarrow \{NAME\}, \{ID\text{-}NUM\} \rightarrow \{ADDR\}, \{ID\text{-}NUM\} \rightarrow \{PH\text{-}NUM\}, \{ID\text{-}NUM\} \rightarrow \{BIRTH\}, \{ID\text{-}NUM\} \rightarrow \{SEX\}\}$.

A relation scheme, RS, *satisfies* the functional dependency $X \rightarrow Y$ iff this FD holds for every instance R of RS. The FD $X \rightarrow Y$ holds in a *database* DB iff for every query of the form:

What are the values of Y_1, Y_2, ..., Y_k for which there exist entities in DB having $X_1 = a_1$, $X_2 = a_2$, ..., and $X_x = a_x$, where $X = \{X_1, X_2,...,X_x\}$ and $Y = \{Y_1, Y_2, ..., Y_k\}$?

the answer contains at most one value for each Y_i.

In the database of Figure 4.1, the FD $\{MAJOR,YREN,SEX\} \rightarrow \{PH\text{-}NUM, ADDR,ID\text{-}NUM\}$ holds. $\{MAJOR,YREN\} \rightarrow \{PH\text{-}NUM\}$ is not a FD of this database as when MAJOR = C.Sci. and YREN = 1976, the PH-NUM could be 291-4192 or 442-9916.

The FD $X \rightarrow Y$ *holds in the database scheme* DBS iff it holds in all instances of DBS.

Functional dependencies are normally provided together with a relational database scheme. They serve as *integrity constraints* on every instance of the DBS. If at any time a purported change to a database will result in a database that violates one of the FDs, then we know that something has gone wrong and the change is not to be permitted. So, for example if the FD $\{ID\text{-}NUM\} \rightarrow \{NAME\}$ is specified for the DBS STUDENT-INFO then the addition of a new record with ID-NUM = 196422 and NAME = Trouble Maker will not be permitted as following this addition the FD $\{ID\text{-}NUM\} \rightarrow \{NAME\}$ will no longer hold. In a real situation, this would mean that an error was made either in the ID-NUM or in the NAME.

The set of FDs of a database scheme DBS is itself a binary relation:

FD(DBS) = $\{(X, Y)| X \rightarrow Y$ holds in DBS$\}$

The binary relations FD(DB), FD(RS), and FD(R) are defined in an analogous way for a database, relation scheme, and relation respectively.

One may verify (see the exercises) that every relation FD(I) where $I \in \{DBS, DB, RS, R\}$ satisfies the following properties (W, X, Y, and Z are subsets of $AT(I)$):

(i)   If $X \subseteq Y$, then $Y \rightarrow X$.

(ii)   If $X \rightarrow Y$ and $Z \subseteq W$, then $X \cup W \rightarrow Y \cup Z$.

(iii) If $X \rightarrow Y$ and $Y \cup W \rightarrow Z$, then $X \cup W \rightarrow Z$.

Normally, one does not fully specify FD(I). For any incomplete specification S of FD(I), FD(I) is the smallest set containing S for which properties (i), (ii), and (iii) hold. FD(I) is the *closure* of S. Note that property (iii) is a generalization of transitivity as when $W = \varnothing$, property (iii) becomes:

If $X \rightarrow Y$ and $Y \rightarrow Z$, then $X \rightarrow Z$.

FDs in themselves are generally not adequate to define all the dependencies that might exist in a relation (or relation scheme, etc.). As an example, consider the relation given in Figure 4.13. In this relation, we have permitted a student to have more than one phone number. The ID-NUM uniquely determines the sex of a student. So, $\{ID\text{-}NUM\} \rightarrow \{SEX\}$ is an FD in this relation. The ID-NUM and YEAR together uniquely determine the MAJOR. Hence, $\{ID\text{-}NUM, YEAR\} \rightarrow \{MAJOR\}$ is also an FD. There is another dependency in Figure 4.13 and this cannot be expressed as an FD. The ID-NUM determines the set of phone numbers corresponding to that student. These phone numbers are not dependent on whether the student is majoring in C.Sci. or Math, or on whether the year is 1978, 1977 or 1980. The set of phone numbers corresponding to a student is a property of the student alone (i.e., of the students ID-NUM). The ID-NUM uniquely determines the set of phone numbers. We shall say that the *multivalued dependency* $\{ID\text{-}NUM\} \Longrightarrow \{PH\text{-}NUM\}$ holds for the relation of Figure 4.13. In our example, it is also the case that $\{ID\text{-}NUM\} \Longrightarrow \{SEX, MAJOR, YEAR\}$.

Let R be a relation and let X, Y, and Z be a partition of $AT(R)$. Let Y_x be the set $\{v | y = VAL(r, Y)$ and $x = VAL(r, X)$ for some $r \in R\}$. Let Y_{xz} be the set $\{y \mid y = VAL(r, Y), x = VAL(r, X),$ and $z = VAL(r, Z)$ for some

$r \in R$. The *multivalued dependency* $X \Longrightarrow Y$ holds in R iff $Y_x = Y_{xz}$ for all x and z for which Y_x and Y_{xz} are nonempty. Intuitively, $X \Longrightarrow Y$ means that the set of values for Y depends only on X and not on Z.

ID-NUM	PH-NUM	SEX	MAJOR	YEAR
91162	271-7341	M	C.Sci.	1978
91162	271-7344	M	C.Sci.	1978
92413	411-2961	F	Math	1977
92143	411-2781	F	Math	1977
92143	429-6111	F	Math	1977
91162	271-7341	M	Math	1980
91162	271-7344	M	Math	1980
92143	411-2961	F	C.Sci.	1980
92143	411-2781	F	C.Sci.	1980
92143	429-6111	F	C.Sci.	1980

Figure 4.13

The notion of a multivalued dependency may be generalized to relation schemes, databases, and database schemes in an analogous manner to that used to generalize the notion of a functional dependency. The following theorem summarizes some of the properties of multivalued dependencies (MVDs).

Theorem 4.3: [Fagin] The following statements are true for every relation R:

(a)   If $X \rightarrow Y$ is an *FD* of R and if $X \cap Y = \emptyset$, then $X \Longrightarrow Y$.
(b)   If X, Y, and Z define a partition of $AT(R)$, then $X \Longrightarrow Y$ iff R is the join of the projections $P(R, X \cup Y)$ and $P(R, X \cup Z)$.
(c)   For every partition X, Y, and Z of $AT(R)$, $X \Longrightarrow Y$ iff $X \Longrightarrow Z$.

Proof: See the exercises. $\square$

A relation R is said to have a *lossless join* with respect to the relation schemes RS1 and RS2 iff R is equal to the join of its projections onto $AT(RS1)$ and $AT(RS2)$. From Theorem 4.3, it follows that if $X \Longrightarrow Z$ is a MVD of R, then R has the lossless join property with respect to the relation schemes for $P(R, X)$ and $P(R, Y)$.

As an example of a relation R that does not have the lossless join property with respect to $AT(RS1)$ and $AT(RS2)$, consider the relation of Figure 4.14(a) and let $AT(RS1) = \{A\}$ and $AT(RS2) = \{B\}$. Figures 4.14(b) and (c) respectively give the projections of R onto $AT(RS1)$ and $AT(RS2)$. Figure 4.14(d) gives the join of these two projections.

The definition of MVDs may be extended to the case where $X \cap Y \neq \varnothing$. This may be done by requiring that $X \Longrightarrow Y$ hold iff R has a lossless join with respect to the relation schemes for $P(R,X)$ and $P(R,Y)$. An alternate (but equivalent) requirement is that $X \Longrightarrow Y$ iff $X \Longrightarrow Y-X$.

A	B
a	b
c	d

A
a
c

B
b
d

A	B
a	b
a	d
c	b
c	d

  (a) R (b) $P(R,\{A\})$ (c) $P(R,\{B\})$ (d) Join of (b) and (c)

Figure 4.14

4.5 Normal Forms

The database literature abounds with normal forms for relations and relation schemes. A relation R is *normalized* (or in 1NF) iff the attribute values of each record in R is a single element (and not a multiset of elements). Figure 4.15(a) shows an unnormalized relation. The relation of Figure 4.15(a) contains two records, one with ID-NUM = 916292 and the other with ID-NUM = 93416. The second component of the first record is the ordered multiset (CSci 5121, CSci 3400, CSci 5122) and the third (A,A,B). An unnormalized relation is easily transformed into an equivalent 1NF relation by introducing new records as in Figure 4.15(b).

ID-NUM	COURSE	GRADE
	C.Sci. 5121	A
91629	C.Sci. 3400	A
	C.Sci. 5122	B
	C.Sci. 8401	A
93416	C.Sci. 5121	B
	C.Sci. 3400	C

ID-NUM	COURSE	GRADE
91629	C.Sci.5121	A
91629	C.Sci. 3400	A
91629	C.Sci. 5122	B
93416	C.Sci. 8401	A
93416	C.Sci. 5121	B
93416	C.Sci. 3400	C

  (a) Unnormalized   (b) Normalized (1NF)

Figure 4.15

In order to introduce the remaining normal forms (2NF,3NF,4NF, and BCNF), we need to define some new terms. A subset X of the attributes of R is a *superkey* iff $X \rightarrow AT(R)$ (i.e., the attributes of R are functionally dependent on the attributes X). For every relation R, $AT(R)$ is a superkey. It is easily seen that if X is a superkey of R and if Y is a set such that $X \subset Y \subseteq AT(X)$, then Y is also a superkey of R. X is a *key* of R iff X is a superkey of R and no proper subset of X is also a superkey R.

The relation of Figure 4.15(a) has two superkeys: {ID-NUM,COURSE} and {ID-NUM,COURSE,GRADE}. Only the first of these is a key.

An attribute of R is a *prime attribute* iff it is in some key of R. ID-NUM and COURSE are prime attributes of R. An attribute that is not prime is *nonprime*.

The above definitions of key, superkey, prime and nonprime attributes are extended to the case of relation schemes with specified functional dependencies by replacing every occurrence of "relation" in the definitions by "relation schemes".

For the relation scheme BIODATA of Figure 4.16(b) and the FDs of Figure 4.16(a), {ID-NUM} is a key. {NAME,ADDRESS} is also a key as (NAME,ADDRESS) $\rightarrow AT(R)$ follows from FDs (i) and (iii) and the transitivity of $\rightarrow$. The prime attributes are ID-NUM, NAME, and ADDRESS. For the relation scheme DEPT, the key is {ID-NUM,MAJOR,DEG}. The prime attributes of DEPT are ID-NUM, MAJOR, and DEG. {ID-NUM,MAJOR,DEG,YREN} is a superkey.

(i) {ID-NUM} $\rightarrow$ {NAME, ADDRESS, SEX, PH-NUM, YREN}
(ii) {NAME} $\rightarrow$ {SEX}
(iii) {NAME, ADDRESS} $\rightarrow$ {ID-NUM}

(a) Some FDs.

BIODATA = (ID-NUM, NAME, ADDRESS, SEX, PH-NUM)
DEPT   = (ID-NUM, NAME, MAJOR, YREN, DEG)
B1   = (ID-NUM, NAME, ADDRESS)
B2   = (ID-NUM, SEX, PH-NUM)

(b) Some relation schemes

Figure 4.16

An attribute A is *fully dependent* on the attribute set X iff $X \rightarrow A$ and there is no Y, $Y \subset X$ for which $Y \rightarrow A$. For the relation scheme BIODATA, SEX is fully dependent on {ID-NUM} but not on {ID-NUM,NAME}.

A relation (relation scheme) is in *2NF (Second Normal Form)* iff every nonprime attribute is fully dependent on each of its keys. The relation of Figure 4.15(a) has only one nonprime attribute: GRADE. This is fully dependent on the key {ID-NUM,COURSE}. So, this relation is in 2NF. The relation scheme BIODATA of Figure 4.16(b) is not in 2NF as the nonprime attribute SEX is not fully dependent on the key {NAME,ADDRESS}. DEPT is not in 2NF as NAME is not fully dependent on the key {ID-NUM,MAJOR,DEG}. If the only FDs specified are (i) and (ii) of Figure 4.16(a), then the relation BIODATA is in 2NF.

A relation (or relation scheme) is in *3NF* iff it is in 2NF and its nonprime attributes are functionally independent (i.e., there is no FD $X \rightarrow Y$ where X and Y are subsets of nonprime attributes). The relation scheme BIODATA of Figure 4.16(b) together with the FDs (i) and (ii) of Figure 4.16(a) is not in 3NF as the nonprime attributes NAME and SEX are not independent. In fact, {NAME} $\rightarrow$ {SEX} is an FD. When the only FDs are (i) and (ii) the relation schemes B1 and B2 of Figure 4.16(b) are in 3NF. The relation of Figure 4.17(a) is in 2NF but not in 3NF. ID-NUM is the only prime attribute and {COURSE} $\rightarrow$ {GRADE}. The relation of Figure 4.17(b) is in 3NF. Its only prime attribute is ID-NUM and the nonprime attributes are independent.

ID-NUM	COURSE	GRADE
91629	5121	A
91610	3400	B
92114	3400	B
61411	5121	A
71164	5121	A

ID-NUM	COURSE	GRADE
91629	5121	A
91610	3400	B
92114	3400	C
61411	5121	A
71164	5121	A

(a) A 2NF relation   (b) A 3NF relation

Figure 4.17

An FD $X \rightarrow Y$ is *trivial* iff $X \subseteq Y$. An MVD $X \Longrightarrow Y$ is *trivial* iff $Y = \varnothing$ or $Y = AT(R) - X$. (We assume that for all MVDS $X \Longrightarrow Y$, $X \cap Y = \varnothing$.) A relation R (or relation scheme RS) is in *4NF* iff whenever $X \Longrightarrow Y$ is a non-trivial MVD then $X \rightarrow AT(R)$ (or $X \rightarrow AT(RS)$) is an FD of R (or RS). A relation R is in *Boyce Codd Normal Form (BCNF)* iff whenever $X \rightarrow Y$ is a nontrivial FD in R then $X \rightarrow AT(R)$ is an FD of R. Similarly, a relation scheme RS, is in BCNF iff whenever $X \rightarrow Y$ is a nontrivial FD in RS then $X \rightarrow AT(RS)$ is an FD of RS (i.e., X is a superkey of RS).

Theorem 4.4: The following statements are true:

(a)   Every 2NF relation (or relation scheme) is also a 1NF relation (or relation scheme).

(b)   Every 3NF relation (or relation scheme) is also a 2NF relation (or relation scheme).

(c)   Every BCNF relation (or relation scheme) is also a 3NF relation (or relation scheme).

(d)   Every 4NF relation (or relation scheme) is also a BCNF relation (or relation scheme).

Proof: Left as an exercise. □

A database is in a given normal form iff all its relations are in that normal form. A database scheme is in a given normal form iff all its relation schemes are in that normal form.

REFERENCES AND SELECTED READINGS

An efficient algorithm to determine the transitive reduction of a transitive relation appears in the paper:

> *The transitive reduction of a directed graph*, by A. Aho, M. Garey, and J. Ullman, SICOMP, 1972.

In some applications, it is desirable to determine the smallest subset T of a binary relation R such that $T^* = R$. This problem is more difficult than the transitive reduction problem. The complexity of this problem is considered in the paper:

> *Computationally related problems*, by S. Sahni, SICOMP, vol. 3, pp. 262-279, 1974.

A topological ordering algorithm that can be easily translated into a computer program can be found in the book:

> *Fundamentals of data structures*, by E. Horowitz and S. Sahni, Computer Science Press, Maryland, 1976.

This book also contains an algorithm to determine the equivalence classes induced by an equivalence relation.

A good general reading on the application of k-ary relations and normal forms to the design of databases and database management languages is:

> *An introduction to database systems*, by C. J. Date, 2nd edition, Addison-Wesley Publishing Company, Massachusetts, 1977.

This book provides a good discussion of the advantages and disadvantages of the various normal forms. Bernstein, and Beeri have studied the problem of obtaining normal form relation schemes from a given set of FDs. The two papers to see are:

> *Synthesizing third normal form relations from functional dependencies*, by P. Bernstein, ACM TODS, Vol.1, No.4, Dec. 1976, pp.277-298.

and

> *Computational problems related to the design of normal form relation schemes*, by C. Beeri and P. Bernstein, ACM TODS, Vol.4, No.1, March 1979, pp.30-59.

Exercises 17 and 18 are from the above paper by Beeri and Bernstein. Our Discussion of MVDs and 4NF is based on the paper:

Multivalued dependencies and a new normal form for relational databases, by R. Fagin, ACM TODS, Vol.2, No.3, Sept.1977, pp.262-278.

EXERCISES

1. Obtain the relation matrix and graphical representation for the following binary relations. In each case, $D = D_1 \cup D_2$.
   (a) $R = \{(0,1), (1,2), (0,0), (2,1), (1,3)\}; D = \{0,1,2,3\}$.
   (b) $R = \{(i,j) | i < j, i, j \in D\}; D = \{0,1,2,3,4\}$.
   (c) $R = \{(i, \lfloor i/2 \rfloor | 0 \leqslant i \leqslant 4\}; D = \{0,1,2,3,4\}$.
   (d) $R = \{(0,0),(2,2),(1,2),(2,1)\}; D = \{0,1,2\}$.

2. Obtain the set representation for the binary relations defined by the following relation matrices:

(i)

	1	2	3
1	1	1	0
2	0	1	1
3	1	0	0

(ii)

	0	1	2	3
0	1	1	1	0
1	1	0	0	1
2	1	0	1	0
3	1	0	1	0

3. Obtain the set representation for the binary relations defined graphically as:

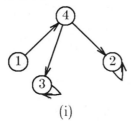

(i)

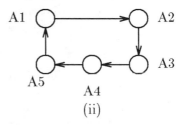

(ii)

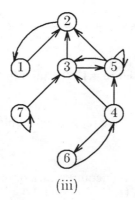

(iii)

Figure 4.18 Relations for Exercise 3.

4. For each of the following fully specified binary relations determine if it is reflexive, irreflexive, symmetric, asymmetric, antisymmetric, transitive.
   (a) $\{(0,0),(1,1),(2,2),(3,3),(4,4),(1,3),(2,3)\}$ $D = \{0,1,2,3,4\}$
   (b) $\{(0,0),(1,1),(2,4),(3,4)\}$, $D = \{0,1,2,3,4\}$
   (c) $\{(2,3),(3,4),(2,4),(1,4)\}$, $D = \{1,2,3,4\}$
   (d) $\{(1,1),(2,4),(4,2),(4,3),(2,3)\}$, $D = \{1,2,3,4\}$

5. Determine the properties of the binary relations $\subseteq$ and $\subset$ over the domain of sets.

6. Show that:
   (a) $R^+ = R.R^* = R^*.R$ for every binary relation R.
   (b) $R = R^+$ for every transitive binary relation R.
   (c) $R = R^*$ for every reflexive transitive binary relation R.

7. Show that if R is a binary relation on the finite set S, then $R^+ = \bigcup_{i=1}^{n} R^i$ where $n = |S|$.

8. Obtain the graphical representation of the relations R^+ and R^* when R has the graphical representation given in Figure 4.19.

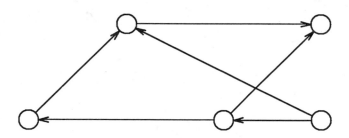

Figure 4.19 Relation for Exercise 8.

9. Show that the transitive reduction of a binary relation isn't necessarily unique.

10. Obtain a transitive reduction of each of the following binary relations:
    (a) $\{(1,2), (1,3), (1,5), (3,1), (1,4)\}^*$
    (b) $\{(1,1), (2,2), (3,3), (1,2), (1,3), (2,3)\}^*$

11. Obtain the equivalence classes resulting from each of the following statements. Assume that all variables that appear in the program also appear in the EQUIVALENCE statement. Determine the number of distinct memory locations needed to handle all the variables.
    (a) EQUIVALENCE (A,F),(B,D,E),(C,G),(C,A)
    (b) EQUIVALENCE (A,R),(B,C),(D,E,F),(E,C),(H,A,T)

12. Let $P = (X_1,X_2,...,X_n)$ be a permutation of the numbers 1,2,3,....,n. $j_1,j_2,...,j_k$, $k > 1$ is a cycle of P iff
    (a) $j_i \neq j_m$ for every i and m, $i \neq m$ unless $i=1$ and $m=k$
    (b) $j_1 = j_k$
    (c) $j_{i+1} = X_{j_i}$, $1 \leqslant i < k$
    As an example, consider the permutation $P = (2,3,6,5,4,1,7)$. 1, 2, 3, 6, 1 is a cycle as it satisfies conditions (a), (b), and (c). The permutation (2,3,6,5,4,1,7) has the three cycles: 1,2,3,6,1; 4,5,4; and 7,7.

    Let R be a binary relation. For any permutation P of $\{1,2,...,n\}$, R is defined as:

    $$R = \{(a,b)|a \text{ and } b \text{ are in the same cycle of } P\}$$

    (i) Show that R is an equivalence relation.
    (ii) Show that a and b are in the same equivalence class iff they are in the same cycle of P.

13. Let R1 and R2 be two equivalence relations.
    (a) Show that the intersection of R1 and R2 is also an equivalence relation.
    (b) Show that $R1 \cup R2$ is not necessarily an equivalence relation.
    (c) What properties does the relation $R1 \cup R2$ satisfy?

14. Under what conditions does the FD $\emptyset \rightarrow X$ hold in a relation R?

15. Prove Theorem 4.3.

16. Let $X \Longrightarrow Y$ and $Y \Longrightarrow Z$ be two MVDs of a relation R.
    (a) Show that if X, Y, and Z are pairwise disjoint then $X \Longrightarrow Z$ must hold in R.
    (b) Show that $X \Longrightarrow Z$ need not hold when X, Y, and Z are not disjoint even though $X \cap Y$ and $Y \cap Z$ are both empty sets (in fact, $X \Longrightarrow Z$ may not even be meaningful).

17. What does the MVD $\emptyset \Longrightarrow Y$ mean? Under what conditions does the MVD $\emptyset \Longrightarrow Y$ hold in a relation R?

18. Show that if $X \cap Y = \emptyset$, $X \cap Z = \emptyset$, $X \Longrightarrow Y$, and $X \Longrightarrow Z$, then each of following holds:
    (a) $X \Longrightarrow Y \cap Z$
    (b) $X \Longrightarrow Y - Z$
    (c) $X \Longrightarrow Z - Y$

19. [Beeri and Bernstein] Consider the FDs:

$$
\begin{aligned}
\{ISBN\} &\rightarrow \{TITLE, AUTHOR\} \\
\{TITLE, AUTHOR\} &\rightarrow \{ISBN\} \\
\{ISBN, TITLE\} &\rightarrow EDITION\text{-}NUM
\end{aligned}
$$

and the relation schemes:

$$
\begin{aligned}
EDITION &= (ISBN, TITLE, EDITION\text{-}NUM) \\
BOOK &= (ISBN, TITLE, AUTHOR)
\end{aligned}
$$

    (a) What are the superkeys, keys, prime and nonprime attributes of EDITION and BOOK?
    (b) Is EDITION in 2NF? Is BOOK in 2NF?

20. [Beeri and Bernstein] Consider the FDs:

$$
\begin{aligned}
\{APT\text{-}TYPE, ADDRESS\} &\rightarrow \{SIZE\} \\
\{ADDRESS\} &\rightarrow \{LANDLORD\} \\
\{ADDRESS, APT\text{-}NUM\} &\rightarrow \{RENT\} \\
\{NAME\} &\rightarrow \{ADDRESS, APT\text{-}NUM\}
\end{aligned}
$$

    (a) What are the superkeys, keys, prime and nonprime attributes of the relation schemes:

$$
\begin{aligned}
DWELLER &= (NAME, ADDRESS, APT\text{-}NUM, RENT) \\
APARTMENT\text{-}TYPE &= (APT\text{-}TYPE, ADDRESS, LANDLORD, SI
\end{aligned}
$$

    (b) Show that DWELLER is in 2NF while APARTMENT-TYPE is not.
    (c) Show that DWELLER is not in 3NF.
    (d) Show that the relation schemes:

$$
\begin{aligned}
TENANT &= (NAME, ADDRESS, APT\text{-}NUM) \\
APARTMENT &= (ADDRESS, APT\text{-}NUM, RENT) \\
APT\text{-}KIND &= (APT\text{-}TYPE, ADDRESS, SIZE) \\
BUILDING &= (ADDRESS, LANDLORD)
\end{aligned}
$$

    are in 3NF.

CHAPTER 5

FUNCTIONS, RECURSION, COMPUTABILITY

5.1 FUNCTIONS

5.1.1 Terminology

Let X and Y be two sets. A *partial function* from X to Y is a mapping of elements of X into elements of Y such that every element of X is mapped onto at most one element of Y. A *total function* from X to Y is a mapping from X to Y in which every element of X is mapped onto exactly one element of Y. X is the *domain* of f and Y is its *range*. We shall use the notation

$$f: X \longrightarrow Y$$

to mean that f is a function (either partial or total) from X to Y. If X is the cartesian product $X_1 \times X_2 \times ... \times X_n$, then we shall often write

$$f: X_1 \times X_2 \times ... \times X_n \longrightarrow Y$$

instead of

$$f: X \longrightarrow Y.$$

Similarly, if $X_i = T$, $1 \leqslant i \leqslant n$, we shall write

$$f: T^n \longrightarrow Y.$$

In either case, f is an *n variable function*. For $x_i \in X_i$, $1 \leqslant i \leqslant n$ and $y \in Y$, we shall write

$$f(x_1, x_2, \cdots, x_n) = y$$

if the function f maps the element $(x_1, x_2, \cdots, x_n) \in X$ onto the element $y \in Y$. y is the *value* of the function f at the point $(x_1, x_2, ..., x_n)$. y is also called the *image* of $(x_1, x_2, ..., x_n)$ under f. If f is a partial function, then its value may not be defined at all points in X. For points $(x_1, x_2, ..., x_n) \in X$ for which there is no $y \in Y$ such that $f(x_1, x_2, ..., x_n) = y$, we shall write

$$f(x_1, x_2, \cdots, x_n) = \uparrow.$$

There are many ways in which a function may be represented. In the tabular representation, an n variable function $f(x_1,x_2,...,x_n)$ is represented by an $n + 1$ column table. The first n of these correspond to $x_1, x_2, ..., x_n$ respectively. The last column corresponds to the value of f. The table contains one row for each tuple $(x_1,x_2, ... , x_n)$ for which $f(x_1,x_2, ... , x_n) = y$ for some $y \in Y$. Figure 5.1(a) gives the tabular representation for a one variable function f_1. If the domain and range of f_1 have been specified to be $X = \{0,1,2,3\}$ and $Y = \{1,2\}$, then f_1 is a total function from X to Y. However, if $X = \{0,1,2,3,4\}$ and $Y = \{1,2\}$, then $f_1(4)$ is undefined. If $Y = \{1\}$ and $X = \{0,1,2,3\}$, then rows three and four of the table of Figure 5.1(a) will have to be deleted. With X and Y defined in this way, f_1 is only a partial function from X to Y as there is no $y \in Y$ for which $f_1(2) = y$ (nor is there a $y \in Y$ for which $f_1(3) = y$).

x	$f_1(x)$	x	$f_2(x)$	x_1	x_2	$f_3(x_1,x_2)$
0	1	a	1	1	1	2
1	1	b	2	1	2	5
2	2	c	3	2	1	5
3	2	d	4	2	2	8
(a)		(b)			(c)	

x	$f_4(x)=x^2$	x	$f_5(x)=\pm\sqrt{x}$	x	$f_6(x)=\sqrt{x}$
0	0	0	0	0	0
1	1	1	1	1	1
2	4	1	-1	4	2
3	9	4	2	9	3
4	16	4	-2	9	3
5	25	9	3	25	5
.	.	.	.	.	.
.	.	.	.	.	.
.	.	.	.	.	.
(d)		(e)		(f)	

Figure 5.1 Some tabular representations.

Figure 5.1(b) gives the tabular representation of another one variable function f_2. This is a total function from X to Y if $X = \{a,b,c,d\}$ and $Y =$

$\{1,2,3,4\}$. If $\{a,b,c,d\} \subset X$ then f_2 is a partial function from X to Y. Figure 5.1(c) gives a two variable function from $X_1 \times X_1$ (or X_1^2) to Y where $X_1 = \{1,2\}$ and $\{2,5,8\} \subseteq Y$. The function $f_4: N \rightarrow N$ (N is the set of natural numbers) is a total function from N to N. It has the property that $f_4(x) = x^2$ for all x, $x \in N$. The table of Figure 5.1(e) does not represent a function from N to I (I is the set of integers) as there are two y's in I for which $f(1) = y$. The term *multivalued function* is used to denote a mapping from X to Y in which an element of X may be mapped onto more than one element in Y. The multivalued function of Figure 5.1(e) is simply the square root function from N to I. f_6 is the square root function from N to N.

One of the shortcomings of the tabular method is that from the table alone, it is not deducible whether or not the function is total. Also, its domain and range are not apparent. We could insist that the table for every function contain a row for every element x in X regardless of whether or not there is a $y \in Y$ for which $f(x) = y$. For every $x \in X$ for which there is no $y \in Y$ for which $f(x) = y$, we can place the symbol $\uparrow$ in the column corresponding to f. Similarly, we could require that for every $y \in Y$ there be at least one row for which the column entry is y. If for some y there is no $x \in X$ for which $f(x) = y$ then the symbol $\uparrow$ is placed in the columns corresponding to the x_is for that y. Figure 5.2 gives a complete tabular representation corresponding to the function f_1 of Figure 5.1(a) when $X = \{0,1,2,3,4\}$ and $Y = \{2,3\}$. Figure 5.1(b) is the complete tabular representation for f_2 when $X = \{a,b,c,d\}$ and $Y = \{1,2,3,4\}$. From a complete tabular representation, the domain and range are easily obtained. It is also easy to determine whether or not the function is total.

x	$f_1(x)$
0	$\uparrow$
1	$\uparrow$
2	2
3	2
4	$\uparrow$
$\uparrow$	3

Figure 5.2 Complete tabular representation.
$X = \{0,1,2,3,4\}$, and $Y = \{2,3\}$.

An alternate representation form for functions is the graphical representation. As in the case of binary relations, the graphical representation of a function from X to Y consists of vertices and directed edges. The vertices are drawn in two columns (see Figure 5.3(a)). In the left column we have one vertex for every $x \in X$. In the right column there is one vertex for every $y \in Y$. There is a directed edge from x to y iff $f(x) = y$. Figures 5.3(a) through (d) give graphical representations for the functions of Figures 5.1 (a)-(d). In each case, it should be apparent what the domain and range are. Figure 5.3(e) gives the graphical representation corresponding to Figure 5.2.

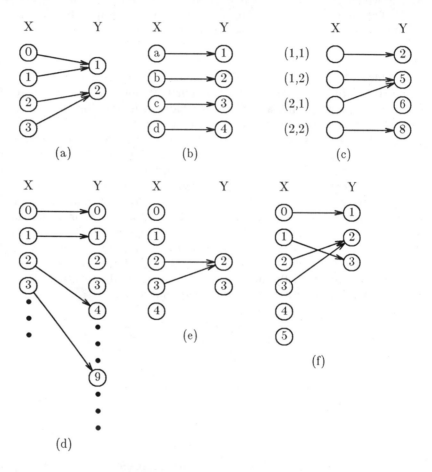

Figure 5.3 Graphical representations.

It should be evident that in the graphical representation of a function there is at most one edge leaving any vertex in the left column of vertices. For a total function there is exactly one edge leaving every vertex in the left column. Both the tabular and graphical representations of a function are explicit ways to specify a function. In a *functional representation* of a function, f, the value $f(x)$ is specified algebraically as in:

(a) $f: N \to N;\ f(x) = x^2$
(b) $f: \{0,1,4,9,...\} \to N;\ f(x) = \sqrt{x}$
(c) $f: I \times I \to N;\ f(x_1,x_2) = x_1^2 + x_2^2$

(d) $f: N \to N;\ f(x) = \begin{cases} x^2 & \text{if x is odd} \\ x^3 & \text{if x is even} \end{cases}$

By now, you should have observed a striking resemblance between the notion of a function and that of a binary relation. Let R be a binary relation defined on the domains D_1 and D_2. Every element of R is a pair (x,y), $x \in D_1$ and $y \in D_2$. If there is no $x \in X$ for which R contains two pairs (x,y_1) and (x,y_2), $y_1 \neq y_2$ then we may regard R as a function from D_1 to D_2. $R(x)$ is undefined for every $x \in D_1$ for which there is no pair (x,y) in R. $R(x)=y$ iff $(x,y) \in R$. Thus, a function is a special case of a binary relation. All functions are binary relations but some binary relations are not functions. In Chapter 7 we shall see another subclass of binary relations, i.e., graphs.

5.1.2 Properties

Let f be a total function from X to Y. f is *onto* (or is a surjective function, or a surjection) iff for every y, $y \in Y$ there exists at least one x, $x \in X$ for which $f(x) = y$. The functions of Figures 5.3(a), and (b) are onto functions while those of Figures 5.3(c), (d), (e), and (f) are not (in fact, the functions of Figures 5.3 (e) and (f) are not even total). f is a *one-to-one* function (or 1-1 function, or injective function, or an injection) iff there do not exist x_1 and x_2 in X, $x_1 \neq x_2$ for which $f(x_1) = f(x_2)$. I.e., for one-to-one functions distinct elements of x are mapped onto distinct elements of Y. The functions of Figures 5.3(b) and (d) are one-to-one. The remaining functions in Figure 5.3 are not one-to-one. A function that is both one-to-one and onto is a *one-to-one onto* (or 1-1 onto, or bijective, or a bijection) function. The function of Figure 5.3(b) is a one-to-one onto function. The remaining functions in this figure are not one-to-one onto.

At this point, it is worth pointing out that two sets A and B can be put into one-to-one correspondence (Chapter 3) iff there exists a one-to-one onto function from A to B (note that only total functions can be one-to-one).

The *composition* of two functions $f: Y \to Z$ and $g: X \to Y$ is the function $f \cdot g: X \to Z;\ f \cdot g(x) = f(y)$ where $y = g(x)$. Clearly, $f \cdot g$ is defined only when g is a total function. The definition of $f \cdot g$ can be extended to non total functions g by stating that $f \cdot g(x)$ is undefined whenever $g(x)$ is undefined in Y. If $f:N \rightarrow N$; $f(x) = x^2$ and $g:N \rightarrow N$; $g(x) = x^5$, then $f \cdot g:N \rightarrow N$; $f \cdot g(x) = x^{10}$. If $g:N \rightarrow N$; $g(x) = 2^x$, then $f \cdot g:N \rightarrow N$; $f \cdot g(x) = 2^{2x}$ and $g \cdot f(x) = g(f(x)) = 2^{x^2}$. So, the operator $\cdot$ is not commutative. In fact, it is quite possible for $g \cdot f$ to be undefined even though $f \cdot g$ is defined. For example, if $f:Y \rightarrow Z$, $g:X \rightarrow Y$, and Z is not a subset of X then $g \cdot f$ is not defined for all $x \in Y$.

The *identity* function $I_X:X \rightarrow X$ is a function such that $I_X(x) = x$ for all $x \in X$. A function $f:X \rightarrow Y$ is said to be *invertible* iff there exists a function $g:Y \rightarrow X$ such that $f \cdot g = I_Y$ and $g \cdot f = I_X$. A function g with the property that $f \cdot g = I_Y$ and $g \cdot f = I_X$ is called the *inverse* of f. The inverse of a function f is denoted f^{-1}. From the definitions, it follows that $f^{-1}(f(x)) = x$ for all $x \in X$ and $f(f^{-1}(y)) = y$ for all $y \in Y$.

It is not too difficult to see that if f is an invertible function, then f must be total. Observe that if $f:X \rightarrow Y$ is not total then there exists an $x \in X$ for which $f(x)$ is undefined. Hence, for every function $g:Y \rightarrow X$, $g(f(x))$ is also undefined. So, $g(f(x)) \neq x$ and f can have no inverse. Furthermore, there exist total functions that are not invertible. For example, the function of Figure 5.3(a) is not invertible. Since $f(2) = f(3) = 2$ there can be no function g for which $g(f(x)) = x$ for all x. So, only one-to-one functions can be invertible. Finally, notice that a function that is not onto is not invertible either. If $f:X \rightarrow Y$ is not onto, then there exists a $y \in Y$ for which $f(x) \neq y$ for all $x \in X$. Hence, there can be no function $g:Y \rightarrow X$; $f(g(y)) = y$ for all y. We therefore conclude that only one-to-one onto functions can be invertible. To see that every such function is invertible, simply define $g:Y \rightarrow X$ to be the function for which $g(y)$ is the x for which $f(x) = y$. Since f is a one-to-one onto function, g is well defined and is also a one-to-one onto function. One may readily verify that g satisfies all the requirements of an inverse for f. Our discussion on the inverse of a function f is summarized by the following theorem.

Theorem 5.1: Let $f:X \rightarrow Y$ and $g:Y \rightarrow X$ be two functions. The following are true:
  (a) f is invertible iff it is a one-to-one onto function.
  (b) g is the inverse of the invertible function f iff $f \cdot g = I_Y$ and $g \cdot f = I_X$.
  (c)   The inverse of an invertible function f is unique.
  (d) f is the inverse of g iff g is the inverse of f. $\square$

Part (b) of Theorem 5.1 deserves further discussion. Are both $f \cdot g = I_Y$ and $g \cdot f = I_X$ necessary requirements? What happens if we replace (b) by (b′)   g is the inverse of the invertible function f iff $f \cdot g = I_Y$

or

(b″)   g is the inverse of the invertible function f iff $g \cdot f = I_X$.

Since (b) is, in fact, the definition of an inverse, one may expect that a change to (b′) or (b″) will alter the properties of an inverse. This does in fact happen. Suppose (b′) is used as the definition of the inverse. The functions f and g given in Figures 5.4(a) and (b) are such that $f \cdot g = I_Y$. However, the function f is not invertible in any intuitive sense as once $f(x)=2$ has been computed, we cannot determine whether x was 1 or 3. Next, suppose (b″) is used as the definition of the inverse. Now, knowing $f(x)$, g can be used to determine x as $g(f(x)) = x$. So, this is more reasonable than (b′). (b″) does however have other difficulties associated with it. For example, intuitively one would expect the inverse of the inverse of f to be f. Consider the functions f_1 and g_1 of Figures 5.4(c) and (d). $g_1 \cdot f_1 = I_X$. So, under (b″) g_1 is the inverse of f_1. However, g_1 has no inverse under (b″). So, it is possible for the inverse of the inverse of f not to exist. To avoid these anomalies, we use (b) as the definition of inverse.

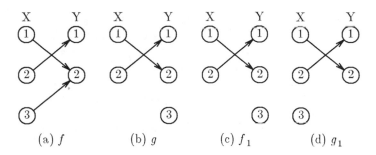

$$(a)\ f \qquad\qquad (b)\ g \qquad\qquad (c)\ f_1 \qquad\qquad (d)\ g_1$$

Figure 5.4

5.1.3. The Pigeon Hole Principle

Let us, for a very brief moment, suppose that we are distributing mail in an office. For each person who is authorised to receive mail in this office, there is a pigeon hole with his or her name on it (see Figure 5.5). In addition, there is a pigeon hole in which mail for unrecognized persons is accumulated (this may, in practice, be the waste paper basket). If we have 15 letters and 10 pigeon holes then we must put two or more letters into at least one pigeon hole. This is so because if we put at most one letter into each hole then there can be no more than 10 letters in all the holes together. If we had 21 letters then there must be at least 3 letters in some hole because if no hole has more than 2 letters in it there can be no more than 20 letters in the 10 pigeon holes.

Bruell	Franta	Hevner	Ibarra	Maly	Munro	Rosen	Sahni	Stein	WASTE

Figure 5.5 Pigeon holes for mail.

Theorem 5.2: [Pigeon Hole Principle I] Let $f:X \rightarrow Y$ be a total function. If $|X| \models n$ and $|Y| \models k$, then there is at least one $y \in Y$ which is the image of at least $\lceil n/k \rceil$ distinct x's in X.

Proof: The proof is by contradiction. Since f is total, for every $x \in X$, $f(x)=y$ for some $y \in Y$. If every y is the image of fewer than $\lceil n/k \rceil$ elements of X then X can have no more than $(\lceil n/k \rceil - 1)k < n$ elements. But, $|X| \models n$. $\square$

In the letter distribution example, f is a function mapping letters into mail boxes (or pigeon holes). X is the set of letters and Y the set of mail boxes.

While the pigeon hole principle may appear to be rather trivial, it is quite useful. For example, it allows us to make the following statements:

(a)   If two objects together weigh 100 lbs then one of these must weigh at least 50 lbs.

(b)   If three objects together weigh 150 lbs then one of these must weigh at least 50 lbs.

(c)   Let x_1, x_2, ..., x_{10} be a sequence of 10 numbers such that $\sum_{1}^{10} x_i =$ 100. At least one of the pairs (x_{2i+1}, x_{2i+2}), $0 \leqslant i \leqslant 4$ must sum to at least 20 (note that there are only five such pairs that together sum to 100).

Theorem 5.3 gives another form of the pigeon hole principle.

Theorem 5.3: [Pigeon hole principle II] Let $f:X \rightarrow Y$ be a total function. If $|X| \models n$ and $|Y| \models k$, then there is at least one $y \in Y$ that is the image of at most $\lceil n/k \rceil$ distinct x's in X.

Proof: See the exercises. $\square$

Example 5.1: Let $X = (x_1, x_2, \ldots, x_n)$ be a sequence of distinct numbers.

n is the length of X. $x_{i_1}, x_{i_2}, ..., x_{i_k}$ is a subsequence of X iff $1 \leqslant i_1 < i_2 < ...$ $< i_k \leqslant n$. x_1,x_2,x_3; x_1,x_5,x_7; x_3,x_4,x_8,x_9,x_{11}; etc. are all subsequences of X provided n is large enough. A sequence $x_1,x_2,...,x_n$ is increasing iff $x_1 < x_2 < ... < x_n$. It is decreasing iff $x_1 > x_2 > ... > x_n$.

We wish to show that every sequence of n distinct numbers contains at least one increasing or decreasing subsequence of length at least $\lceil \sqrt{n} \rceil$. For each i, $1 \leqslant i \leqslant n$ define a_i to be length of the longest increasing subsequence that starts at x_i. Define b_i to be the length of the longest decreasing subsequence starting at x_i. It is not too difficult to see that there can be no i and j such that $(a_i, b_i) = (a_j, b_j)$. To see this, observe that for every i and j, $i \neq j$ either $x_i > x_j$ or $x_i < x_j$. If $x_i > x_j$ and $i > j$, then $a_i < a_j$; if $x_i > x_j$ and $i < j$, then $b_i > b_j$; if $x_i < x_j$ and $i > j$, then $b_j > b_i$; and if $x_i < x_j$ and $i < j$, then $a_j < a_i$.

If each a_i and b_i is less than $\lceil \sqrt{n} \rceil$, then there are fewer than n distinct pairs (a_i,b_i). From Theorem 5.2 it follows that at least two of these n pairs must be identical (here $f:X \rightarrow Y$ is a function from the pairs (a_i,b_i) to the pairs (c_i,d_i) where c_i and d_i are less than $\lceil \sqrt{n} \rceil$. So, the number of letters is n and the number of pigeon holes k, such that $k < n$.). This contradicts our earlier claim that all the pairs (a_i,b_i) were distinct. Hence, at least one a_i or one b_i must be no less than $\lceil \sqrt{n} \rceil$. □

5.2 RECURSION

A *recursive definition* of a function is one in which the function is defined in terms of itself. The following are some examples of recursively defined functions:

(a) $s(n)=\begin{cases} 0 & \text{if } n=0 \\ n+s(n-1) & \text{if } n>0 \end{cases}$; $s:N \rightarrow N$

(b) $p(n)=\begin{cases} 1 & \text{if } n=0 \\ n*p(n-1) & \text{if } n>0 \end{cases}$; $p:N \rightarrow N$

(c) $d(n)=\begin{cases} 100 & \text{if } n=10 \\ d(n+1)-10 & \text{if } n<10 \end{cases}$; $d:\{0, 1,..., 10\} \rightarrow N$

(d) $F(n) = \begin{cases} 0 & \text{if } n=0 \\ 1 & \text{if } n=1 \\ F(n-1)+F(n-2) & \text{if } n>1 \end{cases}$ $;F:N \to N$

(e) $G(b) = \begin{cases} 0 & \text{if } b=1 \\ G(b-1)\|b-1\|G(b-1) & \text{if } b>1 \end{cases}$ $;G:N-\{0\} \to N^+$

(f) $G2(b) = \begin{cases} 0 & \text{if } b=1 \\ 101 & \text{if } b=2 \\ G2(b-1)\|b-1\|G2(b-1) & \text{if } b>2 \end{cases}$ $;G2:N-\{0\} \to N^+$

(g) $A(i,j) = \begin{cases} 2j & \text{if } i=0 \\ 0 & \text{if } i \geq 1 \text{ and } j=0 \\ 2 & \text{if } i \geq 1 \text{ and } j=1 \\ A(i-1,A(i,j-1)) & \text{if } i \geq 1 \text{ and } j \geq 2 \end{cases}$ $;A:N \times N \to N$

(h) $f(i) = \begin{cases} i/2 & \text{if } i \text{ is even} \\ f(f(3i+1)) & \text{if } i \text{ is odd} \end{cases}$ $;f:N \to N$

(i) $g(i) = \begin{cases} 0 & \text{if } i \leq 1 \\ g(i/2) & \text{if } i \text{ is even} \\ g(3i+1) & \text{if } i \text{ is odd} \end{cases}$ $;g:N \to N$

In definitions (e) and (f) the operator "$\|$" denotes string concatenation ($a\|b=ab$). Notice that in each case the function name appears in both the left and right hand side of the equality. When a function is defined recursively, it is not immediately clear whether or not the function is a total function (i.e., is it defined for all values in its domain?). In the case of the example functions (a) - (d) it is not too difficult to see that each is a total function. In fact, for (a) we obtain:

$s(n) = n + s(n-1)$
$\quad\quad = n + (n-1) + s(n-2)$
.
.
.
$\quad\quad = n + (n-1) + \dots + 0$

$$= \sum_{i=0}^{n} i$$
$$= n(n+1)/2, \; n \geq 0.$$

For (b), we obtain:

$$p(n) = n * p(n-1)$$
$$= n * (n-1) * p(n-2)$$

.
.
.

$$= n * (n-1) * (n-2) * \ldots * 1 * 1$$
$$= n!, \; n \geq 0.$$

For (c), we see that

$$d(n) = d(n+1) - 10$$
$$= d(n+2) - 20$$

.
.
.

$$= d(10) - 10 * (10-n)$$
$$= 100 - 100 + 10n$$
$$= 10n, \; 0 \leq n \leq 10.$$

Function (d) defines the Fibonacci numbers. This function was studied in Chapter 2. In that chapter we showed that

$$F(n) = \frac{1}{\sqrt{5}} \left[\left(\frac{1+\sqrt{5}}{2} \right)^n - \left(\frac{1-\sqrt{5}}{2} \right)^n \right], n \geq 0.$$

So, functions (a) − (d) are easily seen to be total. For functions (e) and (f), it is not quite as easy to obtain a formula. However, we can show that both functions are total by using induction on b. Let us look at function (e), i.e., G. For the induction base, we see that when $b=1$, $G(b)=0 \in N$. Now, assume that $G(b) \in N^+$ for $b=m$, $m \in N - \{0\}$. We shall show that $G(m+1) \in N^+$. $G(m+1) = G(m) mG(m)$ which is clearly in N^+ if $G(m)$ is.

The function G is interesting in that it defines a Gray code. A b-bit *Gray code* is a sequence of 2^b different b-bit binary strings such that the ith bit string differs from the $(i+1)$-st in exactly one bit, $1 \leq i < 2^b$. In addition, the first and last bit strings differ in exactly one bit. Figures 5.6(a), (b), and (c) give 1, 2, and 3 bit Gray codes.

$$
\begin{array}{ccc}
0 & 00 & 000 \\
1 & 01 & 001 \\
  & 11 & 011 \\
  & 10 & 010 \\
  &   & 110 \\
  &   & 111 \\
  &   & 101 \\
  &   & 100 \\
\end{array}
$$

(a)   (b)   (c)

Figure 5.6 1,2,and 3 bit Gray codes.

One interesting application of a Gray code is in computing the weight of all possible combinations of n objects. By using an n-bit Gray code, the weight of successive combinations can be found using only one addition or subtraction. Suppose we have 3 objects with weight w_1, w_2, and w_3 respectively. The 3-bit Gray code of Figure 5.6(c) is to be interpreted to mean bit i is 0 iff object i is not in the combination (bits are numbered right to left, 1 to 3). In going from 000 (the empty combination) to 001 object 1 is added to the combination. Its weight becomes w_1. In order to go from the combination 001 to the combination 011, object 2 needs to be added. The new combination weight is $w_1 + w_2$. To go from 011 to 010, object one needs to be removed. The weight of the new combination is w_2.

An examination of Figures 5.6(a), (b), and (c) reveals that the sequences of bit changes in these Gray codes are 0, 010, and 0102010. These correspond exactly to $G(1)$, $G(2)$, and $G(3)$ respectively. One may, in fact, show that $G(b)$ defines a bit change sequence for a b-bit Gray code for all b, $b \geqslant 1$. The proof is a straightforward proof by induction on b. A similar proof can be provided to show that $G2(b)$ is also a bit change sequence for a b-bit Gray code.

Using the properties (obtained in Chapter 2), of Ackermann's function (g) it is possible to prove that $A(i,j)$ is defined for all $(i,j) \in N \times N$ (see the exercises). Function (h) can be shown total as follows. For all even numbers i, $i \in N$, $f(i) = i/2$ and $i/2 \in N$. When i is odd, the binary representation of i is either of the form $...b_4b_301$ or $...b_4b_311$. If it is of the form $...b_4b_301$, then $3i + 1 = 2i + i + 1 = (...b_4b_3010 + ...\ b_4b_301 + 1) = (...b_4b_3100)$. So, $f(3i + 1) = ...b_4b_310$ and $f(f(3i+1)) = ...$

$b_4 b_3 1 = \lfloor i/2 \rfloor + 1$. When i is of the form ... $b_4 b_3 11$, then $3i + 1 = 2i + i + 1 = (...b_4 b_3 110 + ...b_4 b_3 11 + 1) = ...c_4 c_3 010$ and $f(3i + 1) = ...c_4 c_3 01$. From our earlier discussion, it follows that $f(...c_4 c_3 01) = ...c_4 c_3 1$. So, $f(i)$ is defined for all i.

It is not known whether or not function (i) is total. A computer search has shown that $g(i)$ is defined for all i, $0 \leqslant i \leqslant 3 * 10^8$. However, it is not known if there exists an i, $i > 3*10^8$ for which $g(i)$ is undefined.

In each of the nine examples considered, the fact that the function definition is recursive is immediately apparent as the function name appears explicitly on the right hand side of the equality. This isn't always the case. Consider the definitions:

$$f(n) = \begin{cases} 0 & n=0 \\ 1+g(n-1) & n>0 \end{cases}$$

and

$$g(n) = \begin{cases} 1 & n=0 \\ 2+f(n-1) & n>0 \end{cases}$$

Now, f is defined in terms of g, and g in terms of f. So, once again, f is defined in terms of itself (through g). A function f is defined by *indirect recursion* if there exist functions $g_1, g_2, ..., g_k$, $k > 1$, such that f is defined in terms of g_1, g_i is defined in terms of g_{i+1}, $1 \leqslant i < k$ and g_k is defined in terms of f.

While it is possible to come up with recursive definitions from which it may be extremely difficult to determine if the function is total, recursive definitions of functions are usually more concise and easily comprehendible. The Ackermann's function, for instance, would be quite difficult to define non-recursively.

The idea of a recursive definition is readily carried over to the domain of algorithms. A *recursive algorithm* is an algorithm which is defined in terms of itself (either directly or indirectly).

Recursive algorithms for the recursively defined functions (a)-(i) are easy to obtain. Algorithms 5.1 - 5.7 correspond to seven of these nine functions. The similarity between each recursive function definition and the corresponding recursive algorithm is apparent. By specifying a range of permissible values of the input parameters of each algorithm we have assured that the input parameters are assigned only values from the domain of the respective function. So, it is not necessary to explicity check that each variable has a value from the function domain. The procedure corresponding to function g (i.e., (i) above) has not been labeled an algorithm as it is not known whether its value is defined for all i. It is quite possible that for some i, $g(i)$ is not defined.

```
procedure  s(n)
   integer  n,s  range(0:∞)
   if  n = 0 then  return(0)
              else return(n + s(n-1))
   endif
end  s
```

   Algorithm 5.1

```
procedure  p(n)
   integer  n,p  range(0:∞)
   if  n = 0 then  return(1)
              else return(n*p(n−1))
   endif
end  p
```

   Algorithm 5.2

```
procedure  d(n)
   integer  n range(0:10);  integer  d
   if  n = 10 then  return(100)
               else return(d(n+1)−10)
   endif
end  d
```

   Algorithm 5.3

```
procedure  F(n)
   integer  n,F  range (0:∞)
   if  n⩽1 then  return(n)
            else return(F(n−1)+F(n−2))
   endif
end  F
```

   Algorithm 5.4

Since algorithms are usually written for computers, we might wonder how a computer might execute (or evaluate) a recursively defined algorithm. Let's take Algorithm 5.1 as an example. If n happens to be 0, then, of course, our computer should have no difficulty. If $n > 0$, $n + s(n-1)$ cannot be immediately computed as $s(n-1)$ is not known. At this time, the computer will do, more or less, what a human might. It will proceed to evaluate $s(n-1)$. When it has determined the value of $s(n-1)$, then the sum $n+s(n-1)$ will be computed.

```
procedure  G(b)
   integer  b range(1:∞);  string  G
   if  b = 1 then return('0')
              else  return( G(b−1) ‖ b−1 ‖ G(b−1) )
   endif
end  G
```

Algorithm 5.5

```
procedure  A(i,j)
   integer  i,j range(0:∞)
   case
     :i=0:  return(2*j)
     :j=0:  return(0)
     :j=1:  return(2)
     :else:  return(A(i−1,  A(i,j−1)))
   end case
end  A
```

Algorithm 5.6

```
procedure  g(i)
   integer  i range(0:∞)
   case
     :i⩽1:  return  (0)
     :i  even:  return(g(i/2))
     :else:  return(g(3*i + 1))
   end case
end  g
```

Program 5.7

We shall now describe a mechanism by which a computer can handle recursively defined (whether direct or indirect) algorithms (and more generally, programs). A thorough understanding of this mechanism is needed in order to understand how recursively defined algorithms are executed by computers. First, we need to introduce some terms:

(1) A *stack* is a list of items. One end of this list is called the *top* and the other the *bottom*. Additions to a stack are made at the top. Deletions are also made only from the top. Figure 5.7(a) shows a stack containing the three items A, B, C. If a new item, say D, is added to this stack, the new stack configuration will be as in Figure 5.7(b). If an item is to be deleted from the stack of Figure 5.7(a), it will necessarily be item C. The stack configuration following the deletion of an item from the stack of Figure 5.7(a) is given in Figure 5.7(c). In the recur-

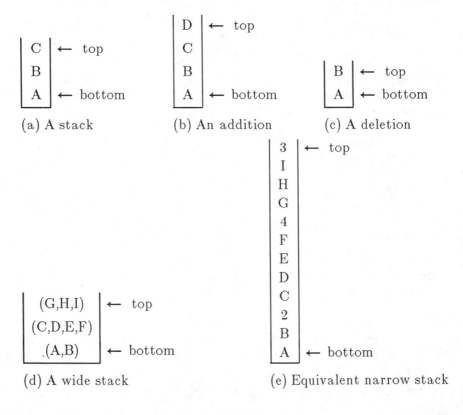

Figure 5.7 Some stacks.

sion mechanism to be described, we shall often be adding k-tuples to the *recursion stack*. For ease in describing the recursion mechanism, we shall allow stack items to be k-tuples as in Figure 5.7(d). In practice, the stack of Figure 5.7(d) will probably look as in Figure 5.7(e). The numbers in this stack tell us that the next (i.e., lower) 'so many elements on the stack form one tuple.

(2) A variable i is *local* to a procedure P if it is not a parameter and it is declared in that procedure. Algorithms 5.1 - 5.7 contain no local variables. In general, recursive algorithms will use local variables (as we shall soon see).

(3) In the procedure definition:

procedure $XYZ(A,B,C,D)$

A, B, C, and D are the *formal parameters* of the procedure. If procedure XYZ is called as in:

call $XYZ(B,C,X,R+Y)$

then the *actual parameters* are B, C, X, and $R+Y$. We shall consider three modes of establishing a correspondence between actual and formal parameters. These modes are: (a) by reference; (b) by value; and (c) by return. Modes (b) and (c) may be combined to get the mode "By value return". Mode (a) is assumed unless otherwise specified. Explicit mode specification can be done in the procedure definition as in:

(i) **procedure** $SUM(n$ **value**$)$
(ii) **procedure** $MULT(A,B,C,n$ **value**$)$
(iii) **procedure** $XY(A$ **value**, B **return**, C **reference**, D **value return**, $E)$

The parameters A, B, and C of (ii) and C and E of (iii) have mode "by reference"; parameters n of (i) and (ii) and A of (iii) have mode "by value". Parameter D of (iii) has mode "by value return".

The different modes are handled differently during a **call** and during a **return**. At the time of a **call** the following happens:

(a) All expressions in the call are evaluated. If the expression corresponds to a formal parameter that has mode "by reference", then the value of the expression is stored in some location in memory.

(b)   An association between formal and actual parameters is made. Formal parameters of mode "reference" get associated with the address of the corresponding actual parameter. Thus, every time a formal parameter of this mode is referred to in the called procedure, a reference to the address of the actual parameter is made. For parameters of mode "value", the formal parameter is assigned as value the current value of the actual parameter.

At the time of return, the values of formal parameters of mode "by return" are assigned to the corresponding actual parameters unless these parameters are expressions or constants.

Our recursion mechanism requires that the following transformations first be made:

(a)   Rewrite all expressions involving implicit calls to algorithms. Each such use of an algorithm is replaced by an explicit call to this algorithm followed by a statement to the type $Ti \leftarrow VALUE$. Ti is a new variable name and $VALUE$ is an algorithm that will extract the value of the function just completed from the recursion stack. Some examples of this transformation are:

(i) $X \leftarrow Y + F(i,j)$ becomes **call** $F(i,j)$
$$T1 \leftarrow VALUE$$
$$X \leftarrow Y + T1$$

(ii) $X \leftarrow Y*F(i) + G(j)*F(i-2)$

becomes	**call** $F(i)$	or	**call** $F(i)$
	$T1 \leftarrow VALUE$		$T1 \leftarrow VALUE$
	call $G(j)$		$T2 \leftarrow Y*VALUE$
	$T2 \leftarrow VALUE$		**call** $G(j)$
	call $F(i-2)$		$T3 \leftarrow VALUE$
	$T3 \leftarrow VALUE$		**call** $F(i-2)$
	$X \leftarrow Y*T1 + T2*T3$		$T4 \leftarrow VALUE$
			$X \leftarrow T2 + T3*T4$

(iii) **return**$(F(n))$ becomes   **call** $F(n)$
$$T1 \leftarrow VALUE$$
$$\textbf{return}(T1)$$

(b)   Assign new and distinct labels (say $L1, L2, ...$) to the first executable statement in each algorithm as well as to all statements that immediately follow a call statement. These labels will be used as return addresses. So, no two statements should be assigned the same label *Li.*

When the transformations (a) and (b) are applied to Algorithms 5.1 and 5.4, Algorithms 5.8 and 5.9 are obtained. In both cases we have changed the definition of the procedures so that their parameters are of mode "by value". The labels assigned in these two algorithms are different as we are assuming both to be sub-programs of some other algorithm. Some simplification is possible. For example, the lines

$$T1 \leftarrow VALUE$$
$$L2: \textbf{return}(n*T1)$$

could be replaced by

$$L2: \textbf{return}(n*VALUE).$$

We shall not be concerned with such simplifications at this time as they do not contribute to our understanding of the recursion mechanism.

```
procedure s(n value)
   integer n range (0:∞)
  L1: if n = 0 then return (0)
               else call s(n−1)
                  L2: T1 ← VALUE
                      return(n + T1)
         endif
end s
```

Algorithm 5.8

```
procedure F(n value)
   integer n,T1 range (0:∞)
  L3: if n⩽1 then return(n)
              else call F(n−1)
                 L4: T1 ← VALUE
                     call F(n−2)
                 L5: T1 ← T1 + VALUE
                     return (T1)
         endif
end F
```

Algorithm 5.9

We are now ready to describe how recursively defined algorithms may be executed. This description will also include algorithms that are not recursively defined (i.e., *iterative algorithms*). The evaluation process is defined by the following:

(I)   At the beginning of execution, initialize the recursion stack (this stack might just as well be called the algorithm stack as we shall be using it for both iterative and recursive algorithms) so that it contains no items.

(II)   Whenever a call statement is reached, during execution, do the following:

IIa.   Store the values (including associations, if any) of all the formal parameters and local variables of the called algorithm (provided it is recursive) in the recursion stack. For example, if we are executing the statement

call $FREC(i,j, 2*t+6)$

where *FREC is* a recursive algorithm defined as:

procedure $FREC$ (a **value**,b **value**,c **value**)

declare f,g

.

.

.

end $FREC$

then at the time the call is to be executed, the current values of the paramaters a, b, and c, and the local variables g, and h are added to the recursion stack.

IIb.   Regardless of whether the called algorithm is recursive or not, the label of the statement immediately following the call statement is added to the stack.

IIc.   Evaluate the actual parameters of the call in case they are expressions (i.e., $2*t+6$ in the above example). These values are stored in memory locations in case they correspond to formal parameters that are of mode "by reference". Formal parameters that are of mode "by value" are now assigned the values of their corresponding actual parameters. In the above example, a, b, and c are now assigned the values of i, j, and $2*t+6$. If *FREC* had instead been defined as:

procedure *FREC* (a,b,c)

then at this time an association is made between i and a, j and b, and the location of the result of $2*t+6$ and c. This is done such that all references to a, b, and c within the called algorithm will actually be references to i, j, and the location of the result of $2*t+6$.

IId. Now proceed to the first executable statement in the called procedure.

(III) Whenever a **return** statement is encountered, do the following:

IIIa. If the return has a value associated with it (as in **return** $(2*b - c*a)$), then compute this value.

IIIb. Delete the return label Li, from the stack. Assign the values of all "by return" formal parameters to their corresponding actual parameters (in case the actual parameters are not expessions). If this is not a return from a recursive procedure, then go to step IIIc. Otherwise, restore all associations of formal parameters of mode "by reference" and restore the values of formal parameters of mode "by value" to what they were immediately preceding the call. These values are available from the top of the stack. Restore the values of all local variables in this procedure.

IIIc. Place the value of the return (if any) onto the stack.

IIId. Proceed to the statement with label Li.

(IV) When the variable *VALUE* is encountered delete an item from the stack and assign it to *VALUE*.

These execution rules should become clear after some examples have been studied. Let us work through Algorithm 5.8. Suppose that this algorithm is called from procedure *MAIN* (Algorithm 5.10). Before starting the execution of *MAIN*, the recursion stack is initialized to be empty (Figure 5.8(a)). Next, the value of i is input. Let us assume it to be 0. The **call** statement is encountered. Since S is a recursive algorithm, the values of its parameters (only n) and local variables (there are none) are put onto the stack. At this time n hasn't been defined. We use '-' to denote an undefined item on the stack. $L4$ is next added to the stack and n assigned the value 0. The stack now looks as in Figure 5.8(b). We now proceed to statement $L1$ of procedure s. Since $n=0$, the **then** clause is entered. It is simply a **return** statement. So, we proceed to do what is required by III. The value associated with the return is 0. The return address L4 is removed from the stack. The value of n is restored to - and the value of the return, 0, is placed on the stack. Figure 5.8(c) gives the new stack configuration. We now proceed to execute statement $L4$ (which is in $MAIN$). In the **print** statement the '0' is removed from the stack when $VALUE$ is encountered. The stack becomes empty. When the **end** statement is reached (it is an implicit return), the stack is empty and there is no place to return to. This signals the end of execution.

```
procedure MAIN
   L3:read (i)
       if i < 0 then print("error", i); stop
                    else call s(i)
              L4: print(i, VALUE)
       endif
end MAIN
```

Algorithm 5.10

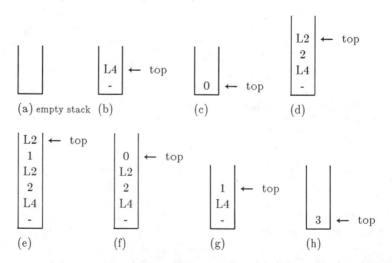

Figure 5.8 Stack configurations.

Let's go through this example once again. This time we shall assume that the input value of i (i.e., the value read in the statement $L3$ is 2. Following the call from *MAIN*, the stack is as in Figure 5.8(b) and n has the value 2. As a result, another call is made to *s* from within *s*. At this time the current value of *s*'s parameters and local variables are placed on the stack. The return address, $L2$, is also stacked and *n* assigned its new value, 1. The stack is as in Figure 5.8(d) and we proceed to statement $L1$. Now, $n=1$ and another call is made to *s*. The stack configuration changes to that of Figure 5.8(e) and *n* is assigned the value 0. We again proceed to statement $L1$. Since $n=0$, the **return**(0) statement is reached. $L2$ and 1 are removed from the stack. *n* is reassigned its previous value of 1; the value of the return (i.e.,0) is placed on the stack; and we proceed to executed statement $L2$. The stack now looks as in Figure 5.8(f). $T1$ is assigned the value at the top of the stack (i.e., 0). This item is removed from the stack and the configuration of Figure 5.8(d) is obtained.

The next statement encountered is **return**$(n+T1)$. $n+T1=1$ is computed. $L2$ and 2 are removed from the stack and the value of the return is placed on it (see Figure 5.8(g)). *n* is assigned the value 2 and we proceed to execute statement $L2$. This results in $T1$ getting assigned the value 1. $n+T1=3$ is computed. $L4$ and - are removed from the stack. 3 is put onto the stack. The stack now has the configuration of Figure 5.8(h). n is assigned the value - and execution is resumed at statement $L4$. The 3 is removed from the stack in statement $L4$ and an empty stack results. Execution terminates when the statement **end** *MAIN* is reached as the stack is empty and the implicit return cannot be carried out.

From the preceding discussion, it might appear that we can do away with the stacking of formal parameters and local variables for calls made from outside the recursive algorithm. For example, when procedure s is called from *MAIN*, the current values of the parameters and local variables in s are of no importance and need not be saved for later restoration. So, we can certainly do away with the initial stacking of parameters and local variables in this case. However, when indirect recursion exists, it is necessary to stack the values of parameters and local variables even when the call is made from outside the algorithm being called.

Consider the two procedures *A* and *B* (Algorithms 5.11 and 5.12). If parameter and local variable stacking is not done for calls made to *B* from outside *B* and for calls made to *A* from outside *A*, then there will be no saving of previous values of *X* and *Y* in *A* or of *X* in *B*. Clearly, the result will be different from when such a saving of values followed by their restoration on return is done. Since the intent of the expression $B(X) + X$ is to add the present value of *X* to $B(X)$ and not to add the value of *X* that might result from the call to $B(X)$, it is necessary in the case of indirect recursion to stack parameter and local variable values even for calls from outside the algorithm.

procedure A (X **value**, Y **value**)
   integer A,X,Y
   if $X \leqslant 2$ **then return endif**
   $X \leftarrow B(X) + X$
   return $(X + Y)$
end A

procedure B (X **value**)
   integer B,X,C
   $C \leftarrow A(X/2, X/4)$
   return $(C+X)$
end B

Algorithm 5.11   **Algorithm 5.12**

The recursion mechanism just described can also be used to mechanically transform recursive algorithms into iterative ones. This transformation is needed when a recursively defined algorithm is to be programmed in a programming language that doesn't support recursion (e.g., FORTRAN). The transformation from recursive to iterative algorithms simply requires the introduction of code corresponding to steps I, II, and III. We assume that the transformations given in (i) and (ii) have already been performed. We also assume that a stack is available to us. Associated with this stack are the three algorithms:

> ADD_STACK (X) ... adds X to the stack.
> $DELETE\_STACK$ ... deletes top most item in stack and
>   returns this item as value.
> $VALUE$ ... this is identical to $DELETE\_STACK$.

Since this stack is to be used by all algorithms, we shall assume that the initialization of the stack is done by the computer system itself at the start of execution. So, we need only be concerned with steps II and III. Every **call** statement, **call** REC (e_1, e_2, ...) that invokes a recursive algorithm from outside this algorithm (e.g. the call from $MAIN$) is replaced by the following code.

(a)   Code to stack the values (or associations) of all parameters and local variables in the called algorithm.

(b)   Code to stack the return address.

(c)   **call** $REC(e_1, e_2, \cdots)$

Calls to a recursive algorithm from within itself (e.g., the call to s appearing in procedure s) are replaced by the code:

(a)   Code to stack the values (or associations) of all parameters and local variables in this algorithm.

(b)   Code to stack the return address.
(c)   Code to assign the new parameter values.
(d)   An unconditional branch to the first executable statement in this
algorithm.

Return statements appearing within recursive algorithms are replaced
by the following code:

(a)   Code to unstack the return address and the parameter and local variable values.
(b)   Code to restore the parameter and local variable values (or associations).
(c)   Code to stack the value (if any) of this return.
(d)   Code to determine if the return address is within this algorithm. If
so an unconditional branch is made to this address. If not, a normal
return is made to the calling algorithm.

Algorithms 5.13 and 5.14 are the result of applying these transformations to Algorithms 5.8 and 5.9 respectively.

```
procedure s (n )
  integer n,NEW range(0:∞); address ADDR
  L1: if n=0 then ADDR ← DELETE_STACK
                  n ← DELETE_STACK
                  call ADD_STACK(0)
                  if ADDR ∈ {L1,L2} then go to ADDR
                                            else return
              endif
         else call ADD_STACK (n )
              call ADD_STACK ('L2')
              n ← n−1
              go to L1
         L2: T1 ← VALUE
              NEW ← n + T1
              ADDR ← DELETE_STACK
              n ← DELETE_STACK
              call ADD_STACK (NEW )
              if ADDR ∈ {L1,L2} then go to ADDR
                                        else return
              endif
      endif
  end s (n )
```

Algorithm 5.13

```
procedure F(n)
  integer n,T1,NEW range(0:∞); address ADDR
  L3: if n ⩽ 1 then NEW ← n
                   ADDR ← DELETE_STACK
                   (n,T1) ← DELETE_STACK
                   call ADD_STACK (NEW)
                   if ADDR ∈ {L3,L4,L5} then go to ADDR
                                          else return
                   endif
            else call ADD_STACK ((n,T1))
                 call ADD_STACK ('L4')
                 n ← n−1
                 go to L3
            L4: T1 ← VALUE
                 call ADD_STACK ((n,T1))
                 call ADD_STACK ('L5')
                 n ← n−2
                 go to L3
            L5: T1 ← T1 + VALUE
                 ADDR ← DELETE_STACK
                 (n,T1) ← DELETE_STACK
                 call ADD_STACK (T1)
                 if ADDR ∈ {L3,L4,L5} then go to ADDR
                                        else return
                 endif
       endif
end F
```

Algorithm 5.14

It is important to realize that Algorithms 5.13 and 5.14 do not involve any more work than the recursive versions given in Algorithms 5.1 and 5.4. In Algorithms 5.1 and 5.4 the additional book keeping work (i.e., the stacking) is hidden. Still, it has to be done during execution. Thus, while Algorithm 5.1 and 5.4 may appear very simple, their execution could entail a fair amount of work. In practice, of course, one would not write procedure $s(n)$ as in Algorithm 5.1. As remarked earlier, $s(n)$ just computes the sum of the first $n + 1$ natural numbers ($\sum_{i=0}^{n} i$) which is just $n(n+1)/2$. So, the algorithm for $s(n)$ would be as in Algorithm 5.15. Since, $p(n) = n!$, we would use Algorithm 5.16 rather than Algorithm 5.2. Algorithm 5.17 is a cheaper way to compute $F(n)$ than Algorithm 5.4. Algorithms 5.15 through 5.17 are cheaper than their recursive counterparts both in terms of the amount of computing time needed to compute the respective functions (as there is no stacking or procedure calls) and in terms of the space needed (as additional stack space is not used).

procedure $s(n)$
    integer n **range**$(0:\infty)$
    return$(n*(n+1)/2)$
end s

Algorithm 5.15

procedure $p(n)$
    integer n,i,t **range** $(0:\infty)$
    $t \leftarrow 1$
    for $i \leftarrow 1$ **to** n **do**
        $t \leftarrow t * i$
    end for
    return(t)
end p

Algorithm 5.16

procedure $F(n)$
    integer f,g,h,i,n **range**$(0:\infty)$
    if $n \leqslant 1$ **then return**(n) **endif**
    $f \leftarrow 0; \ g \leftarrow 1$
    for $i \leftarrow 2$ **to** n **do** //compute $h = F(i)$//
        $h \leftarrow f+g, \ f \leftarrow g, \ g \leftarrow h$
    end for
    return(h)
end F

Algorithm 5.17

The point we wish to make is that recursively defined functions are trivially translated into recursive algorithms that can be directly run on a computer. However, for many recursively defined functions, there exist iterative algorithms that are also easy to obtain. These iterative algorithms can be expected to run faster and to use less space than their recursive counterparts. Using the recursive algorithms in such cases is an abuse of recursion. This is not to say that recursive algorithms should never be used. Recursion is a very elegant and powerful tool that can often be used to describe rather complex processes in a very understandable way. The resulting algorithms are often easily proved correct. These considerations will often outweigh the possible loss in computer time and space. In the remainder of this section we shall look at several examples employing recursion. For each of these it will be apparent that the iterative versions of the corresponding algorithms are much harder to obtain and prove correct.

Example 5.2: [Insert Sort] In the sorting problem, we start with n numbers $X(1), X(2), ..., X(n)$. We are required to rearrange these numbers so that following the sort, the sequence of numbers is in nondecreasing order. I.e. $X(1) \leqslant X(2) \leqslant ... \leqslant X(n)$. A common way to arrive at a recursive algorithm for any problem is to proceed as one does in a proof by induction. When $n=1$, the sorting problem is trivial. If we know how to sort $n-1$ numbers, then n numbers can be sorted by first sorting the first $n-1$ numbers and then inserting $X(n)$ into its correct place. Algorithm 5.18 is the resulting algorithm.

The insertion of $X(n)$ into the sorted sequence $X(1), ..., X(n-1)$ can also be done recursively. If $n=1$, or $X(n-1) \leqslant X(n)$, then nothing has to be done. Otherwise, $X(n)$ is to be inserted into $X(1) ... X(n-2)$. The resulting insertion algorithm is procedure *INSERT* (Algorithm 5.19). This algorithm isn't any easier to understand than the **for** loop used in procedure *INSORT* and it is certainly less efficient. So, we shall not use this recursive version.

Actually, the recursive call to *INSORT* in Algorithm 5.18 can be replaced by a **for** loop without sacrificing too much the readability (or comprehensibility) of the algorithm. When this is done, we obtain Algorithm 5.20. Notice that the correctness of Algorithm 5.18 was an immediate consequence of our recursive formulation of the solution to the sorting problem. The correctness of Algorithm 5.20 follows from that of Algorithm 5.18. So, even though we have decided not to use the recursive algorithm, it was useful in arriving at a correct iterative algorithm. □

```
procedure INSORT(X,n)
  //sort X(1), ..., X(n)//
  integer i,n; declare T,X(n)
  if n ≤ 1 then return endif
  call INSORT(X,n−1) //sort first n−1 numbers//
  //insert X(n) into correct place//
  T ← X(n)
  for i ← n−1 downto 1 do
    if X(i) ≤ T then exit endif
    X(i+1) ← X(i)
  end for
  X(i+1) ← T //i=0 on normal exit from for loop//
end INSORT
```

Algorithm 5.18 Recursive insertion sort.

```
procedure INSERT(X,n)
   //insert X(n) into X(1) ... X(n-1)//
   integer n; declare T, X(n)
   if n ≤ 1 then return endif
   if X(n-1) ≤ X(n) then return endif
   T ← X(n)
   X(n) ← X(n-1)
   X(n-1) ← T
   call INSERT (X,n-1)
end INSERT
```

Algorithm 5.19 Recursive insert.

```
line    procedure INSORT(X,n)
           //iterative version of Algorithm 5.18//
0          integer i,j,n; declare X(n),T
1          if n ≤ 1 then return endif
2          for j ← 2 to n do//sort X(1),X(2),...,X(j)//
              //insert X(j) into sorted X(1)...X(j-1)//
3             T ← X(j)
4             for i ← j-1 down to 1 do
5               if X(i) ≤ T then exit endif
6               X(i+1) ← X(i)
7             end for
8             X(i+1) ← T
9          end for
10      end INSORT
```

Algorithm 5.20 Final iterative version of insertion sort.

Example 5.3: [Merge Sort] By thinking recursively, one can arrive at at least two other sorting strategies: merge sort and quicksort. Suppose we have two sorted sequences: $x_1, x_2,...,x_a$ and $y_1, y_2, ..., y_b$. These two sequences can be merged together to obtain a sorted sequence of $a+b$ numbers. Algorithm 5.21 describes one possible merging strategy. So, if we can sort two sequences of $n/2$ numbers each then a sequence of n numbers can be sorted by recursively sorting each half of this sequence and then merging together the resulting two sorted sequences. Procedure MSORT is the resulting algorithm. □

procedure $MERGE(X,a,Y,b,Z)$
  //merge X and Y into a sorted sequence Z//
  //$X=x_1,x_2,...,x_a$; $Y=y_1,y_2,...,y_b$//
  sequence X,Y,Z; **integer** a,b,i,j,k
  $i \leftarrow j \leftarrow k \leftarrow 1$
  while $i \leqslant a$ **and** $j \leqslant b$ **do**
   if $x_i \leqslant y_j$ **then** $z_k \leftarrow x_i$; $i \leftarrow i + 1$
        else $z_k \leftarrow y_j$; $j \leftarrow j + 1$
   endif
   $k \leftarrow k + 1$
  end while
  if $i >$ a **then** $(z_k,...,z_{a+b}) \leftarrow (y_j,...,y_b)$
      else $(z_k,...,z_{a+b}) \leftarrow (x_i,...,x_a)$
  endif
end $MERGE$

Algorithm 5.21 Merging two sequences.

procedure $MSORT(X,n)$
  sequence X,Z; **integer** n
  if $n \leqslant 1$ **then return endif**
  call MSORT$((x_1,x_2, \ldots , x_{\lceil n/2 \rceil}) , \lceil n/2 \rceil)$
  call MSORT$((x_{\lceil n/2 \rceil+1},...,x_n), \lfloor n/2 \rfloor)$
  call MERGE$((x_1,x_2,...,x_{\lceil n/2 \rceil}), \lceil n/2 \rceil, (x_{\lceil n/2 \rceil+1},...,x_n), \lfloor n/2 \rfloor, Z)$
  $X \leftarrow Z$
  return
end $MSORT$

Algorithm 5.22 Recursive merge sort algorithm.

Example 5.4: [Quicksort] In insertion sort we first sort $n-1$ numbers and then insert the nth number into its correct spot. If we reverse the argument we will have a new sorting method: quicksort. In quicksort, one picks an arbitrary element y from the sequence x_1, x_2,..., x_n and positions it in its correct position. This done by rearranging the n elements x_1, x_2, ..., x_n such that all numbers appearing to the left of y are less than or equal to y, and all those appearing to the right of y are greater than or equal to y. Let $S1$ be the ordered multiset of numbers on the left of y and $S2$ the ordered multiset of numbers on the right of y. Since y is in its correct position, $\forall z \in S1 (z \leqslant y)$, and $\forall z \in S2$ $(z \geqslant y)$, we need only to sort $S1$ and $S2$ independently. The resulting sorting algorithm is procedure $QUICKSORT$ (Algorithm 5.23). Its correctness follows from the above discussion. $\square$

procedure $QUICKSORT(X,n)$
  sequence $X,S1,S2$; **integer** n,i,k;
  if $n \leqslant 1$ **then return endif**
  $i \leftarrow RANDOM(1,n)$ //randomly select element in X//
  $y \leftarrow x_i$.
  partition the elements in $X-\{x_i\}$ into
  two multisets $S1$ and $S2$ such that
  $\forall z \in S1(z \leqslant x_i)$ and $\forall z \in S2(z \geqslant x_i)$
  call $QUICKSORT(S1,|S1|)$
  call $QUICKSORT(S2,|S2|)$
  $X \leftarrow S1\|y\|S2$
  return
end $QUICKSORT$

Algorithm 5.23

Example 5.5: [Selection] In this problem, we are given a multiset X of n numbers and a number k, $1 \leqslant k \leqslant n$. We are required to find the kth smallest number in X. One way to do this is to first sort X. Following this, $X(k)$ is the kth smallest number in X. Another way to solve this problem is to follow along the lines of QUICKSORT. Randomly select a partitioning element y and partition X into three sets $S1$, $S2$, and $S3$ with the properties:

(a) $\forall z \in S1(z < y)$
(b) $\forall z \in S2(z = y)$
(c) $\forall z \in S3(z > y)$

Now, if $k \leqslant |S1|$ then the kth smallest element of X is also the kth smallest element of $S1$ and we may proceed recursively to find this element in $S1$. Note that $|S1| < |X|$. If $k \leqslant |S1|+|S2|$ and $k > |S1|$ then the kth smallest element of X is in S2 and it must be equal to y. If neither of these two cases hold, then $k > |S1|+|S2|$. In this case, the kth smallest element of X is in $S3$ and it is the $(k-|S1|-|S2|)$th element in S3. Since $|S3| < |X|$, we can proceed recursively to find this element in $S3$. Procedure SEL (Algorithm 5.24) is the resulting algorithm. This algorithm assumes that on the initial call, $0 < k \leqslant |X|$. □

Example 5.6: [Permutations] In this example, we shall consider the problem of generating all $n!$ permutations of the set $\{1,2,...,n\}$. We wish to generate these permutations in such a way that successive permutations differ from each other only in that two adjacent elements have been interchanged (e.g., 1234 and 1324 differ in this way). It should be apparent that this is the

procedure $SEL(X,k)$
　//Find the kth smallest element of X//
　multiset $X,S1,S2,S3$; **integer** k,i
　if $|X| = 1$ **then return**(x_1) **endif**
　$i \leftarrow RANDOM(1,|X|)$ //randomly select an element of X//
　Partition X into the three multisets $S1,S2$, and $S3$
　such that all elements in $S1$ are less than x_i;
　all elements in $S2$ equal x_i; and all elements
　in $S3$ are greater than x_i.
　case
　　:$k < |S1|$: **return**$(SEL(S1,k))$
　　:$k < |S1| + |S2|$: **return**(x_i)
　　:else: **return**$(SEL(S3,k-|S1|-|S2|))$
　end case
end SEL

Agorithm 5.24 Recursive selection algorithm.

minimal difference that can exist between two different permutations of the same set. Figure 5.9 lists the $3!=6$ permutations of the set $\{1,2,3\}$ in 4 different orders each of which satisfy the minimal difference property.

(a)　123,　132,　312,　321,　231,　213
(b)　123,　213,　231,　321,　312,　132
(c)　312,　132,　123,　213,　231,　321
(d)　321,　312,　132,　123,　213,　231

Figure 5.9 Some minimal difference permuation sequences.

　　A recursive algorithm to generate a minimal difference permutation sequence of $\{1,2, ..., n\}$ is not too hard to arrive at. Once again, we shall proceed along the lines of a proof by induction. When $n=1$, the only permutation in the sequence is 1. Next, suppose we have a way to generate a minimal difference permutation sequence for $n-1$ numbers for some arbitrary n, $n>2$. How can we use the sequence to obtain a minimal change sequence for n numbers? In order to get some ideas on how this might be accomplished, let us examine one of the sequences of Figure 5.9 and see how it may be obtained from the permutation sequence 12, 21 for two numbers. Let us look at sequence (a). 123 is obtained from 12 by placing the 3 at the right end; 132 is obtained by moving the 3 left; 312 is obtained by moving the 3 left again; 321 is obtained from 21 by placing the three at the left; 231 is obtained by moving the 3 right; 213 is obtained by moving the 3 right again.

Let $P_1, P_2, ..., P_m$ be a minimal difference sequence for the n-1 numbers $\{1, 2, ..., n-1\}$. A minimal difference sequence for $\{1, ..., n\}$ can be obtained as follows:

(a) $P_1\, n$ is the first permutation
(b)   The next $n-1$ permutations are obtained by moving n one position left each time.
(c) $n\, P_2$ is the $(n+1)$st permutation
(d)   The next $n-1$ permutations are obtained by moving n one position right each time.
(e) $P_3\, n$ is the $(2n+1)$st permutation.
(f)   The next $n-1$ permutations are obtained by moving n one position left each time.

.
.
.

We can prove that the new sequence is a minimal difference sequence if $P_1, ..., P_m$ is. The permutations generated in (a) and (b) are clearly in minimal difference order. Those in (c) and (d) are; those in (e) and (f) are; etc. So, we need only be concerned about the switch over from (b) to (c); from (d) to (e); from (f) to (g) etc. If the switch over is to a permutation of type $n\, P_j$ then the permutation preceding it is nP_{j-1}. P_j and P_{j-1} are minimal difference, so nP_j and nP_{j-1} are. If the switch over is to $P_j n$ then the preceding permutation is P_{j-1}. Once again the minimal difference property is preserved.

Algorithm 5.25 is the resulting recursive algorithm. Its correctness has already been established. While this algorithm was fairly easy to come by, it is not very practical as all $(n-1)!$ permutations of $\{1, 2, ..., n-1\}$ need to be computed before the first permutation of $\{1, 2, ..., n\}$ is generated. This of course means that $(n-1)!$ units of storage are needed by the algorithm. Even for relatively small values of n, this amount of storage is impractical. So, we are interested in obtaining an iterative version which generates the $n!$ permutations one by one using much less storage. An algorithm with this property is much harder to arrive at and prove correct. The exercises examine one such algorithm. □

5.3 COMPUTABILITY

For any given function $f: X \rightarrow Y$ one might well ask the question: Is f computable? Before attempting to answer this question, we should perhaps try to get a feel for what one means by the word computable. We would agree that the functions (a)-(h) defined in Section 5.2 are computable. We know

procedure $PERM(n)$
//generate a minimal difference permutation//
//sequence for $\{1,2,...,n\}$//
integer n,i **range**$(1:\infty)$; **ordered set** Q,R
if $n=1$ **then return**$((1))$ **endif**
if $n=2$ **then return**$((12,21))$**endif**
$Q \leftarrow PERM(n-1)$
$R \leftarrow 0$
for $i \leftarrow 1$ **to** $(n-1)!$ **by** 2 **do**
  // $Q=Q_1,Q_2,...,Q_{(n-1)!}$
  Compute the permutations $R_{(i-1)n+1}, R_{(i-1)n+2},...,R_{in}$)
  starting with $R_{(i-1)n+1} = Q_in$ and then moving n one
  position left each time.
  Compute the permutations $R_{in+1}, ..., R_{(i+1)n}$
  starting with $R_{in+1} = nQ_{i+1}$ and then moving n
  one position right each time.
end for
return(R)
end $PERM$

Algorithm 5.25 A recursive permutation algorithm.

that the algorithms corresponding to these functions can be run on a computer (and even by hand by a human) and the value of the function obtained (in a finite amount of time) for any desired element in the function domain. For function (i), we don't know if the corresponding program (Program 5.7) is in fact an algorithm (i.e., does it always terminate?). However, it is not too difficult to see that this program will (in a finite amount of time) produce the value of $g(i)$ for every i at which $g(i)$ is defined. For values of i for which $g(i)$ is not defined (there may not be any such i), the program will not terminate and so its output is also not defined. For partial functions this is perhaps a sufficient requirement for computabilty. A reasonable definition for computability then is:

> A function $f:X \rightarrow Y$ is *computable* iff there exists a computer program that (a) computes $f(x)$ in a finite amount of time, for every $x \in X$ for which $f(x)$ is defined and (b) for every $x \in X$ for which $f(x)$ is undefined, the program runs forever (i.e., it does not terminate).

Notice that our definition doesn't say anything about the programming language used. It is not necessary to do so as a program written in any reasonable language can be translated into an equivalent program in another language. If this is not convincing, we could fix the programming language to be some machine language. Since all programs can be translated into equivalent machine language programs, this language is clearly sufficiently powerful to permit all programs one might conceive of. Since every program that can be written can contain only a finite number of instructions, we might as well add this to our definition of computability:

> A function $f: X \rightarrow Y$ is *computable* iff there exists a computer program, containing a finite number of instructions, that (a) computes $f(n)$, in a finite amount of time, for every $x \in X$ for which $f(x)$ is defined and (b) for every $x \in X$ for which $f(x)$ is undefined, the program runs forever.

Since there is no way to formally establish this as a correct definition of computability, we justify this definition simply by stating that no one has been able to come up with an intuitively satisfying definition of computability under which the set of computable functions properly contains the set of functions computable under the definition we have just given.

Now that we have a definition of the term "computable", we can return to our earlier question: Is f computable? This question is trivially answerable if it turns out that every single function is computable. Unfortunately, not all functions are computable. One may arrive at this conclusion using the following argument. The set of all finite length computer programs is countably infinite (see Chapter 3). From Theorem 5.4 we know that the set of all total functions is uncountable. So, there are more total functions than programs. Consequently, there must exist total functions for which there are no programs. Hence there exist non-computable total functions.

Theorem 5.4: The set of all total functions is uncountable.

Proof: We shall show that the set F of all total functions of the type $f: N \rightarrow \{0, 1\}$ is uncountable. Since this is a proper subset of the set of all total functions, this set must also be uncountable. The proof is a diagonalization proof. If F is countably infinite, then it can be put into one-to-one correspondence with N. Consider any such one-to-one correspondence. Let f_i be the total function in F that corresponds to i in this one-to-one correspondence. Define the total function $g: N \rightarrow \{0, 1\}$ as below:

$$g(i) = 1 - f_i(i)$$

The function g has the property that it differs from the ith function in F (i.e., the function that corresponds to i in the one-to-one correspondence) at at least one x, $x \in N$ (i.e., $x = i$). Since g is clearly a total function from N to

$\{0,1\}$, it is an element of F. However, g cannot be in correspondence with any $j \in N$ in the one-to-correspondence. To see this, suppose $g = f_j$ for some j. By definition of g, $g(j) = f_j(j) = 1 - f_j(j)$. Since $f_j(j) \in \{0,1\}$, $f_j(j) \neq 1 - f_j(j)$. If g is not in correspondence with any j, then the supposed one-to-one correspespondence is not a one-to-one correspondence.

The assumption that a one-to-one correspondence between F and N exists leads to the conclusion that no such correspondence exists. So, the assumption is false and F cannot be put into one-to-one correspondence with N. Hence, F is an uncountable set. □

Theorem 5.5: There exist non-computable total functions.□

It is difficult, at first, to accept the conclusion that some functions are not computable. Our first reaction is to find fault with our definition of the term computable. As remarked earlier, this definition has withstood the test of time. All alternate and reasonable proposals for the definition of computability have proved equivalent to the one given earlier. So long as computability is defined in terms of a countable set of computing devices (programs, Turing machines, Markov algorithms, etc) there will remain functions that are not computable.

Let us turn our attention now to examining the computability of certain specific (and possibly useful) functions. As mentioned in the discussion preceding Theorem 5.4, the set of programs is countably infinite. So, this set can be put into one-to-one correspondence with N. Let's choose some one-to-one correspondence and let P_i denote the program that corresponds to i. i is the index of P_i. We may assume that every program computes a function f from some domain X to some range Y. So, every program has exactly one input and one output (which may be undefined for some inputs). While one customarily thinks of programs as having several inputs and outputs, all the inputs can be lumped together into one ordered multiset and all the outputs lumped together as another ordered multiset. Since every input and output is necessarily of finite length and composed of a finite number of different symbols, the set, IO, of ordered multisets representing possible inputs and outputs is countably infinite and can be put into one-to-one correspondence with N. So, by selecting one of the many one-to-one correspondences between IO and N, we can simply use elements of N to refer to members of the set IO. Thus, each program P_i may be viewed as computing a function from N to N.

Some specific functions whose computability is of interest to us are:

1. *Halting Problem*: Does program P_i terminate (halt) on all inputs? I.e., is program P_i an algorithm? We may define the function *HP* as below:

$$HP(i) = \begin{cases} 1 & \text{if } P_i(j) \text{ halts for all } j \in N \\ 0 & \text{otherwise} \end{cases}$$

2. *Correctness Problem*: Does program P_i compute the function f?

$$CP((i,f)) = \begin{cases} 1 & \text{if } P_i(j) = f(j) \text{ for all } j \\ 0 & \text{otherwise} \end{cases}$$

3. *Length Problem*: Let the length of an instruction be the number of operators and operands in it. The length, $|P_i|$, of program P_i is the sum of the lengths of the instructions in P_i. In the length problem we wish to determine the shortest program that is equivalent to P_i.

$LP(i) = j$ such that P_j is the shortest program equivalent to P_i

4. *Speed Problem*: Find the fastest program equivalent to P_i.

$SP(i) = j$ such that P_j is the fastest program equivalent to P_i

Are the total functions *HP*, *CP*, *LP*, and *SP* computable? Since these functions are total, they are computable iff there exists some algorithm that computes them. Let us first consider the halting problem. The halting problem is of interest because we often write programs for total functions. Unfortunately the programs one comes up with often contain infinite loops. These are detected only during execution of the programs. This is not a very desirable way to detect the occurrence of such loops as we could spend a lot of time and money debugging a particular program before all the infinite loops have been removed. It would therefore be nice if we had an algorithm that would accept a program (or its index) and determine if it terminates on all inputs. Before tackling the problem of whether or not HP is computable, we shall look at some apparently simple versions of the halting problem: halting problem on index and halting problem on zero.

In the *halting problem on index*, we are interested in determining if P_i halts on the input i. The corresponding total function that we wish to compute is:

$$HPI(i) = \begin{cases} 1 & \text{if } P_i(i) \text{ halts} \\ 0 & \text{otherwise} \end{cases}$$

Theorem 5.6: *HPI* is not computable.

Proof: If *HPI* is computable then there exists a program (which is also an algorithm), P_j, that computes *HPI* (i.e., $P_j(i) = HPI(i)$ for all $i \in N$).

From P_j we can construct another program R (Program 5.26). Since R is a program, it must be one of the P_is. Let $R=P_k$. Clearly, $P_k(k)$ either halts or it doesn't (there is no middle ground). If $P_k(k)$ halts then $P_j(k) = 1$ and from the definition of R (which is actually P_k) it is clear that $R_k(k) = P_k(k)$ doesn't halt. On the other hand, if $P_k(k)$ doesn't halt, then $P_j(k) = 0$ and from the definition of P_k it follows that $P_k(k)$ does halt. So, $P_k(k)$ halts iff it doesn't. This is clearly a contradiction ($S \iff \bar{S}$). Thus the only assumption made by us (i.e. the existence of an algorithm for *HPI*) must be incorrect. Hence, HPI is not computable. □

procedure $R(i)$
  $L1$: **if** $P_j(\text{i}) = 1$ **then go to** $L1$ **endif**
  return (0)
end R

Program 5.26

The proof of Theorem 5.6 should be studied carefully. It is a proof involving diagonalization. Diagonalization has been used in the construction of program R. The halting properties of R differ from those of every program P_i for at least one input (i.e., the input i). If $P_i(i)$ halts (in this case $P_i(i) = 1$) then $R(i)$ doesn't. If $P_i(i)$ doesn't halt then $R(i)$ does.

In the *halting problem on zero*, we wish to compute the function *HP0*:

$$HP0(i) = \begin{cases} 1 & \text{if } P_i(0) \text{ halts} \\ 0 & \text{otherwise} \end{cases}$$

Theorem 5.7: *HP0* is not computable.

Proof: This theorem follows from Theorem 5.6. For every program P_j, we can construct a new program $P_{g(j)}$ which is identical to P_j except that the statement $i \leftarrow j$ is inserted at the front (see Figure 5.10). Given j, the index $g(j)$ is computable from j. We can determine the program P_j it corresponds to. From P_j, $P_{g(j)}$ can be constructed and then by examining the one-to-one correspondence between programs and N, we can determine $g(j)$. The new program $P_{g(j)}$ disregards its input and computes $P_{g(j)}(i) = P_j(j)$ for all i.

Suppose there is an algorithm P_k that computes *HP0*. From P_k we can obtain an algorithm for *HPI* (see Program 5.27). From Theorem 5.6, we know that such an algorithm doesn't exist. So, no P_k computing *HP0* exists. □

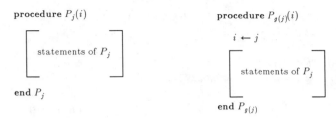

Figure 5.10

procedure $HPI(j)$
  Determine P_j
  Insert $i \leftarrow j$ at the front of P_j
  to get a new program $P_{g(j)}$
  Determine $g(j)$. Let $k = g(j)$.
  return$(HP0(k))$
end HPI

Program 5.27

The construction used in the proof of Theorem 5.8 also proves that the halting problem on all inputs is not computable. To see this just notice that $HP(g(j)) = HPI(j)$ for all j.

Theorem 5.8: HP is not computable. $\square$

The functions $CP((i,f))$, $LP(i)$, and $SP(i)$ are also not computable.

Theorem 5.9: CP is not computable.

Proof: We shall show that the simpler version CPZ:

$$CPZ(i) = \begin{cases} 1 & \text{if } P_i(j) = 0 \text{ for all } j \\ 0 & \text{otherwise} \end{cases}$$

is not computable. This may be proved as follows: From any given program index j, we can obtain the corresponding program P_j. From P_j we can construct another program $P_{g(j)}$ which simulates the execution of $P_j(j)$. $P_{g(j)}(i)$ simulates $P_j(j)$ for upto i steps (each step of $P_j(j)$ is a simple operation like add, subtract, **go to**, compare etc.). This simulation is easily done by using either a clock or counter that keeps track of the number of basic operations performed by $P_j(j)$ so far. The simulation stops whenever one the following two events occurs:

(a) $P_j(j)$ stops.
(b)   The counter (or clock) reaches i.

If $P_j(j)$ stops in i steps then $P_{g(j)}(i)$ has value 1, otherwise it has value 0. The algorithm corresponding to $P_{g(j)}$ is Algorithm 5.28. So, $P_{g(j)}$ computes the constant function zero iff $P_j(j)$ does not halt.

Now, suppose CPZ is computable. Using the algorithm for CPZ and the above simulator construction, we can construct an algorithm (Algorithm 5.29) for HPI. This however contradicts Theorem 5.7 which states that HPI is not computable. So, CPZ is not computable. $\square$

procedure $P_{g(j)}(i)$
   Simulate $P_j(j)$ for i steps.
   if $P_j(j)$ stopped in i steps **then return**(1)
                                   else return(0)
   endif
end $P_{g(j)}$

Algorithm 5.28

procedure HPI(j)
   Determine P_j.
   Construct the simulator $P_{g(j)}$ for P_j(j).
   Determine $g(j)$. Let $k = g(j)$.
   return$(1 - CPZ(k))$
end HPI

Algorithm 5.29

Theorem 5.10:   LP and SP are not computable.

Proof: This is easily proved using the fact that CPZ (see Theorem 5.9) is not computable. The proof is left as an exercise. $\square$

REFERENCES AND SELECTED READINGS

Martin Gardner's article:

> Mathematical games: What unifies dinner guests, strolling school girls, and handcuffed prisoners, *Scientific American*, Vol. 242, no. 5, May 1980, pp. 16-28.

contains an interesting discussion of principles that are related to the pigeon hole principle. The minimal difference permutation sequence algorithm (Algorithm 5.25) is due to H. Trotter. A FORTRAN code corresponding to the iterative version of this algorithm appears in the paper:

> *Algorithm 115: Perm*, by H. Trotter, CACM, Vol. 5, 1964, pp.430-435

Efficient algorithms to generate minimal difference permutation sequences and other combinatorial objects can be found in the papers:

> *Algorithm 433: Four combinatorial algorithms*, by G. Ehrlich, CACM, Vol. 16, 1973, pp. 690-691.

and

> *Loopless algorithms for generating permutations, combinations, and other combinatorial configurations*, by G. Ehrlich, JACM, Vol. 3, 1973, pp.500-513.

The paper

> *Parallel matrix and graph algorithms*, by E. Dekel, D. Nassimi, and S. Sahni, Proceedings Annual Allerton Conference, 1979.

contains an interesting application of Gray codes to the computation of matrix products on parallel computers.

A more advanced treatment of iterative and recursive algorithms can be found in the texts:

> *Fundamentals of data structures*, by E. Horowitz and S. Sahni, Computer Science Press Inc., Maryland, 1976.

and

> *Fundamentals of computer algorithms*, by E. Horowitz and S. Sahni, Computer Science Press Inc., Maryland, 1978.

The latter text contains versions of our Algorithms 5.20-5.24, that are more readily translated into computer programs. A good reference on the subject of computability is the text:

Introduction to computability, by F. Hennie, Addison-Wesley Publishing Co., Massachusetts, 1977.

EXERCISES

1. For each of the following functions, state whether it is onto, into, or one-to-one onto. Also, state whether it is a partial or total function from X to Y. For those functions that are invertible, obtain the inverse.

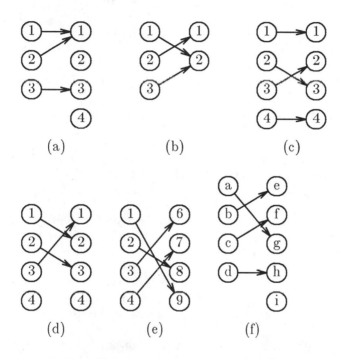

      (a)   (b)   (c)

      (d)   (e)   (f)

2. (a) Show that if $f: X \to Y$ and $g: Y \to Z$ are onto functions then $g \cdot f: X \to Z$ is an onto function.
   (b) Are f and g necessarily onto functions if $g \cdot f$ is?

3. (a) Show that if $f: X \to Y$ and $g: Y \to Z$ are one-to-one functions then $g \cdot f: X \to Z$ is also a one-to-one function.
   (b) Are f and g necessarily one-to-one functions if $g \cdot f$ is?

4. (a) Show that if $f: X \to Y$ and $g: Y \to Z$ are one-to-one onto functions then $g \cdot f: X \to Z$ is also a one-to-one onto function.
   (b) Are f and g necessarily one-to-one onto functions if $\cdot f$ is?

5. (a) Does Theorem 5.2 neccessarily hold when f is a partial function from X to Y?
   (b) Prove Theorem 5.3.
   (c) Does Theorem 5.3 neccessarily hold when f is a partial function?

6. (a) What is the 3-bit Gray code that results from the bit change sequence $G1(3)$ (See Section 5.2)?
   (b) Show that the functions $G(b)$ and $G1(b)$ both define bit change sequences for b-bit gray codes, $b \leqslant 1$.

7. (a) Is the Ackermann's function a total function from $N \times N$ to N?
   (b) Is it an onto function?

8. (a) Show that function (i) (i.e. g) of Section 5.2 is defined for all i of the form $i = 2^k$ for some $k \in N$.
   (b) Can you think of any other classes of i for which you can show that $g(i)$ is defined?

9. Use the recursion mechanism of Section 5.2 to evaluate $p(3)$ (Algorithm 5.2). Show the stack configuration following each call and return.

10. Use the recursion mechanism of Section 5.2 to evaluate $F(3)$ (Algorithm 5.4). Show the stack configuration following each call and return.

11. Write a recursive algorithm for function (h) of Section 5.2. Use the recursion mechanism of that section to compute $f(11)$. Show the stack configuration following each call and return.

12. Use the transformation rules given in Section 5.2 to transform Algorithm 5.3 into an equivalent iterative algorithm that uses the recursion stack.

13. Do Exercise 12 for Algorithm 5.5.

14. Do Exercise 12 for Program 5.7.

15. Show that if $P_1, P_2, ..., P_{n!}$ is a minimal difference permutation sequence generated by Algorithm 5.25 then P_1 can be obtained from $P_{n!}$ by interchanging two adjacent elements.

16. Algorithm 5.30 is an iterative algorithm to generate a minimal change permutation sequence. Prove that this algorithm is correct and that it generates the same sequence as generated by Algorithm 5.25.

```
procedure PERM(n)
  integer P(0:n+1),I(1:n),D(1:n),i,j,k
  for i ← 1 to n do
    P(i) ← I(i) ← i; D(i) ← −1
  end for
  j ← n + 1
  P(0) ← P(n+1) ← j
  while j ≠ 1 do
    print(P(1),P(2),...,P(n)) //output a permutation//
    j ← n
    while P(I(j) + D(j)) > j do
      D(j) ← −D(j)
      j ← j − 1
    end while
    P(I(j)) ⟷ P(I(j)+D(j)) //interchange//
    I(j) ⟷ I(P(I(j)))
  end while
end PERM
```

Algorithm 5.30

17. A permutation $P=(P(1), P(2),...,P(n))$ of $\{1,2,...,n\}$ is a *derangement* iff $P(i) \neq i$, $1 \leq i \leq n$.
   (a) Let $d(n)$ be the number of derangements of $\{1,2,...,n\}$. Show that $d(1)=0$, $d(2)=1$, and $d(n)=(n-1)(d(n-1) + d(n-2))$, $n > 2$.
   (b) Write a recursive algorithm to compute $d(n)$, $n \geq 1$.
   (c) Write a recursive algorithm to list all the derangements of $\{1,2,...,n\}$.

18. Let F be a recursively defined algorithm. Assume that the last statement appearing in F (i.e., the one just preceding **end** F) is a call to F. Can this call to F be replaced by code assigning the parameters of F their values and an unconditional branch to the first statement in F (the previous parameter values, local variable values and return address are not stacked)?

19. The halting problem on some input is given by the function:

$$HPS(i) = \begin{cases} 1 & \text{if there is at least one } x \in N \text{ for} \\ & \text{which } P_i(x) \text{ halts} \\ 0 & \text{otherwise} \end{cases}$$

Is HPS a computable function?

20. Prove that the functions SP and LP of Section 5.3 are not computable.

21. Are the following functions computable? Prove your answers.

(a) $BB(i) = \begin{cases} 1 & \text{if } P_i(i) \text{ executes at least} \\ & \text{one instruction two or more} \\ & \text{times} \\ 0 & \text{otherwise} \end{cases}$

(b) $BBM(i) = \begin{cases} 1 & \text{if } P_i(i) \text{ executes at least one of} \\ & \text{the instructions in its main program} \\ & \text{two or more times} \\ 0 & \text{otherwise} \end{cases}$

CHAPTER 6

ANALYSIS OF ALGORITHMS

6.1 COMPLEXITY

We have already seen a large number of algorithms in this text. So far, we have been concerned only with establishing the correctness of an algorithm. I.e., does it compute the function it was intended to compute? Once correctness has been established, we will be interested in knowing the amount of computer time and memory needed to run this algorithm (or to run programs resulting from this algorithm). *Complexity analysis* is concerned with determining these two quantities. The *space complexity* of an algorithm is the amount of memory it needs to run to completion. The *time complexity* of an algorithm is the amount of computer time it needs to run to completion.

Space Complexity

Procedure *ABC* (Algorithm 6.1) computes the expression $A+B+B*C+(A+B-C)/(A+B)+4$; procedure *SUM* (Algorithm 6.2) computes the sum $\sum_{i=1}^{n} A(i)$; and procedure *RSUM* is a recursive algorithm that computes $\sum_{i=1}^{n} A(i)$. The space needed by each of these algorithms is seen to be the sum of the following components:

(a) Space needed to store the compiled version of the algorithm (this is the code that can actually be executed on the computer of interest).

(b) Space needed to store the values of simple (i.e., non-array) variables and constants (e.g., 4, 1, etc.).

(c) Space needed for arrays.

(d) Space needed by the recursion stack.

procedure $ABC(A,B,C)$
  real A,B,C
  return$(A+B+B*C+(A+B-C)/(A+B)+4)$
end ABC

Algorithm 6.1

The space needed by the compiled version of the algorithm is itself dependent upon such factors as:

(a)   The programming language in which the algorithm is written.

(b)   The experience of the person who translates the algorithm from its present form into the target programming language.

(c)   The compiler used to compile the resulting program into machine code.

The programming language used is of importance especially in the case of recursive algorithms. If a programming language that does not support recursion (e.g., FORTRAN) is used, then the programmer will have to eliminate the recursion (possibly using the method described in Section 5.2). The resulting iterative algorithm will generally contain considerably more statements than the original recursive algorithm did. Hence we would expect the machine code for the iterative algorithm to be longer than that for the original recursive algorithm.

line	
	procedure $SUM(A,n)$
0	**integer**$,n,S,A(n)$
1	$S \leftarrow 0$
2	**for** $i \leftarrow 1$ **to** n **do**
3	$S \leftarrow S+A(i)$
4	**end for**
5	**return**(S)
6	**end** SUM

Algorithm 6.2

line	
	procedure $RSUM(A,n)$
0	**integer** $n,A(n)$
1	**if** $n \leqslant 0$ **then return**(0) **endif**
2	**return**$(RSUM(A,n-1)+A(n))$
3	**end** $RSUM$

Algorithm 6.3

Programmer experience is also an important factor. An experienced programmer might make the observation that $\sum_{i=1}^{n} A(i)$ is easily computed without using recursion and will opt to save the overhead (stacking of parameter values etc.) incurred when recursion is used. In this case, when given Algorithm 6.3, the programmer might in fact program the equivalent of Algorithm 6.2.

The compiler used is a very important factor determining the space needed by the resulting code. Figure 6.1 shows two possible codes for the evaluation of $A+B+B*C+(A+B-C)/(A+B)+4$. Both of these codes perform exactly the same arithmetic operations (i.e., every operator has the same operands) but each needs a different amount of space. The compiler in use determines exactly which code will be generated.

LOAD	A	LOAD	A
ADD	B	ADD	B
STORE	$T1$	STORE	$T1$
LOAD	B	SUB	C
MULT	C	DIV	$T1$
STORE	$T2$	STORE	$T2$
LOAD	$T1$	LOAD	B
ADD	$T2$	MUL	C
STORE	$T3$	STORE	$T3$
LOAD	A	LOAD	$T1$
ADD	B	ADD	$T3$
SUB	C	ADD	$T2$
STORE	$T4$	ADD	4
LOAD	A		
ADD	B		
STORE	$T5$		
LOAD	$T4$		
DIV	$T5$		
STORE	$T6$		
LOAD	$T3$		
ADD	$T6$		
ADD	4		

(a)   (b)

Figure 6.1 Two equivalent codes.

For simple variables and constants, the space requirements are a function of the computer used and the size of the numbers involved. The reason for this is that we will normally be concerned with the number of words of memory required. Since the number of bits per word varies from computer to computer, the number of words needed per variable also varies. Also, it takes more bits to store the number 2^{100} than it does to store 2^3. The space needed for arrays also depends on the above two factors. It, of course, also depends on the array size (i.e., the number of elements in the array). The array of Algorithm 6.2 is of size n. So, the space it needs varies with the size, n, of the problem instance being solved.

The space needed by the recursion stack is often neglected by beginning complexity analysts. This is so because one must first understand how recursion is implemented in languages that support recursion. It is only after one has become aware of the recursion mechanism that one realizes that a recursion stack will be created and this stack needs space. The amount of space needed by the recursion stack will depend on the number of local variables to be stacked; the number of parameters; location of the recursive function within the program; and the maximum depth of recursion (this is the maximum number of nested recursive calls). For Algorithm 6.3 recursive calls get nested until $n=0$. At this time the nesting looks as in Figure 6.2. The maximum depth of recursion for this algorithm is therefore $n+1$.

call $RSUM(n)$
call $RSUM(n-1)$
call $RSUM(n-2)$

.

.

.

call $RSUM(1)$
call $RSUM(0)$

Figure 6.2 Nesting of recursive calls for Algorithm 6.3.

In summary, the space needed by an algorithm depends upon many factors. Many of these are not determined at the time the algorithm is conceived or written (e.g., the programming language; the computer to be used; who is going to write the final computer program, etc.). Until these factors have been determined, we cannot make an accurate analysis of the space requirements of an algorithm. We can, however, determine the contribution of those components which depend upon such factors as the size of the problem instance (i.e., the number of inputs and outputs, magnitude of the numbers involved, etc.) being solved.

Of the four components ((a)-(d)) listed earlier, (a) and (b) are relatively insensitive to the particular problem instance being solved (unless of course the magnitude of the numbers involved becomes too large for a word. But then, we would probably have to rewrite the algorithm using multiprecision arithmetic.). The space needed by some of the arrays may also be independent of the problem size (e.g., an array declared as **integer** $B(6)$). Some recursive calls may result in a depth of nesting that too is independent of the specific problem instance being solved. And, provided each such call stacks a fixed number of values, the stack space needed by such a call to a recursive algorithm will also be independent of the size of the problem instance being solved.

We may divide the total space needed by an algorithm into two parts:

(A)   A fixed part which is independent of the characteristics of the inputs and outputs. This part typically includes the program space (i.e., space for the code); space for simple variables and fixed size arrays (i.e., **integer** $A(4)$ etc.); space for constants, etc.

(b)   A variable part which consists of the space needed by arrays whose size is dependent on the particular problem instance being solved and the space needed by the recursion stack (in so far as this space depends on the instance characteristics). The space requirement $S(A)$, of any algorithm A may therefore be written as $S(A) = c + S_A$ (instance characteristics) where c is a constant.

When analyzing the space complexity of an algorithm, we shall concentrate solely on estimating S_A (instance characteristics). Making the assumption that one word is adequate to store the values of each of A, B, and C in Algorithm 6.1, we obtain $S_{ABC}( ) = 0$. Under a similar assumption, $S_{SUM}(n) = n$. If calls to $RSUM$ do not result in the stacking of $A(1),...,A(n)$ (this is often the case; often, only simple parameters, simple local variables, and array names (not the value of all elements) get stacked), then $c_1 n \leqslant S_{RSUM(n)} \leqslant c_2 n$ for some constants c_1 and c_2. If calls to $RSUM$ result in stacking $A(1),...,A(n)$ each time, then $c_3 n^2 \leqslant S_{RSUM}(n) \leqslant c_4 n^2$ for some constants c_3 and c_4. As far as Algorithm 6.3 is concerned, its outcome is independent of whether or not array elements get stacked on calls to recursive algorithms. In our further discussions, we shall assume that array elements do not get stacked but that array names do. With this added assumption, the variable contribution, S_{RSUM}, to $S(RSUM)$ is such that $c_1 n \leqslant S_{RSUM}(n) \leqslant c_2 n$ for some constants c_1 and c_2.

For most algorithms, the instance characteristics with respect to which the space $S(A)$ is defined are the number of inputs or the number of outputs or both or some function of these. For Algorithms 6.2 and 6.3 we used only the number of elements in the array A.

Time Complexity

The time complexity of an algorithm depends upon all the factors that the space complexity depends on. An algorithm will run faster on a computer capable of executing 10^8 instructions per second than on one that can execute only 10^6 instructions per second. The code of Figure 6.1(b) will require less execution time than the code of Figure 6.1(a). Some compilers will take less time than others to generate the corresponding computer code. Compilers for FORTRAN may be faster than those for COBOL. Smaller problem instances will probably take less time than larger instances.

As in the case of space complexity, the time complexity $T(A)$ of algorithm A can be written as the sum $T(A) = c + t_A$ (instance characteristics) where c is a constant (i.e., it is independent of the instance characteristics with respect to which t_A is defined). The constant c normally includes only the time needed to compile the program. t_A includes all the execution time (i.e. the time to actually compute the function that A is supposed to compute). When analyzing the time complexity of an algorithm one is concerned mainly with determining t_A. This lack of concern for c is often justified by the argument that a program need be compiled only once. The compiled version can however be executed many times (using different data).

Because many of the factors t_A depends on are not known at the time an algorithm is conceived, it is reasonable to attempt to only estimate t_A. If we knew the characteristics of the compiler to be used and the algorithm was specified as a computer program, we could proceed to determine the number of additions, subtractions, multiplications, divisions, compares, loads, stores, etc. that would be made by the code for A. Having done this, we could present a formula for t_A. Letting n denote the instance characteristics, we might then have an expression for $t_A(n)$ of the form:

$$t_A(n) = c_a ADD(n) + c_s SUB(n) + c_m MUL(n) + c_d DIV(n) + \ldots$$

where c_a, c_s, c_m, c_d, etc., respectively, denote the time needed for an addition, subtraction, multiplication, division, etc. and ADD, SUB, MUL, DIV, etc. are functions whose value is the number of additions, subtractions, multiplications, divisions, etc. that will be performed when the code for A is used on an instance with characteristic n.

Obtaining such a formula is in itself an impossible task as the time needed for an addition, subtraction, multiplication, etc. often depends on the actual numbers being added, subtracted, multiplied, etc. In reality then, the true value of $t_A(n)$ for any given n can be obtained only experimentally. The algorithm is programmed, compiled, and run on a particular machine. The execution time is physically clocked and $t_A(n)$ obtained. Even with this experimental approach, one could face difficulties. The execution time will depend upon system load: How many other programs are on the computer at the time algorithm A is run? What are the characteristics of these other programs? etc.

Given the minimal utility of determining the exact number of additions, subtractions, etc. that are needed to solve a problem instance with characteristics given by n, we might as well lump all the operations together (provided that the time required by each is relatively independent of n) and obtain a count for the total number of operations. We can go one step further and count only the number of program steps.

A *program step* is loosely defined to be a semantically meaningful segment of a program that has an execution time which is independent of the instance characteristics. For example, the statement **return**$(A+B+B*C \cdots)$ of Algorithm 6.1 could be regarded as a step as its execution time is independent of the instance characteristics (this statement isn't strictly true as the time for a multiply and divide will generally depend on the actual numbers involved in the operation).

The number of steps any program statement is to be assigned depends on the nature of that statement. The following discussion considers the various statement types that can appear in a program and states the complexity of each in terms of the number of steps:

(a) *Declarative Statements*
These are statements that specify array sizes and variable types (e.g. **integer, real**, etc.). These statements are not executable statements and so count as zero steps.

(b) *Assignment Statements*
Almost every statement of the form $<$variable$> \leftarrow <$expression$>$ has a step count of one. The exceptions are:

(i) If $<$expression$>$ uses the exponentiation operator, $\uparrow$, then the step count of the assignment statement is as below:
I If all uses of $\uparrow$ in $<$expression$>$ are of the form $x \uparrow c$ where

c is independent of the instance characteristics, then the step count is one. So, the step count for the statement:

$$X \leftarrow A{\uparrow}2 + B{\uparrow}4 + C{\uparrow}8.2 + D{*}E{*}F$$

. is 1.

II If $X{\uparrow}Y$ is computed as **alog**$(Y$ **log** $X)$ where **alog** is the inverse of **log**, then again the step count is 1.

III Let $X{\uparrow}n_1$, $X{\uparrow}n_2$, ..., $X{\uparrow}n_k$ be all the uses of ${\uparrow}$ in which the exponent is a natural number that is a function of the instance characteristics. If $X{\uparrow}n_i$ is computed as $X{*}X{*}X{*}...{*}X$ the step count is $\sum_{i=1}^{k} n_i$. If it is computed using the binary expansion of the n_is then the step count is $\sum_{i=1}^{k} \log_2 n_i$.

(ii) If <expression> contains implied calls to subalgorithms where the stacking needed to effect the call takes an amount of time that is dependent on the instance characteristics, the step count assignable to this assignment statement is the sum of the number of items stacked by each of the implied calls contained in this statement. Note that this step count does not include the number of steps needed to execute the called subalgorithm. As an example, consider the statement:

$$X \leftarrow SUM(A,m)$$

The step count assignable to this line is 1 as the time needed to effect the implied call to SUM is a constant.

(iii) If <expression> contains both ${\uparrow}$s and implied calls, then the step count is the sum of the counts for cases (i) and (ii) above.

(c) *Iteration Statements*
This class of statements includes the **for, while,** and **until** statements of the form:

> **for** $a \leftarrow$ <expr> **to** <expr1> **by** <expr2> **do**
> **while** <expr> **do**
> **until** <expr>

Each execution of a **while**, and **until** statement will be given a step count equal to the number of step counts assignable to $<$expr$>$. The step count for the first execution of the **for** statement equals the step count assignable to $<$expr$>$, $<$expr1$>$, and $<$expr2$>$ together (note that these expressions are computed only when the loop is started). Generally, these expressions are simply constants or variable names. Hence, the step count for the first execution is generally 1. Remaining executions of the **for** statement have a step count of 1.

(d) *Case Statement*
This can always be given a step count of 1.

(e) *If-then-else Statement*
The **if-then-else** statement may be broken down into four parts as in:

> **if** $<$expr$>$
>   **then** $<$statements1$>$
>   **else** $<$statements2$>$
> **endif**

Each of the first three parts is assigned the appropriate number of steps corresponding to $<$expr$>$, $<$statements1$>$, and $<$statements2$>$. The step count for **endif** is zero.

(f) *Call Statement*
All calls to algorithms count as one step unless the call involves stacking a number of items that depends on the instance characteristics. In this latter case, the count equals the number of items stacked.

(g) *End Statement*
Each **end**, **end for**, **end while**, and **end case** counts as one step.

(h) *Procedure Statement*
This counts as zero steps as its cost has already been assigned to a **call** statement.

(i) *Return Statement*
A **return** statement is one step unless it is of the type **return**($<$expr$>$). In this case it counts as as many steps as does $<$exp$>$ (see the discussion for assignment statements).

With the above assignment of step counts to statements, we can proceed to determine the number of steps needed by an algorithm to solve a particular problem instance. We can go about this in one of two ways. In the first method, we introduce a new variable, *COUNT*, into the algorithm. This is a global variable with initial value 0. Every assignment, **call**, **for**, and **while** statement is followed by a statement that increases the value of *COUNT* by the number of steps assignable to that statement. Every **end**, **end for**, **end while**, **case**, and **return** statement is preceded by a statement that increases *COUNT* by the appropriate number of steps. If any of these latter statements are labeled, the label is moved to the statement just added. Every **end for** and **end while** statement is followed by a statement incrementing *COUNT* by the number of steps attributable to the first execution of the **for** and the last execution of the **while** statement, respectively. Every **if** <expr> **then** is preceded by a statement that increase *COUNT* by the number of steps attributable to <expr>. If the **if** statement is labeled, the label is moved to the statement just added. Every **end case** results in a statement incrementing *COUNT* by 1 at the end of each of the cases contained in the **case-end case** block.

When all the statements incrementing *COUNT* are introduced into Algorithms 6.2 and 6.3 the result is Algorithms 6.4 and 6.5. The value of *COUNT* at the time the algorithms terminate is the number of steps executed by Algorithms 6.2 and 6.3 respectively.

```
procedure SUM (A,n)
  integer i,n,S,A (n); global integer COUNT
  S ← 0
  COUNT ← COUNT+1
  for i ← 1 to n do
    COUNT ← COUNT+1 //for for//
    S ← S+A (i)
    COUNT ← COUNT+1 //for assignment//
    COUNT ← COUNT+1 //for end for//
  end for
  COUNT ← COUNT+1 //for for//
  COUNT ← COUNT+1 //for return//
  return(S)
  COUNT ← COUNT+1 //for end SUM//
end SUM
```

Algorithm 6.4

Since we are interested in determining only the value of *COUNT* when the algorithm terminates, Algorithm 6.4 may be simplified to Algorithm 6.6. It should be easy to see that both Algorithms 6.4 and 6.6 compute the same value for *COUNT*. It is easy to see that in the **for** loop the value of *COUNT* will increase by a total of $3n$. So, if *COUNT* is zero to start with, then it will be $3n+3$ on termination. Therefore, a call to procedure $SUM(A,n)$ (Algorithm 6.2) will execute a total of $3n+3$ steps.

procedure $RSUM(A,n)$
  integer n, $A(n)$; **global integer** *COUNT*
  $COUNT \leftarrow COUNT+1$ //for **if**//
  if $n \leqslant 0$ **then** $COUNT \leftarrow COUNT+1$ //for **return**//
  return(0); **end if**
  $COUNT \leftarrow COUNT+1$ //for **return**//
  return$(RSUM(A,n-1)+A(n))$
  $COUNT \leftarrow COUNT+1$ //for **end** *RSUM*//
end *RSUM*

Algorithm 6.5

procedure $SUM(A,n)$
  integer i,n; **global integer** *COUNT*
  $COUNT \leftarrow COUNT+1$
  for $i \leftarrow 1$ **to** n **do**
    $COUNT \leftarrow COUNT+3$
  end for
  $COUNT \leftarrow COUNT+2$
  return
  $COUNT \leftarrow COUNT+1$
end *SUM*

Algorithm 6.6

Let $t_{RSUM}(n)$ be the value of *COUNT* when procedure *RSUM* (Algorithm 6.5) terminates. We can see that $t_{RSUM}(0)=2$. When $n>0$, *COUNT* increases by 2 plus whatever increase results from the call to *RSUM* from within the **return** statement. From the definition of t_{RSUM}, it follows that *COUNT* increases by $t_{RSUM}(n-1)$ as a result of the call from within the **return**. So, the value of *COUNT* at the time of termination is $2+t_{RSUM}(n-1)$, $n>0$.

When analyzing a recursive algorithm for its step count, we shall obtain a recursive formula for the step count (i.e., say $t_{RSUM}(n) =$

$2+t_{RSUM}(n-1)$, $n>0$ and $t_{RSUM}(0)=2)$. These recursive formulas will be referred to as *recurrence relations*. In a later chapter, we shall study how to solve recurrence relations. For now, it is enough to know that the solution to the recurrence relation for t_{RSUM} is $t_{RSUM}(n) = 2n+2$, $n\geqslant 0$. So, the step count for procedure *RSUM* (Algorithm 6.3) is $2n+2$.

Comparing the step counts of Algorithms 6.2 and 6.3, we see that the count for Algorithm 6.3 is less than that for Algorithm 6.2. This doesn't tell us that Algorithm 6.2 is necessarily slower than Algorithm 6.3 because a step doesn't correspond to a definite time unit. Each step of procedure *SUM* may be shorter than every step of procedure *RSUM*. So, it might well be (in fact, we expect it to be so) that *SUM* is faster than *RSUM*. The step count is useful in that it tells us how the computing time for *SUM* and *RSUM* vary with changes in *n*. If *n* is doubled, we expect the time to double; if *n* increases by a factor of 10, we expect the computing time to increase by a factor of 10; etc. So, we expect the computing time to grow *linearly* in *n*. We shall say that *SUM* and *RSUM* are *linear* algorithms (their time complexity is linear in the instance characteristic n).

Algorithm 6.7 is an algorithm to add two $m \times n$ matrices A and B together. Introducing the *COUNT* incrementing statements leads to Algorithm 6.8. Algorithm 6.9 is a simplified version of Algorithm 6.8 which computes the same value for *COUNT*. Examining Algorithm 6.9, we see that line 2 is executed a total of *m* times; line 4 is executed *n* times for each value of *i* or a total of *mn* times; line 6 is executed m times; and line 8 is executed once. If *COUNT* is zero to begin with, it will be $3mn+3m+2$ when Algorithm 6.9 terminates. From this analysis we see that if $m > n$ then it would be better to interchange the two **for** statements in Algorithm 6.7. If this were done, the step count would be $3mn+3n+2$. Note that in this example the instance characteristics are given by *m* and *n*.

```
line  procedure MADD (A,B,C,m,n)
 0      declare A (m,n),B (m,n),C (m,n); integer i,j,m,n
 1      for i ← 1 to m do
 2        for j ← 1 to n do
 3          C (i,j) ← A (i,j)+B (i,j)
 4        end for
 5      end for
 6    end MADD
```

Algorithm 6.7

```
procedure MADD (A,B,C,m,n)
  declare A (m,n),B (m,n),C (m,n); integer i,j,m,n
   global integer COUNT
   for i ← 1 to m do
     COUNT ← COUNT+1
     for j ← 1 to n do
       COUNT ← COUNT+1
       C (i,j) ← A (i,j)+B (i,j)
       COUNT ← COUNT+1
       COUNT ← COUNT+1
     end for
     COUNT ← COUNT+1
     COUNT ← COUNT+1
   end for
   COUNT ← COUNT+1
   COUNT ← COUNT+1
 end MADD
```

Algorithm 6.8

```
line    procedure MADD (m,n)
 0        integer i,j,m,n; global integer COUNT
 1        for i ← 1 to m do
 2          COUNT ← COUNT+1
 3          for j ← 1 to n do
 4            COUNT ← COUNT+3
 5          end for
 6          COUNT ← COUNT+2
 7        end for
 8        COUNT ← COUNT+2
 9      end MADD
```

Algorithm 6.9 Step count for matrix addition.

The second method to determine the step count of an algorithm is to build a table in which we list the total number of steps contributed by each statement. This figure is often arrived at by first determining the number of steps per execution of the statement and the total number of times (i.e., frequency) each statement is executed. By combining these two quantities, the total contribution of each statement can be obtained. By adding up the contributions of all statements, the step count for the entire algorithm can be obtained.

There is an important difference between the step count of a statement and its steps per execution (s/e). The step count does not necessarily reflect the complexity of the statement. For example, the statement:

$$X \leftarrow SUM(A,m)$$

has a step count of 1 while the total change in $COUNT$ resulting from the execution of this statement is actually 1 plus the change resulting from the implied call to SUM (i.e., $3m+3$). The steps per execution of the above statement is $1+3m+3 = 3m+4$. *The s/e of a statement is the amount by which COUNT changes as a result of the execution of that statement.*

Consider the following **case** statement:

case
 $:X=SUM(A,n)$: **return**
 $:X=SUM(B,m)$: $i \leftarrow 2$; $j \leftarrow SUM(C,p)$
 :else : $j \leftarrow 2$
end case

The steps per execution of the statements **case** and **end case** is 1. The s/e for the first part of the case $(:X=SUM(A,n):$ **return**$)$ is $3n+5$ as it takes $3n+3$ steps to compute $SUM(A,n)$, one to check if $X = SUM(A,n)$, and another to execute the return. The s/e for $:X=SUM(B,m)$: $i \leftarrow 2$; $j \leftarrow SUM(C,p)$ is $3(n+m+p)+13$ as it takes $3n+4$ steps to determine that $X \neq SUM(A,n)$, $3m+4$ to determine that $X = SUM(B,m)$; 1 to execute $i \leftarrow 2$; and $3p+4$ to execute $j \leftarrow SUM(C,p)$. The s/e of the **else** statement is $3(n+m)+9$.

In Table 6.1 the number of steps per execution and frequency of each of the statements in procedure SUM (Algorithm 6.2) have been listed. The total number of steps required by the algorithm is determined to be $3n+3$.

line	s/e	frequency	total steps
0	0	0	0
1	1	1	1
2	1	$n+1$	$n+1$
3	1	n	n
4	1	n	n
5	1	1	1
6	1	0	0

Total number of steps $=$ $\quad 3n+3$

Table 6.1 Step table for Algorithm 6.2.

Table 6.2 gives the step count for procedure *RSUM* (Algorithm 6.3). Line 1.1 refers to the **if** clause of line 1 of Algorithm 6.3 and line 1.2 refers to the statement in the **then** clause of the **if**. Notice that under the s/e (steps per execution) column line 2 has been given a count of $1 + t_{RSUM}(n-1)$. This is the total cost of line 2 each time it is executed. It includes all the steps that get executed as a result of the call to RSUM from line 2. The frequency and total steps columns have been split into two parts: one for the case $n = 0$ and the other for the case $n > 0$. This is necessary as the frequency (and hence total steps) for some statements is different for each of these cases.

line	s/e	frequency		total steps	
		$n=0$	$n>0$	$n=0$	$n>0$
0	0	0	0	0	0
1.1	1	1	1	1	1
1.2	1	1	0	1	0
2	$1 + t_{RSUM}(n-1)$	0	1	0	$1 + t_{RSUM}(n-1)$
3	1	0	0	0	0
	Total number of steps			2	$2 + t_{RSUM}(n-1)$

Table 6.2 Step table for Algorithm 6.3.

Table 6.3 corresponds to procedure *MADD* (Algorithm 6.7). It is important to note that the frequency of line 1 is $m+1$ and not m. This is so as i needs to be incremented up to $m+1$ before the loop can terminate. Similarly, the frequency for line 2 is $m \ n+1$).

line	s/e	frequency	total steps
0	0	0	0
1	1	$m+1$	$m+1$
2	1	$m(n+1)$	$mn+m$
3	1	mn	mn
4	1	mn	mn
5	1	m	m
6	1	1	1
		Total	$3mn+3m+2+m$

Table 6.3 Step table for Algorithm 6.7.

Algorithm 6.10 transposes an $n \times n$ matrix A. Recall that B is the transpose of A iff $B(i,j) = A(j,i)$ for all i and j. The steps per execution are fairly easy to obtain (see Table 6.4). The frequency for each of lines 0, 1, 5, and 6 is relatively easy to determine. Let us examine line 2. For each value of i, line 2 is executed $n-i+1$ times. So, the frequency for this line is

$$\sum_{i=1}^{n-1} (n-i+1) = (n+1)(n-1) - \sum_{1}^{n-1} i = (n+1)(n-1) - n(n-1)/2 =$$

$(n-1)(n+2)/2$. The frequency for line 3 is $\sum_{i=1}^{n-1} (n-i) = n(n-1)/2$.

The frequency for line 4 is the same as that for line 3.

line	procedure $TRANSPOSE(A,n)$
0	**declare** $A(n,n)$; **integer** n,i,j
1	**for** $i \leftarrow 1$ **to** $n-1$ **do**
2	**for** $j \leftarrow i+1$ **to** n **do**
3	$A(i,j) \Longleftrightarrow A(j,i)$
4	**end for**
5	**end for**
6	**end** $TRANSPOSE$

Algorithm 6.10

line	s/e	frequency	total steps
0	0	0	0
1	1	n	n
2	1	$(n-1)(n+2)/2$	$(n-1)(n+2)/2$
3	3	$n(n-1)/2$	$3n(n-1)/2$
4	1	$n(n-1)/2$	$n(n-1)/2$
5	1	$n-1$	$n-1$
6	1	1	1
		Total	$(n-1)(5n+2)/2+2n$

Table 6.4 Step table for Algorithm 6.10.

In some cases, the number of steps per execution of a statement is not fixed. This is the case, for example, for line 2 of procedure $INEF$ (Algorithm 6.11). Procedure $INEF$ is a very inefficient way to compute the sums $\sum_{i=1}^{j} A(i)$ for $j = 1, 2, ..., n$. SUM is Algorithm 6.2. The step count of a call to $SUM(A,n)$ has already been determined to be $3n+3$. The number of steps per execution of line 2 of procedure $INEF$ depends on the value of j. $3j+3$ steps result from the call to SUM and another from the assignment statement of line 2 itself. So, the number of steps per execution of line 2 is $3j+4$. The frequency of this line is n. But, the total number of steps resulting from this line is not $(3j+4)n$. Instead, it is $\sum_{j=1}^{n}(3j+4) = 3n(n+1)/2+4n$. Table 6.5 gives the complete step analysis for this algorithm.

line	procedure $INEF(A,B,n)$
0	declare $A(n),B(n)$; integer n,j
1	for $j \leftarrow 1$ to n do
2	$B(j) \leftarrow SUM(A,j)$
3	end for
4	end $INEF$

Algorithm 6.11

line	s/e	frequency	total steps
0	0	0	0
1	1	$n+1$	$n+1$
2	$3j+4$	n	$3n(n+1)/2+4n$
3	1	n	n
4	1	1	1
		Total	$3n(n+1)/2+6n+2$

Table 6.5 Step table for Algorithm 6.11.

In summary, the time complexity of an algorithm is given by the number of steps taken by the algorithm to compute the function it was written for. The number of steps is itself a function of the instance characteristics. While any specific instance may have several characteristics (e.g., the number of inputs, the number of outputs, the magnitudes of the inputs and outputs, etc.), the number of steps is computed as a function of some subset of these. Usually, we choose those characteristics that are of importance to us. For example, we might wish to know only how the computing time (i.e., time complexity) increases as the number of inputs increase. In this case the number of steps will be computed as a function of the number of inputs alone. For a different algorithm, we might be interested in determining how the computing time increases as the magnitude of one of the inputs increases. In this case the number of steps will be computed as a function of this inputs magnitude. Thus, before the step count of an algorithm can be determined, we need to know exactly which characteristics of the problem instance are to be used. These define the variables in the expression for the step count. In the case of procedure *SUM*, we chose to measure the time complexity as a function of the number, n, of elements being added. For procedure *MADD* the choice of characteristics was the number m of rows and the number n of columns in the matrices being added.

Once the relevant characteristics $(n,m,p,q,r,...)$ have been selected, we can define what a step is. A step is any computation unit that is independent of the characteristics $(n,m,p,q,r,...)$. Thus 10 additions could be one step; 100 multiplications could also be one step; but n additions could not. Nor could $m/2$ additions, $p+q$ subtractions etc. be counted as one step. A systematic way to assign step counts was also discussed. Once this has been done, the time complexity (i.e., the total step count) of an algorithm can be obtained using either of the two methods discussed.

The examples we have looked at so far were sufficiently simple that the time complexities were nice functions of fairly simple characteristics like the number of elements, and the number of rows and columns. For many algorithms, the time complexity is not dependent solely on the number of inputs or outputs or some other easily specified parameter. Consider procedure *INSORT* (Algorithm 5.20) for example. This algorithm sorts the n numbers $X(1)$, $X(2)$,...,$X(n)$ into non-decreasing order. A natural parameter with respect to which one might wish to determine the step count is the number, n, of elements to be sorted. I.e., we would like to know how the computing time changes as we change the number of elements n.

line	s/e	frequency	total steps
0	0	0	0
1.1	1	1	1
1.2	1	0	0
2	1	n	n
3	1	$n-1$	$n-1$
4	1	$n-1$	$n-1$
5.1	1	1	1
5.2	1	1	1
6	1	0	0
7	1	0	0
8	1	$n-1$	$n-1$
9	1	$n-1$	$n-1$
10	1	1	1
		Total	$5n+1$

Table 6.6   Step count for Algorithm 5.20 when $n>1$ and $X(1) \leqslant X(2) \leqslant ... \leqslant X(n)$.

The parameter n, however, is not adequate. For the same n, the step count varies with the relative order of the elements in X. For example, when $n=4$ and $X(1{:}4) = (2,3,6,10)$ line 5 is executed only once for each value of j (as $X(j-1) \leqslant T = X(j)$ for all j). When the elements are already sorted, the step count for procedure INSORT is $5n+1$ (see Table 6.6). In this table line 5.1 refers to the $X(i) \leqslant T$ part of the **if** statement of line 5 of procedure INSORT. Line 5.2 refers to the **then** clause of this **if**. Note that even though the **for** statement of line 4 is written as:

for $i \leftarrow j-1$ **down to** 1 **do** ;

it is executed only once for each value of j (as the loop is exited immediately from line 5.2). So, its frequency is j. Lines 6 and 7 do not get executed even once.

If the elements to be sorted are initially in the order $X(1) > X(2) > ... > X(n)$, then it will never be the case that $X(i) \leqslant T$ in line 5. In this case the frequency of line 5.2 will be 0. Line 4 will get executed j times for each j, or a total of $\sum_2^n j = n(n+1)/2 - 1$ times. Lines 5.1, 6 and 7 will get executed $n-1$ times less than line 4, or a total of $n(n-1)/2$ times. The step count for procedure $INSORT$ when $X(1) > X(2) > ... > X(n)$ is $2n^2 + 3n - 2$ (see Table 6.7).

line	s/e	frequency	total steps
0	0	0	0
1.1	1	1	1
1.2	1	0	0
2	1	n	n
3	1	$n-1$	$n-1$
4	1	$n(n+1)/2-1$	$n(n+1)/2-1$
5.1	1	$n(n-1)/2$	$n(n-1)/2$
5.2	1	0	0
6	1	$n(n-1)/2$	$n(n-1)/2$
7	1	$n(n-1)/2$	$n(n-1)/2$
8	1	$n-1$	$n-1$
9	1	$n-1$	$n-1$
10	1	1	1
		Total	$2n^2+3n-2$

Table 6.7 Step count for Algorithm 5.20 when $n>1$ and $X(1) > X(2) > ... > X(n)$.

We can extricate ourselves from the difficulties resulting from situations when the chosen parameters are not adequate to determine the step count uniquely by defining two kinds of steps counts: worst case and average.

Let A be an algorithm. Suppose we wish to determine its step count $t_A(n_1,n_2,...,n_k)$ as a function of the parameters n_1, n_2, ..., n_k. Let $f:X \rightarrow Y$ be the total function computed by A. For any $I \in X$ (I is an instance of the problem solved by A) let $STEP_A(I)$ be the number of steps needed by A to compute $f(I)$. Let $S(n_1,n_2,...,n_k)$ be the set

$$S(n_1, n_2, \ldots, n_k) = \{I \,|\, I \in X \text{ and } I \text{ has the characteristics } n_1,n_2,...,n_k\}.$$

The *worst case* step count of A, $t_A^{WC}(n_1,n_2,...,n_k)$, is a function of $n_1,n_2,...,n_k$ such that:

$$t_A^{WC}(n_1,n_2, \cdots ,n_k) = \max\{STEP_A(I)\,|\,I \in S(n_1, n_2, ..., n_k)\}.$$

The *average* step count of A, $t_A^{AVG}(n_1,n_2,...,n_k)$ is a function of $n_1,n_2, \ldots, n_k$ such that:

$$t_A^{AVG}(n_1,n_2, \cdots ,n_k) = \frac{1}{|S(n_1,n_2,...,n_k)|} \sum_{I \in S(n_1,...,n_k)} STEP_A(I)$$

Both t_A^{WC} and t_A^{AVG} are meaningful quantities to obtain as t_A^{WC} tells us the worst that can happen on a problem instance with characteristics n_1, $n_2,...$, n_k and t_A^{AVG} tells us how much time we would expect to spend on the average (or on a randomly chosen instance). The formulation for t_A^{AVG} assumes that all $I \in S$ are equally likely instances. If this is not the case then the equation needs to be modified to

$$t_A^{AVG}(n_1,n_2, \cdots ,n_k) = \sum_{I \in S(n_1,...,n_k)} p(I)STEP_A(I)$$

where p(I) is the normalized frequency (or probability, frequency/$|S(n_1,n_2, \ldots, n_k)|$) with which instance I will be solved.

It is not too difficult to see that the worst case step count t_{INSORT}^{WC} is $2n^2+3n-2$. Figuring out the average step count is somewhat harder. First, we shall determine the average frequency of each of the lines in Algorithm 5.19. The frequencies for lines 0 to 3 and 8 to 10 depend only on n (and not on any other characteristics of the ordered set $(X(1),...,X(n))$). So, the average frequencies for these lines, are the same as those given in Tables 6.6 and 6.7. The frequencies of lines 4 through 7 however depend on the ordering of elements in $(X(1),...,X(n))$. For any given j, we are trying to determine where $T=X(j)$ should be placed. There are j possibilities (see Figure 6.1). Let the j possibilities be indexed $1,2,...,j$ as in Figure 6.1. If the actual position for insertion is k then the frequency count for lines 4 to 7 is as given in Tables 6.8(a) and (b). Since each of the lines 4, 5.1,5.2,6, and 7 has a step count of 1, the total step count attributable to lines 4 to 7 is $4k-1$ when

$k < j$ and $4k+1$ when $k=j$. So, assuming each value of k is equally likely, the average step count attributable to these lines is $(\sum_{k=1}^{j-1}(4k-1)+4j+1)/j =$ $(2j^2+j+2)/j$. Thus the average total contribution of lines 4 to 7 to the step count is $\sum_{j=2}^{n}(2j^2+j+2)/j = \sum_{j=2}^{n}(2j+1+2/j) = n^2+2n-3-\sum_{2}2/j$. The average total contribution of the remaining lines (when $n>1$) is $4n-1$. So,

$$t_{INSORT}^{AVG}(n)=n^2+6n-4-\sum_{2}^{n}1/j.$$

$$
\begin{array}{cccccc}
X(1) & X(2) & & \cdots & X(j-2) & X(j-1) \\
\uparrow & \uparrow & \uparrow & \uparrow & \uparrow & \uparrow \\
j & j-1 & & 3 & 2 & 1
\end{array}
$$

Figure 6.1 The j possible positions to insert $T - X(j)$.

line	frequency	line	frequency
4	k	4	$k+1$
5.1	k	5.1	k
5.2	1	5.2	0
6	$k-1$	6	k
7	$k-1$	7	k
total	$4k-1$		$4k+1$
	(a) $k<j$		(b) $k = j$

Table 6.8 Frequencies for lines 4 to 7 of *INSORT*.

Before concluding this section, we shall look at an algorithm for which the worst case analysis is a little more involved than simply determining worst case frequencies and the number of steps per execution of each statement. Procedure *REV* (Algorithm 6.12) takes a sequence $A(1),...,A(n)$ of n numbers and prints them out reversing subsequences contained within zeroes. The zeroes are not printed out. For example, the sequence 2,3,4,0,1,6,5,9,0,8,4,2, will be output as 4,3,2,9,5,6,1,2,4,8. The worst case frequency for each of the lines of Algorithm 6.12 is given in Table 6.9.

Some of the entries in Table 6.9 need explanation. The worst case frequency for line 4 is n. This happens when $A(i) = 0$, $1 \leqslant i \leqslant n$. It also happens when $A(1:n) = (1, 0, 1, 0, 1, 0, 1, 0, ..., 1, 0)$ for n even. It is possible to execute line 5 $i-1$ times for any given i. The sequence $A(1:i)=(1,2,3,...i-1,0)$ causes this to happen (for example). However, it is incorrect to conclude from this that the frequency of line 5 is $\sum_{1}^{n}(i-1) = n(n-1)/2$. Each time line 5 is executed, a $B(\,)$ gets printed out. B values

```
line   procedure REV(A,n)
 0        declare A(n),B(n+1); integer i,k,n
 1        B(1) ← 0; k ← 1
 2        for i ← 1 to n do
 3          if A(i) = 0 then
 4                          while B(k) ≠ 0 do
 5                            print (B(k)); k ←k−1
 6                          end while
 7                       else k ← k+1; B(k) ←A(i)
 8          endif
 9        end for
10        for i ← k down to 2 do
11          print(B(k))
12        end for
13     end REV
```

Algorithm 6.12

line	s/e	frequency	total steps
0	0	0	0
1	2	1	2
2	1	$n+1$	$n+1$
3	1	n	n
4	1	n	n
5	2	$n-1$	$2n-2$
6	1	$n-1$	$n-1$
7	2	n	$2n$
8	0	0	0
9	1	n	n
10	1	$n+1$	$n+1$
11	1	n	n
12	1	n	n
13	1	1	1
		Total	$12n+2$

Table 6.9 Worst case counts for Algorithm 6.12.

are collected in line 7. This line can be executed at most n times. It does get executed n times when none of the $A(i)$s equals zero. But in this case, one cannot get to line 5 at all. So, if line 5 is ever reached then line 7 can be reached at most $n-1$ times. So, at most $n-1$ items can get into B and hence line 5 can be executed at most $n-1$ times. It is executed exactly $n-1$ times when $A(i) \neq 0$, $1 \leqslant i < n$ and $A(n) = 0$. So, the worst case frequency of line 5 is $n-1$. The worst case frequency for line 10 cannot be any more than $n+1$. Its frequency is $n+1$ when $A(i) \neq 0$, $1 \leqslant i \leqslant n$.

Now that we have determined the worst case frequency for each of the statements in Algorithm 6.12, we can compute the worst case total steps contributed by each statement. Adding these together yields the sum $12n+2$. This unfortunately is not the correct value for $t_{REV}^{WC}(n)$. This is so because if line 11 is executed n times then line 5 cannot be executed even once (otherwise we will have more than n items printed out). Also, if line 7 is executed n times then lines 4 to 6 cannot be executed even once. In other words, there is no input sequence $A(1:n)$ for which each statement in Algorithm 6.12 executes as many times as its worst case frequency. It is however true that $t_{REV}^{WC}(n) < 12n+2$. Determining the exact expression for $t_{REV}^{WC}(n)$ is quite difficult. We shall not do this as there is little more information to be obtained from the exact expression.

Whether the exact expression is $10n+14$, or $11n+20$, or $7n+16$, or $11n+18$ is not important. This is so because the notion of a step itself is not exact. Two statements, each counted as one step, need not take the same amount of computer time to execute (e.g., $A \leftarrow X$ is one step and so also is $A \leftarrow Y(i,j) + D*B*C - R \uparrow 2 + G*H/P$. The important information to be extracted from the analysis of procedure REV is that its worst case computing time grows linearly with n.

6.2 ASYMPTOTIC NOTATION (O, Ω, Θ, o)

Determining the exact step count (either worst case or average) of an algorithm could prove to be an exceedingly difficult task (try this for Algorithm 6.12 for example). Expending immense effort to determine the step count exactly isn't a very worthwhile proposition as the notion of a step is itself inexact. One motivation for determining step counts is to be able to compare the time complexities of two algorithms that compute the same function. Procedures SUM and $RSUM$ compute the same functions and their step counts are $3n+3$ and $2n+2$ respectively. What can we say about the relative efficiency of the two algorithms? About all we can say is that the two algorithms are almost equally efficient. If we examine the relative complexities of steps in the two algorithms, we would probably decide that the steps of $RSUM$ are more complex than those of SUM, so SUM is expected to have a smaller time complexity than $RSUM$.

Because of the inexactness of what a step stands for, the exact step count isn't very useful for comparative purposes. An exception to this is when the difference is very large as in $3n+3$ vs $100n+10$. We might feel quite safe in predicting that the algorithm with step count $3n+3$ will run in less time than the one with step count $100n+10$. But even in this case, it isn't necessary to know that the exact step count is $100n+10$. Something like "it's about $80n$, or $85n$, or $75n$" would have been adequate to arrive at the same conclusion.

For most situations, it is adequate to be able to make a statement like $c_1 n^2 \leqslant t_A^{WC}(n) \leqslant c_2 n^2$ or $t_B^{WC}(n,m) = c_1 n + c_2 m$ where c_1 and c_2 are nonnegative constants. This is so because if we have two algorithms with a complexity of $c_1 n^2 + c_2 n$ and $c_3 n$, respectively, then we know that the one with complexity $c_3 n$ will be faster than the one with complexity $c_1 n^2 + c_2 n$ for sufficiently large values of n. For small values of n, either algorithm could be faster (depending on c_1, c_2, and c_3). If $c_1 = 1$, $c_2 = 2$, and $c_3 = 100$ then $c_1 n^2 + c_2 n \leqslant c_3 n$ for $n \leqslant 98$ and $c_1 n^2 + c_2 n > c_3 n$ for $n > 98$. If $c_1 = 1$, $c_2 = 2$, and $c_3 = 1000$ then $c_1 n^2 + c_2 n \leqslant c_3 n$ for $n \leqslant 998$. No matter what the values of c_1, c_2, and c_3, there will be an n beyond which the algorithm with complexity $c_3 n$ will be faster than the one with complexity $c_1 n^2 + c_2 n$. This value of n will be called the *break even point*. If the break even point is 0 then the algorithm with complexity $c_3 n$ is always faster (or at least as fast). The exact break even point cannot be determined analytically. The algorithms have to be programmed and run on a computer in order to determine the break even point. To know that there is a break even point it is adequate to know that one algorithm has complexity $c_1 n^2 + c_2 n$ and the other $c_3 n$ for some constants c_1, c_2, and c_3. There is little advantage to determining the exact values of c_1, c_2, and c_3.

With the previous discussion as motivation, we introduce some terminology that will enable us to make meaningful (but inexact) statements about the complexity (both time and space) of an algorithm.

Definition: [Asymptotic notation] $f(n) = O(g(n))$ (read as "f of n is big oh of g of n") iff there exist positive constants c and n_0 such that $f(n) \leqslant cg(n)$ for all n, $n \geqslant n_0$. $f(n) = \Omega(g(n))$ (read as "f of n is omega of g of n") iff there exist positive constants c and n_0 such that $f(n) \geqslant cg(n)$ for all n, $n \geqslant n_0$. $f(n) = \Theta(g(n))$ (read as "f of n is theta of g of n") iff there exist positive constants c_1, c_2, and n_0 such that $c_1 g(n) \leqslant f(n) \leqslant c_2 g(n)$ for all n, $n \geqslant n_0$. $f(n) = o(g(n))$ (read as "f of n is little oh of g of n") iff $\displaystyle\lim_{n \to \infty} f(n)/g(n) = 1$. $\square$

The definitions of O, Ω, Θ, and o are easily extended to include functions of more than one variable. For example, $f(n,m) = O(g(n,m))$ iff there exist positive constants c, n_0, and m_0 such that $f(n,m) \leqslant cg(n,m)$ for all $n \geqslant n_0$ and all $m \geqslant m_0$.

Example 6.1: $3n+2 = O(n)$ as $3n+2 \leqslant 4n$ for all $n \geqslant 2$. $3n+3 = O(n)$ as $3n+3 \leqslant 4 n$ for all $n \geqslant 3$. $100n+6 = O(n)$ as $100n+6 \leqslant 101n$ for $n \geqslant 6$. $10n^2+4n+2 = O(n^2)$ as $10n^2+4n+2 \leqslant 11n^2$ for $n \geqslant 5$. $1000n^2+100n-6 = O(n^2)$ as $1000n^2+ 100n-6 \leqslant 1001n^2$ for $n \geqslant 100$. $6*2^n+n^2 = O(2^n)$ as $6*2^n+n^2 \leqslant 7 *2^n$ for $n \geqslant 4$. $3n+3 = O(n^2)$ as $3n+3 \leqslant 3n^2$ for $n \geqslant 2$. $10n^2+4n+2 = O(n^4)$ as $10n^2+4n+2 \leqslant 10n^4$ for $n \geqslant 2$. $3n+2 \neq O(1)$ as

$3n+2$ is not less than or equal to c for any constant c and all n, $n \geq n_0$. $10n^2+4n+2 \neq O(n)$. $\square$

As illustrated by the previous example, the statement $f(n) = O(g(n))$ only states that $g(n)$ is an upper bound on the value of $f(n)$ for all n, $n \geq n_0$. It doesn't say anything about how good this bound is. Notice that $n = O(n^2)$, $n = O(n^{2.5})$, $n = O(n^3)$, $n = O(2^n)$, etc. In order for the statement $f(n) = O(g(n))$ to be informative, $g(n)$ should be as small a function of n as one can come up with for which $f(n) = O(g(n))$. So, while we shall often say $3n+3 = O(n)$, we shall almost never say $3n+3 = O(n^2)$ even though the latter statement is correct.

From the definition of O, it should be clear that $f(n) = O(g(n))$ is not the same as $O(g(n)) = f(n)$. In fact, it is meaningless to say that $O(g(n)) = f(n)$, or that $\Omega(g(n)) = f(n)$, or that $\Theta(g(n)) = f(n)$. The use of the '$=$' symbol is unfortunate as this symbol commonly denotes an equivalence relation. The use of '$=$' in connection with O, Ω, and Θ is irreflexive. Some of the confusion that might result from the use of this symbol (which is standard terminology) can be avoided by reading the symbol '$=$' as 'is' and not as 'equals'.

Theorem 6.1 obtains a very useful result concerning the order of $f(n)$ (i.e., the $g(n)$ in $f(n) = O(g(n))$) when $f(n)$ is a polynomial in n.

Theorem 6.1: If $f(n) = a_m n^m + ... + a_1 n + a_0$, then $f(n) = O(n^m)$.

Proof: $f(n) \leq \sum_{i=0}^{m} |a_i| n^i$

$$\leq n^m \sum_0^m |a_i| n^{i-m}$$

$$\leq n^m \sum_0^m |a_i|, \text{ for } n \geq 1.$$

So, $f(n) = O(n^m)$. $\square$

Example 6.2: $3n+2 = \Omega(n)$ as $3n+2 \geq 3n$ for $n \geq 1$ (actually the inequality holds for $n \geq 0$ but the definition of Ω requires an $n_0 > 0$). $3n+3 = \Omega(n)$ as $3n+3 \geq 3n$ for $n \geq 1$. $100n+6 = \Omega(n)$ as $100n+6 \geq 100n$ for $n \geq 1$. $10n^2+4n+2 = \Omega(n^2)$ as $10n^2+4n+2 \geq n^2$ for $n \geq 1$. $6*2^n+n^2 = \Omega(2^n)$ as $6*2^n+n^2 \geq 2^n$ for $n \geq 1$. Observe also that $3n+3 = \Omega(1)$; $10n^2+4n+2 = \Omega(n)$; $10n^2+4n+2 = \Omega(1)$; $6*2^n+n^2 = \Omega(n^{100})$; $6*2^n+n^2 = \Omega(n^{50.2})$; $6*2^n+n^2 = \Omega(n^2)$; $6*2^n+n^2 = \Omega(n)$; and $6*2^n+n^2 = \Omega(1)$. $\square$

As in the case of the "big oh" notation, there are several functions $g(n)$ for which $f(n) = \Omega(g(n))$. $g(n)$ is only a lower bound on $f(n)$. For the statement $f(n) = \Omega(g(n))$ to be informative, $g(n)$ should be as large a function of n as possible for which the statement $f(n) = \Omega(g(n))$ is true. So, while we shall say that $3n+3 = \Omega(n)$ and that $6*2^n + n^2 = \Omega(2^n)$, we shall almost never say that $3n+3 = \Omega(1)$ or that $6*2^n + n^2 = \Omega(1)$ even though both these statements are correct.

Theorem 6.2 is the analogue of Theorem 6.1 for the omega notation.

Theorem 6.2: If $f(n) = a_m n^m + ... + a_1 n + a_0$ and $a_m > 0$, then $f(n) = \Omega(n^m)$.

Proof: Left as an exercise. $\square$

Example 6.3: $3n+2 = \Theta(n)$; $3n+3 = \Theta(n)$; $10n^2 + 4n + 2 = \Theta(n^2)$; $6*2^n + n^2 = \Theta(2^n)$; and $10*\log n + 4 = \Theta(\log n)$. $3n+2 \neq \Theta(1)$; $3n+3 \neq \Theta(n^2)$; $10n^2 + 4n + 2 \neq \Theta(n)$; $10n^2 + 4n + 2 \neq \Theta(1)$; $6*2^n + n^2 \neq \Theta(n^2)$; $6*2^n + n^2 \neq \Theta(n^{100})$; and $6*2^n + n^2 \neq \Theta(1)$. $\square$

The theta notation is more precise than both the "big oh" and omega notations. $f(n) = \Theta(g(n))$ iff $g(n)$ is both an upper and lower bound on $f(n)$.

Notice that the coefficients in all of the $g(n)$s used in the preceding three examples have been 1. This is in accordance with practice. We shall almost never find ourselves saying that $3n+3 = O(3n)$, or that $10 = O(100)$, or that $10n^2 + 4n + 2 = \Omega(4n^2)$, or that $6*2^n + n^2 = \Omega(6*2^n)$, or that $6*2^n + n^2 = \Theta(4*2^n)$, even though each of these statements is true.

Theorem 6.3: If $f(n) = a_m n^m + ... + a_1 n + a_0$ and $a_m > 0$, then $f(n) = \Theta(n^m)$.

Proof: Left as an exercise. $\square$

Example 6.4: $3n+2 = o(3n)$; $4n+6 = o(4n)$; $3n+2 \neq o(2n)$; $4n+6 \neq o(8n)$; $10n^2 + 4n + 2 = o(10n^2)$; $10n^2 + 4n + 2 \neq o(12n^2)$; $6*2^n + n^2 = o(6*2^n)$; $6*2^n + n^2 \neq o(8*2^n)$. $\square$

Theorem 6.4: If $f(n) = a_m n^m + ... + a_1 n + a_0$ and $a_m > 0$, then $f(n) = o(a_m n^m)$.

Proof: Left as an exercise. $\square$

The following theorem is useful in computations involving asymptotic notation.

Theorem 6.5: The following are true for every real number x, $x > 0$ and every real ϵ, $\epsilon > 0$:

(1) There exists an n_0 such that $(\log n)^x < (\log n)^{x+\epsilon}$ for every n, $n \geq n_0$.

(2) There exists an n_0 such that $(\log n)^x < n$ for every n, $n \geq n_0$.

(3) There exists an n_0 such that $n^x < n^{x+\epsilon}$ for every n, $n \geq n_0$.

(4) For every real y, there exists an n_0 such that $n^y (\log n)^y < n^{x+\epsilon}$ for every n, $n \geq n_0$.

(5) There exists an n_0 such that $n^x < 2^n$ for every n, $n \geq n_0$.

Proof: Follows from the definition of the individual functions. $\square$

Example 6.5: From Theorem 6.5, we obtain the following: $n^3 + n^2 \log n = \Theta(n^3)$; $2^n/n^2 = \Omega(n^k)$ for every natural number k ; $n^4 + n^{2.5} \log^{20} n = \Theta(n^4)$; $2^n n^4 \log^3 n + 2^n n^4/\log n = \Theta(2^n n^4 \log^3 n)$. $\square$

Table 6.10 lists some of the more useful identities involving the "big oh", omega, and theta notations. In this table, all symbols other than n are positive constants. Some of these identities will be proved correct in the exercises. The correctness of the others should be immediately apparent. Table 6.11 lists some useful inference rules for sums and products.

If you have studied Tables 6.10 and 6.11, then you are ready to see how asymptotic notation can be used to describe the time complexity (or step count) of an algorithm. Let us restudy the time complexity analyses of the previous section. For procedure SUM (Algorithm 6.2) we had determined that $t_{SUM}(n) = 3n + 3$. So, $t_{SUM}(n) = \Theta(n)$; $t_{RSUM}(n) = 2n + 2 = \Theta(n)$; $t_{MADD}(m,n) = 3mn + 3n + 2 = \Theta(mn)$; $t_{TRANSPOSE}(n) = (n-1)(5n+2)/2 + 2n = \Theta(n^2)$; and $t_{INEF}(n) = 3n(n+1)/2 + 6n + 2 = \Theta(n^2)$.

For procedure $INSORT$, we had determined that $t_{INSORT}^{WC}(n) = 2n^2 + 3n - 2 = \Theta(n^2)$. From the expression for t_{INSORT}^{WC} it also follows that $t_{INSORT}(n) = O(n^2)$. This last equality states that the computing time of procedure $INSORT$ is bounded by cn^2 for some positive constants c and n_0, and all inputs of size n, $n \geq n_0$. Note that $t_{INSORT}(n)$ is really a multivalued function as the value of $t(n)$ is different for different instances of n elements. Also, $t_{INSORT}(n) = \Omega(n)$ as the best situation for $INSORT$ is when $X(1) \leq ... \leq X(n)$. For this case, $t_{INSORT}(n) = 5n + 1$. $t_{INSORT}^{AVG}(n) = n^2 + 6n - 4 - \sum_{2}^{n} 1/j = \Theta(n^2)$.

	f(n)	Asymptotic
E1	c	$\oplus (1)$
E2	$\sum_{0}^{k} c_i n^i$	$\oplus (n^k)$
E3	$\sum_{1}^{n} i$	$\oplus (n^2)$
E4	$\sum_{1}^{n} i^2$	$\oplus (n^3)$
E5	$\sum_{1}^{n} i^k, \ k>0$	$\oplus (n^{k+1})$
E6	$\sum_{0}^{n} r^i$	$\oplus (r^n)$
E7	$n!$	$\oplus (\sqrt{n}(n/e)^n)$
E8	$\sum_{i=1}^{n} 1/i$	$\oplus (\log n)$

$\oplus$ can be any one of O, Ω, and Θ

Table 6.10

I1 $\{f(n) = \oplus (g(n))\} \models \sum_{n=a}^{b} f(n) = \oplus (\sum_{n=a}^{b} g(n))$

I2 $\{f_i(n) = \oplus (g_i(n)), \ 1 \leqslant i \leqslant k\} \models \sum_{1}^{k} f_i(n) = \oplus (\max_{1 \leqslant i \leqslant k} \{g_i(n)\})$

I3 $\{f_i(n) = \oplus (g_i(n)), \ 1 \leqslant i \leqslant k\} \models \prod_{1}^{k} f_i(n) = \oplus (\prod_{1}^{k} g_i(n))$

I4 $\{f_1(n) = O(g_1(n)), \ f_2(n) = \Theta(g_2(n))\}$
$\models f_1(n) + f_2(n) = O(g_1(n) + g_2(n))$

I5 $\{f_1(n) = \Theta(g_1(n)), \ f_2(n) = \Omega(g_2(n))\}$
$\models f_1(n) + f_2(n) = \Omega(g_1(n) + g_2(n))$

I6 $\{f_1(n) = O(g(n)), \ f_2(n) = \Theta(g(n))\}$
$\models f_1(n) + f_2(n) = \Theta(g(n))$

Table 6.11 Inference rules for $\oplus \in \{O, \Omega, \Theta\}$

While we might all see that the O, Ω, and Θ notations have been used correctly in the preceding paragraph, we are still left with the question: Of what use are these notations if one has to first determine the step count exactly? The answer to this question is that the asymptotic complexity (i.e., the complexity in terms of O, Ω, and Θ) can be determined quite easily without determining the exact step count. This is usually done by first determining the asymptotic complexity of each statement (or group of statements) in the algorithm and then adding up these complexities. Tables 6.12 to 6.18 do just this for procedures *SUM, RSUM, MADD, TRANSPOSE, INEF, INSORT*, and *REV*.

line	s/e	frequency	total steps
0	0	0	$\Theta(0)$
1	1	1	$\Theta(1)$
2	1	$n+1$	$\Theta(n)$
3	1	n	$\Theta(n)$
4	1	n	$\Theta(n)$
5	1	1	$\Theta(1)$
6	1	0	$\Theta(0)$

$$t_{SUM}(n) = \Theta\left(\max_{0 \le i \le 6} \{g_i(n)\}\right) = \Theta(n)$$

Table 6.12 Asymptotic complexity of *SUM*.

line	s/e	frequency $n=0$	frequency $n>0$	total steps $n=0$	total steps $n>0$
0	0	0	0	0	$\Theta(0)$
1.1	1	1	1	1	$\Theta(1)$
1.2	1	1	0	1	$\Theta(0)$
2	$2+t_{RSUM}(n-1)$	0	1	0	$\Theta(2+t_{RSUM}(n-1))$
3	1	0	0	0	$\Theta(0)$

$$t_{RSUM}(n) = \begin{cases} c + t_{RSUM}(n-1) & n>0 \\ 2 & n=0 \end{cases}$$

Table 6.13 Asymptotic complexity of *RSUM*.

line	s/e	frequency	total steps
0	0	0	$\Theta(0)$
1 and 5	1	$\Theta(m)$	$\Theta(m)$
2 to 4	1	$\Theta(mn)$	$\Theta(mn)$
6	1	1	$\Theta(1)$
		$t_{MADD}(m,n) =$	$\Theta(mn)$

Table 6.14 Asymptotic complexity of *MADD*.

line(s)	s/e	frequency	total steps
0	0	0	$\Theta(0)$
1 and 5	1	$\Theta(n)$	$\Theta(n)$
2 to 4	1	$\Theta(\sum_{i=1}^{n-1}(n-1)) = \Theta(n^2)$	$\Theta(n^2)$
6	1	1	$\Theta(1)$
		$t_{TRANSPOSE}(n) =$	$\Theta(n^2)$

Table 6.15 Asymptotic complexity of *TRANSPOSE*.

line	s/e	frequency	total steps
0	0	0	$\Theta(0)$
1 and 3	1	$\Theta(n)$	$\Theta(n)$
2	$3j+4$	n	$\Theta(\sum_{1}^{n}(3j+4)) = \Theta(n^2)$
4	1	1	$\Theta(1)$

$$t_{inef}(n) = \Theta(n^2)$$

Table 6.16 Asymptotic complexity of INEF

In the table for *RSUM*, *c* is a constant. Note that in the table for *MADD*, lines 1 and 5 have been lumped together even though they have different frequencies. This lumping together of these two lines is possible because their frequencies are of the same order. For the same reason, it is possible to lump together lines 2 through 4.

line	s/e	frequency	total steps
0	0	0	$\Theta(0)$
1.1	1	1	$\Theta(1)$
1.2	0	0	$\Theta(0)$
2,3,8,9	$\Theta(1)$	$\Theta(n)$	$\Theta(n)$
4 to 7	$\Theta(1)$	$\Omega(n), O(\sum_{2}^{n} j = O(n^2)$	$\Omega(n), O(n^2)$
10	1	1	$\Theta(1)$

$$t_{INSORT}(n) = \Omega(n)$$
$$t_{INSORT}(n) = O(n^2)$$

Table 6.17 Asymptotic complexity of $INSORT$ $(n > 1)$.

line(s)	s/e	frequency	total steps
0 and 8	0	0	$\Theta(0)$
1 and 13	$\Theta(1)$	1	$\Theta(1)$
2,3, and 9	$\Theta(1)$	$\Theta(n)$	$\Theta(n)$
4 to 6	$\Theta(1)$	$O(n)$	$O(n)$
7	$\Theta(1)$	$O(n)$	$O(n)$
10 to 12	$\Theta(1)$	$O(n)$	$O(n)$
		$t_{REV}(n) =$	$\Theta(n)$

Table 6.18 Asymptotic complexity of REV.

In Table 6.15 there is an apparent misuse of notation. This is actually not the case because by the equality $\Theta(\sum_{i=1}^{n}(n-i) = \Theta(n^2))$, we mean something different from when we say $f(n) = \Theta(g(n))$. Indeed, $\Theta(\sum_{1}^{n}(n-i))$ is not a function. By the equality $\Theta(g_1(n)) = \Theta(g_2(n))$ we mean that both $\Theta(g_1(n))$ and $\Theta(g_2(n))$ give the asymptotic frequency of lines 2 to 4 (i.e., $f(n) = \Theta(g_1(n))$ iff $f(n) = \Theta(g_2(n))$). Another explanation can be provided by interpreting $O(g(n))$, $\Omega(g(n))$, and $\Theta(g(n))$ as being the sets defined below:

$$O(g(n))=\{f(n)\,|\,f(n)=O(g(n))\}$$
$$\Omega(g(n))=\{f(n)\,|\,f(n)=\Omega(g(n))\}$$
$$\Theta(g(n))=\{f(n)\,|\,f(n)=\Theta(g(n))\}.$$

Under this interpretation, it is meaningful to make statements such as $O(g_1(n))=O(g_2(n))$; $\Theta(g_1(n))=\Theta(g_2(n))$; etc. When using this interpretation, it is also convenient to read $f(n) = O(g(n))$ as "f of n is in (or is a member of) big oh of g of n", etc.

The asymptotic complexity table for procedure *INSORT* (Table 6.17) contains both the upper and lower bounds (O and Ω) on the frequencies. This enables us to obtain both an upper and a lower bound on $t_{INSORT}(n)$. In Table 6.18 we have used inference rule I7 to obtain $t_{REV}(n) = O(n)$. Note that the asymptotic analysis of procedure *REV* is much simpler than its exact analysis.

While all the analyses of Tables 6.12 to 6.18 were actually carried out in terms of step counts, it is correct to interpret $t_A(n) = \Theta(g(n))$, or $t_A(n) = O(g(n)$ or $t_A(n) = \Omega(g(n))$ as a statement about the computing time of algorithm A. This is so because each step takes only $\Theta(1)$ time to execute.

The asymptotic notation "o" is missing from our examples on asymptotic analysis. This notation is seldom used to analyze the time complexity (or even to obtain the step count) of an algorithm as it is too precise a statement. The statement $t_A(n) = o(10n)$ means that algorithm A needs $o(10n)$ time units to execute. Such a statement, of course, cannot be made until A is programmed and run on a machine and we determine the function that represents its computing time exactly. Even if this were done, we would have to say something like $t_A(n) = o(10n)$ when A is programmed by Marge Jablonskii in Pascal and run on a CDC 6400 computer using version 4.6 of the Pascal compiler! The "little oh" notation could be used for step counts. The step count for procedure *SUM* is $3n+3 = o(3n)$. It is wasteful to use a precise notation such as "little oh" when counting an imprecise quantity such as a step.

6.3 PRACTICAL COMPLEXITIES

We have seen that the time complexity of an algorithm is generally some function of the instance characteristics. This function is very useful in determining how the time requirements vary as the instance characteristics change. The complexity function may also be used to compare two algorithms A and B that perform the same task. Assume that algorithm A has complexity $\Theta(n)$ and algorithm B is of complexity $\Theta(n^2)$. We can assert that algorithm A is faster than algorithm B for "sufficiently large" n. To see the validity of this assertion, observe that the actual computing time of A is bounded from above by n for some constant c and for all n, $n \geqslant n_1$ while that of B is bounded from below by dn^2 for some constant d and all n, $n \geqslant n_2$. Since $cn \leqslant dn^2$ for $n \geqslant c/d$, algorithm A is faster than algorithm B whenever $n \geqslant \max\{n_1, n_2, c/d\}$.

One should always be cautiously aware of the presence of the phrase "sufficiently large" in the assertion of the preceding discussion. When deciding which of the two algorithms to use, we must know whether the n we are dealing with is in fact "sufficiently large". If algorithm A actually runs in $10^6 n$ milliseconds while algorithm B runs in n^2 milliseconds and if we always have $n \leqslant 10^6$, then algorithm B is the one to use.

To get a feel for how the various functions grow with n, you are advised to study Table 6.19 and Figure 6.3 very closely. As is evident from the table and the figure, the function 2^n grows very rapidly with n. In fact, if an algorithm needs 2^n steps for execution, then when $n=40$ the number of steps needed is approximately $1.1*10^{12}$. On a computer performing one billion steps per second, this would require about 18.3 minutes. If $n=50$, the same algorithm would run for about 13 days on this computer. When $n=60$, about 36.56 years will be required to execute the algorithm and when $n=100$, about $4*10^{13}$ years will be needed. So, we may conclude that the utility of algorithms with exponential complexity is limited to small n (typically $n \leqslant 40$).

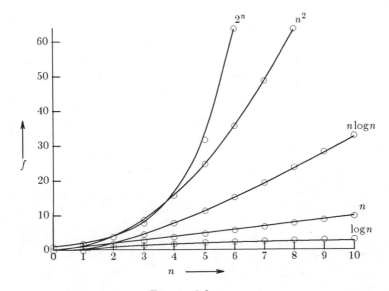

Figure 6.3

$\log n$	n	$n \log n$	n^2	n^3	2^n
0	1	0	1	1	2
1	2	2	4	8	4
2	4	8	16	64	16
3	8	24	64	512	256
4	16	64	256	4096	65536
5	32	160	1024	32768	4294967296

Table 6.19

Algorithms that have a complexity that is a polynomial of high degree are also of limited utility. For example, if an algorithm needs n^{10} steps, then using our one billion steps per second computer we will need 10 seconds when $n=10$; 3,171 years when $n=100$; and $3.17*10^{13}$ years when $n=1000$. If the algorithm's complexity had been n^3 steps instead, then we would need one second when $n=1000$, 16.67 minutes when $n=10,000$; and 11.57 days when $n=100,000$.

Table 6.20 gives the time needed by a one billion instructions per second computer to execute a program of complexity $f(n)$ instructions. One should note that currently only the fastest computers can execute about one billion instructions per second. From a practical standpoint, it is evident that for reasonably large n (say $n > 100$) only algorithms of small complexity (such as n, $n \log n$, n^2, n^3, etc.) are feasible. Further, this is the case even if one could build a computer capable of executing 10^{12} instructions per second. In this case, the computing times of Table 6.20 would decrease by a factor of 1000. Now, when $n = 100$ it would take 3.17 years to execute n^{10} instructions, and $4*10^{10}$ years to execute 2^n instructions.

Time for f(n) instr. on a 10^9 instr/sec computer

n	$f(n)=n$	$f(n)=n\log_2 n$	$f(n)=n^2$	$f(n)=n^3$	$f(n)=n^4$	$f(n)=n^{10}$	$f(n)=2^n$
10	$.01\mu s$	$.03\mu s$	$.1\mu s$	$1\mu s$	$10\mu s$	10sec	$1\mu s$
20	$.02\mu s$	$.09\mu s$	$.4\mu s$	$8\mu s$	$160\mu s$	2.84hr	1ms
30	$.03\mu s$	$.15\mu s$	$.9\mu s$	$27\mu s$	$810\mu s$	6.83d	1sec
40	$.04\mu s$	$.21\mu s$	$1.6\mu s$	$64\mu s$	2.56ms	121.36d	18.3min
50	$.05\mu s$	$.28\mu s$	$2.5\mu s$	$125\mu s$	6.25ms	3.1yr	13d
100	$.10\mu s$	$.66\mu s$	$10\mu s$	1ms	100ms	3171yr	$4*10^{13}$yr
1,000	$1.00\mu s$	$9.96\mu s$	1ms	1sec	16.67min	$3.17*10^{13}$yr	$32*10^{283}$yr
10,000	$10.00\mu s$	$130.3\mu s$	100ms	16.67min	115.7d	$3.17*10^{23}$yr	
100,000	$100.00\mu s$	1.66ms	10sec	11.57d	3171yr	$3.17*10^{33}$yr	
1,000,000	1.00ms	19.92ms	16.67min	31.71yr	$3.17*10^7$yr	$3.17*10^{43}$yr	

μs = microsecond = 10^{-6}seconds
ms = millisecond = 10^{-3}seconds
sec = seconds
min = minutes
hr = hours
d = days
yr = years

Table 6.20

REFERENCES AND SELECTED READINGS

The study of algorithms and their mathematical analysis was given great impetus by Donald E. Knuth, most specifically in his projected seven volume work:

The art of computer programming
Volume 1, Fundamental Algorithms (1968)
Volume 2, Seminumerical Algorithms (1969)
Volume 3, Sorting and Searching (1973)
Volume 4, Combinatorial Search and Recursion (to appear)
Volume 5, Syntactical Algorithms (to appear)
Volume 6, Theory of languages (to appear)
Volume 7, Compilers (to appear)
Addison-Wesley, Massachusetts.

Some entertaining articles on the analysis of algorithms are:

Algorithms, by D. Knuth, Scientific American, April 1977.

Computer science and its relation to mathematics, by D. Knuth, Amer. Math. Monthly, April, 1974.

Our definitions of O, Ω, and Θ are due to D. Knuth and appear in

Big omichron, big omega, and big theta, by D. Knuth, SIGACT News, ACM, April 1976.

Additional ideas on obtaining step counts appear in

On the analysis of algorithms by W. Paul, Fakultat für Mathematik der Universitat Bielefeld, Germany, 1978.

The following books contain several algorithms for which asymptotic analyses are also provided:

Fundamentals of computer algorithms, by E. Horowitz and S. Sahni, Computer Science Press, Inc., Maryland, 1978.

Fundamentals of data structures, by E. Horowitz and S. Sahni, Computer Science Press, Inc., Maryland, 1976.

The design and analysis of computer algorithms, by A. Aho, J. Hopcroft, and J. Ullman, Addison-Wesley, Massachusetts, 1974.

Combinatorial algorithms, by E. Reingold, J. Nievergelt, and N. Deo, Prentice Hall, New Jersey, 1977.

Several additional analyses can be found in papers appearing in J.ACM and SICOMP.

EXERCISES

1. (a) Introduce statements to increment COUNT at all appropriate points in Algorithm 6.13.

> **procedure** $D(X,n)$
> **integer** $X(n),n,i$
> **for** $i \leftarrow 1$ **to** n **by** 2 **do**
> $X(i) \leftarrow X(i)+2$
> **end for**
> $i \leftarrow 1$
> **while** $i \leqslant n/2$ **do**
> $X(i) \leftarrow X(i)+X(i+1)$
> $i \leftarrow i+1$
> **end while**
> **end** D

Algorithm 6.13

(b) Simplify the resulting algorithm by eliminating statements. The simplified algorithm should compute the same value for *COUNT* as computed by the algorithm of (a).

(c) What is the exact value of *COUNT* when the algorithm terminates?

(d) Obtain the step count for Algorithm 6.13 using the frequency method. Clearly show the step count table.

2. Do Exercise 1(d) for procedure *TRANSPOSE* (Algorithm 6.10).

3. Do Exercise 1(d) for procedure *INEF* (Algorithm 6.11)

4. Do Exercise 1 for Algorithm 6.14. This algorithm multiplies two $n \times n$ matrices A and B.

> **procedure** $MMUL(A,B,C,n)$
>   **declare** $A(n,n),B(n,n),C(n,n)$; **integer** i,j,k,n
>   **for** $i \leftarrow 1$ **to** n **do**
>   **for** $j \leftarrow 1$ **to** n **do**
> $C(i,j) \leftarrow 0$
>   **for** $k \leftarrow 1$ **to** n **do**
> $C(i,j) \leftarrow C(i,j)+A(i,k)*B(k,j)$
>   **end for**
>   **end for**
>   **end for**
> **end** $MMUL$

Algorithm 6.14

(a) Do Exercise 1 for Algorithm 6.15. This algorithm multiplies two matrices A and B where A is an $m \times n$ matrix and B is an $n \times p$ matrix.

```
line procedure MPROD (A,B,C,m,n,p)
  0    declare A (m,n),B (n,p),C (m,p)
  1    integer m,n,p,i,j,k
  2    for i ← 1 to m do
  3      for j ← 1 to p do
  4        C (i,j) ← 0
  5        for k ← 1 to n do
  6          C (i,j) ← C (i,j)+A (i,k)*B (k,j)
  7        end for
  8      end for
  9    end for
 10  end MPROD
```

Algorithm 6.15

(b) Under what conditions will it be profitable to interchange lines 2 and 3?

6. Show that the following equalities are correct (do not use Theorems 6.1, 6.2, 6.3 or Table 6.10).

(a) $5n^2-6n = \Theta(n^2)$
(b) $n! = O(n^n)$
(c) $2n^2 2^n + n\log n = \Theta(n^2 2^n)$
(d) $\displaystyle\sum_{i=0}^{n} i^2 = \Theta(n^3)$
(e) $\displaystyle\sum_{i=0}^{n} i^3 = \Theta(n^4)$
(f) $2^{2^n}+6*2^n = \Theta(2^{2^n})$
(g) $n^3+10^6 n^2 = \Theta(n^3)$
(h) $6n^3/(\log n+1) = O(n^3)$
(i) $n^{1.001}+n\log n = \Theta(n^{1.001})$
(j) $n^{k+\epsilon}+n^k \log n = \Theta(n^{k+\epsilon})$ for all k and ϵ, $k \geqslant 0$, and $\epsilon > 0$.
(k) $10n^3+15n^4+100n^2 2^n = o(100n^2 2^n)$

7. Show that the following equalities are incorrect:

(a) $10n^2+9 = O(n)$
(b) $n^2 \log n = \Theta(n^2)$
(c) $n^2/\log n = \Theta(n^2)$
(d) $n^3 2^n + 6n^2 3^n = O(n^3 2^n)$

8. Prove Theorems 6.2, 6.3, and 6.4.

9. Prove that equivalences E5 through E8 (Table 6.10) are correct.

10. Prove the correctness of inference rules I1 to I6 (Table 6.11).

11. Obtain the asymptotic time complexity of Algorithm 6.13. Set up a frequency table similar to Tables 6.12 to 6.17.

12. Do Exercise 11 for Algorithm 6.14.

13. Do Exercise 11 for Algorithm 6.15.

14. (a) Show that procedure *QUICKSORT* (Algorithm 5.22) exhibits its worst case behavior (for both time and space) when the input set S is such that one of $S1$ and $S2$ is empty at all levels of the recursion.
    (b) What is the worst case space complexity of procedure *QUICKSORT*?
    (c) Show that procedure *QUICKSORT* takes minimum time when $|S_1| \simeq |S_2|$ at all levels of the recursion.
    (d) Is the space complexity of *QUICKSORT* affected by the nature of the input? If so, how?
    (e) How would you obtain an iterative version of procedure *QUICKSORT* that has a worst case space complexity of $O(\log n)$ where $|S| = n$?

15. (a) Show that procedure *SEL* (Algorithm 5.23) exhibits its worst case time behavior when $k = 1$ and $|S_2| = 0$ at all levels of the recursion.
    (b) Under what conditions does *SEL* exhibit its best case behavior?
    (c) What is the worst case space complexity of procedure *SEL*?
    (d) What is its best case space complexity?

16. Is it necessary for an algorithm to exhibit its worst case time behavior and worst case space behavior at the same time (i.e., for the same input)? Prove your answer.

CHAPTER 7

RECURRENCE RELATIONS

7.1 INTRODUCTION

The computing time of an algorithm (particularly a recursive algorithm) is often easily expressed recursively (i.e., in terms of itself). This was the case, for instance, for procedure *RSUM*. We had determined that $t_{RSUM}(n) = c + t_{RSUM}(n-1)$ where c is some constant. Let $t()$ denote the worst case time complexity of procedure *MERGE* (Algorithm 5.21). The worst case computing time, $t_M^w(n)$, of Algorithm 5.22 is easily seen to satisfy the inequality:

$$t_M^w(n) \leqslant \begin{cases} c_1 & n=1 \\ t_M^w(\lceil n/2 \rceil) + t_M^w(\lfloor n/2 \rfloor) & n>1 \\ + t(\lceil n/2 \rceil, \lfloor n/2 \rfloor) + c_2 \end{cases}$$

An analysis of procedure *MERGE* reveals that:

$$t(a,b) = \Theta(a+b)$$

Substituting this into the inequality for t_M^w, we get:

$$t_M^w(n) \leqslant \begin{cases} c_1 & n=1 \\ t_M^w(\lceil n/2 \rceil) + t_M^w(\lfloor n/2 \rfloor) + c_4 n & n>1 \end{cases} \tag{7.1}$$

We expect the recurrence (7.1) to be difficult to solve because of the presence of the ceiling and floor functions. If we attempt to solve (7.1) only for values of n that are a power of 2 ($n=2^k$), then (7.1) becomes:

$$t_M^w(n) \leqslant \begin{cases} c_1 & n=1 \\ 2t_M^w(n/2) + c_4 n & n>1 \text{ and a power of 2} \end{cases} \tag{7.2}$$

If the inequality of (7.2) is converted to the equality:

$$t_M'(n) = \begin{cases} c_1 & n=1 \\ 2t_M'(n/2)+c_4n & n>1 \text{ and a power of } 2 \end{cases} \tag{7.3}$$

then $t_M'(n)$ is an upper bound on $t_M^W(n)$. So, if $t_M'(n) = f(n)$ then $t_M^w(n) = O(f(n))$. Since it is also the case that there exist constants c_5 and c_6 such that:

$$t_M^w(n) \geqslant \begin{cases} c_5 & n=1 \\ 2t_M^w(n/2)+c_6n & n>1 \text{ and } n \text{ a power of } 2 \end{cases}$$

it follows that $t_M^w(n) = \Omega(f(n))$. Hence, $t_M^w(n) = \Theta(f(n))$.

The entire discussion concerning the worst case complexity t_M^w can be repeated with respect to the best case complexity (i.e. the minimum time spent on any input of n numbers). The conclusion is that $t_M^b(n) = \Theta(f(n))$. Since both the best and worst case complexities are $\Theta(f(n))$, it follows that $t_M^a(n) = \Theta(f(n))$ and $t_M(n) = \Theta(f(n))$.

Let us analyze procedure QUICKSORT (Algorithm 5.23). The selection of a random number in the range 1 to n and the partitioning of S into $S1$ and $S2$ can be done in $\Theta(n)$ time. The assignment $X \leftarrow S1|y|S2$ can be done in $\Theta(1)$ time. So,

$$t_Q(n) = \begin{cases} c_1 & n \leqslant 1 \\ c_2n + t_Q(|S1|) + t_Q(|S2|) & n>1 \end{cases} \tag{7.4}$$

In (7.4), Q has been used as an abbreviation for QUICKSORT. $|S1|$ can be any number in the range 0 to $n-1$. Since y is randomly selected, $|S1|$ equals each of 0, 1, ... $n-1$ with equal probability. So, for the average complexity of QUICKSORT, we obtain:

$$t_Q^a(n) = \begin{cases} c_1 & n \leqslant 1 \\ c_2n + \dfrac{1}{n}(\sum\limits_{i=1}^{n-1}[t_Q^a(i)+t_Q^a(n-i)]) & n>1 \end{cases} \tag{7.5}$$

$$= \begin{cases} c_1 & n \leqslant 1 \\ c_2n + \dfrac{2}{n}\sum\limits_{i=1}^{n-1}t_Q^a(i) & n>1 \end{cases}$$

The worst case for *QUICKSORT* is when one of $S1$ and $S2$ is empty at all levels of the recursion (see the exercises). In this case, we obtain the recursion:

$$t_Q^w(n) = \begin{cases} c_1 & n \leqslant 1 \\ c_2 n + t_Q^w(n-1) & n > 1 \end{cases} \tag{7.6}$$

The best case for *QUICKSORT* is when $|S1| \simeq |S2|$ at all levels of the recursion (see the exercises). The recurrence for this case is:

$$t_Q^b(n) = \begin{cases} c_1 & n \leqslant 1 \\ c_2 n + t_Q^b(\lceil \frac{n-1}{2} \rceil) + t_Q^b(n-1-\lceil \frac{n-1}{2} \rceil) & n > 1 \end{cases} \tag{7.7}$$

A function $g(n)$ such that $t_Q^b(n) = \Theta(g(n))$ for n a power of 2 can be obtained by solving the recurrence:

$$g(n) = t_Q'(n) = \begin{cases} c_1 & n \leqslant 1 \\ c_2 n + 2 t_Q'(n/2) & n > 1 \text{ and a power of 2} \end{cases} \tag{7.8}$$

For procedure *SEL* (Algorithm 5.24), it can be shown (see the exercises) that the worst case is when $k=1$ and $|S2| = 0$ at all levels of the recursion. So, the worst case computing time of *SEL* is given by the recurrence:

$$t_{SEL}^w(n) = \begin{cases} c_1 & n=1 \\ c_2 n + t_{SEL}^w(n-1) & n > 1 \end{cases} \tag{7.9}$$

To obtain the recurrence for the average computing time of *SEL*, we need to introduce some new functions. First, we shall assume that all the elements in X are distinct. Let $t^k(n)$ be the average time to find the kth smallest element in $X, |X| = n$. This average is taken over all $n!$ permutations of the elements in X. The average computing time of *SEL* is given by:

$$t_{SEL}^a(n) = \frac{1}{n} \sum_{k=1}^{n} t^k(n)$$

Define $R(n)$ to be the largest $t^k(n)$. I.e.,

$$R(n) = \max_{1 \leqslant k \leqslant n} \{t^k(n)\}$$

It is easy to see that $t_{SEL}^a(n) \leqslant R(n)$.

    With these definitions in mind, let us proceed to analyze procedure *SEL* for the case when all elements in X are distinct. Since x_i is chosen at random, there is an equal probability that $|S1| = j = 0, 1, 2, ..., n-1$. This leads to the following inequality for $t^k(n)$:

$$t^k(n) \leqslant \begin{cases} c & n=1 \\ cn + \dfrac{1}{n}[\displaystyle\sum_{k<j\leqslant n} t^k(j-1) + \sum_{1\leqslant j<k} t^{k-j}(n-j)] & n\geqslant 2 \end{cases}$$

From this, we conclude that:

$$R(n) \leqslant cn + \frac{1}{n}\max_k\{\sum_{k<j\leqslant n} R(j-1) + \sum_{1\leqslant j<k} R(n-j)\}, n\geqslant 2$$

$$= cn + \frac{1}{n}\max_k\{\sum_{k}^{n-1} R(i) + \sum_{n-k+1}^{n-1} R(i)\}, n\geqslant 2$$

Since R is an increasing function of n, it can be shown (see the exercises) that:

$$R(n) \leqslant \begin{cases} c & n=1 \\ 2.5c & n=2 \\ cn + \dfrac{2}{n}\displaystyle\sum_{i=n/2}^{n-1} R(i) & n \text{ even and } n>2 \\ cn + \dfrac{2}{n}\displaystyle\sum_{i=(n+1)/2}^{n-1} R(i) \text{ else} \end{cases} \qquad (7.10)$$

    If $R(n) = \Theta(f(n))$, then it follows from our earlier observation $(t_{SEL}^a(n) \leqslant R(n))$ that $t_{SEL}^a(n) = O(f(n))$. We shall later see that $R(n) = \Theta(n)$. This together with the observation $t_{SEL}^a(n) = \Omega(n)$ will lead to the conclusion that $t_{SEL}^a(n) = \Theta(n)$.

Even though procedure *BINSRCH* (Algorithm 2.2) is not a recursive algorithm, its worst case time complexity is best described by a recurrence relation. It is not too difficult to see that the following recurrence is correct:

$$
t_B^w(n) = \begin{cases} c_1 & n=1 \\ c_2 + t_B^w(\lceil \dfrac{n-1}{2} \rceil) & n>1 \end{cases}
\tag{7.11}
$$

When n is a power of 2, (7.11) simplifies to:

$$
t_B^w(n) = \begin{cases} c_1 & n=1 \\ c_2 + t_B^w(n/2) & n>1 \text{ and a power of 2} \end{cases}
\tag{7.12}
$$

Hopefully, these examples have convinced you that recurrence relations are indeed useful in describing the time complexity of both iterative and recursive algorithms. In each of the above examples, the recurrence relations themselves were easily obtained. Having obtained the recurrence, we must now solve it to determine the asymptotic growth rate of the time complexity. We shall consider four methods of solving recurrence relations:

(a) substitution
(b) induction
(c) characteristic roots
(d) generating functions.

7.2 SUBSTITUTION

In the substitution method of solving a recurrence relation for $f(n)$, the recurrence for $f(n)$ is repeatedly used to eliminate all occurrences of $f()$ from the right hand side of the recurrence. Once this has been done, the terms in the right hand side are collected together to obtain a compact expression for $f(n)$. The mechanics of this method are best described by means of examples.

Example 7.1: Consider the recurrence:

$$
t(n) = \begin{cases} c_1 & n=0 \\ c_2 + t(n-1) & n \geq 1 \end{cases}
\tag{7.13}
$$

When $c_1 = c_2 = 2$, $t(n)$ is the recurrence for the step count of procedure *RSUM*. If $n > 2$ then $t(n-1) = c_2 + t(n-2)$. If $n > 3$ then $t(n-2) = c_2 + t(n-3)$ etc. These equalities are immediate consequences of (7.13) and are used in the following derivation of a nonrecursive expression for $t(n)$:

$$t(n) = c_2 + t(n-1)$$

$$= c_2 + c_2 + t(n-2)$$

$$= c_2 + c_2 + c_2 + t(n-3)$$

$$\cdot$$
$$\cdot$$
$$\cdot$$

$$= c_2 n + t(0)$$

$$= c_2 n + c_1, \; n \geqslant 0$$

So, we see that $t(n) = c_2 n + c_1$, $n \geqslant 0$. From this, we obtain $t_{RSUM}(n) = 2n + 2$. $\square$

Example 7.2: Recurrence (7.12) may be solved by substitution. Observe that (7.12) defines $t_B^w(n)$ only for values of n that are a power of 2. If n is not a power of 2, then the value of $t(n)$ is not defined by (7.12). For example, if $n = 5$ then $t_B^w(2.5)$ appears on the right hand side of (7.12). But t_B^w is a function whose domain is the natural numbers. If $n = 6$, then from (7.12) we obtain (t is used as an abbreviation for t_B^w):

$$t(6) = c_2 + t(3)$$

$$= c_2 + c_2 + t(1.5)$$

But, $t(1.5)$ is undefined. When n is a power of 2, $t(n)$ is always defined (i.e., using the recurrence 7.12). $t(n)$ is of course defined for all $n \in N - \{1\}$ when (7.11) is used. Assuming n is a power of 2 (say, $n = 2^k$), the substitution method leads to the following series of equalities:

$$t(n) = c_2 + t(n/2)$$

$$= c_2 + c_2 + t(n/4)$$

$$= c_2 + c_2 + c_2 + t(n/8)$$

$$\cdot$$
$$\cdot$$
$$\cdot$$

$$= kc_2 + t(n/2^k)$$

$$= kc_2 + t(1)$$

$$= c_1 + kc_2$$

$$= c_1 + c_2 \log n, \; n \text{ a power of } 2$$

Unless otherwise specified, all logarithms in this chapter are base 2. At this point, we only have an expression for $t_B^w(n)$ for values of n that are a power of 2. If n is between 2^k and 2^{k+1}, then $t_B^w(n)$ will be between $t(2^k)$ and $t(2^{k+1})$. So, $c_1 + c_2 \lfloor \log n \rfloor \leqslant t_B^w \leqslant c_1 + c_2 \lceil \log n \rceil$ for all n. This implies that $t_B^w(n) = \Theta(\log n)$. $\square$

Example 7.3: Consider the recurrence:

$$t(n) = \begin{cases} c & n=1 \\ a*t(n/b)+cn & n \geqslant 2 \end{cases} \tag{7.14}$$

This recurrence defines $t(n)$ only for values of n that are a power of b. The recurrence (7.3) is the same as (7.14) when $c_1 = c_4 = c$, and $a = b = 2$. Even though (7.3) is not an instance of (7.14) (as $c_1 \neq c_4$ in general), the solution to (7.14) with $a = 2$ and $b = 2$ does give us a function $g(n)$ such that $t_M'(n) = O(g(n))$. And, from the discussion following (7.3) it should be clear that $t_M^w(n) = \Theta(g(n))$. When $c = c_1$, and $a = b = 2$, (7.14) becomes the same as (7.8).

Assume that $n = b^k$ for some natural number k. Solving (7.14) by the substitution method yields:

$$t(n) = a*t(n/b)+cn$$

$$= a[a*t(n/b^2)+c(n/b)]+cn$$

$$= a^2 t(n/b^2)+cn(a/b)+cn$$

$$= a^2[a*t(n/b^3)+c(n/b^2)]+cn[a/b+1]$$

$$= a^3 t(n/b^3)+cn(a^2/b^2)+cn[a/b+1]$$

$$= a^3[a*t(n/b^4)+c(n/b^3)]+cn[a^2/b^2+a/b+1]$$

$$= a^4 t(n/b^4)+cn[a^3/b^3+a^2/b^2+a/b+1]$$

.

.

.

$$= a^k t(n/b^k) + cn \sum_{i=0}^{k-1} (a/b)^i$$

$$= a^k t(1) + cn \sum_{i=0}^{k-1} (a/b)^i$$

$$= a^k c + cn \sum_{i=0}^{k-1} (a/b)^i$$

$$= (a/b)^k cn + cn \sum_{i=0}^{k-1} (a/b)^i, \ (b^k = b^{\log_b n} = n)$$

$$= cn \sum_{i=0}^{k} (a/b)^i.$$

When $a = b$, $\sum_{0}^{k} (a/b)^i = k+1$. When $a \neq b$, $\sum_{i=0}^{k} (a/b)^i = ((a/b)^{k+1} - 1)/(a/b - 1)$. If $a < b$ then $a/b < 1$ and $((a/b)^{k+1} - 1)/(a/b - 1) = (1 - (a/b)^{k+1})/(1 - a/b) < 1/(1 - a/b)$. So, $\sum_{0}^{k} (a/b)^i = \Theta(1)$. When $a > b$, $((a/b)^{k+1} - 1)/(a/b - 1) = \Theta((a/b)^k) = \Theta(a^k/b^{\log_b n}) = \Theta(a^{\log_b n}/n) = \Theta(n^{\log_b a}/n)$. So, we obtain:

$$t(n) = \begin{cases} \Theta(n) & a < b \\ \Theta(n \log n) & a = b \\ \Theta(n^{\log_b a}) & a > b \end{cases}$$

From this and our earlier discussion, we conclude that $t_M^w(n) = \Theta(n \log n)$ and $t_Q^b(n) = \Theta(n \log n)$. $\square$

Recurrence (7.14) is a very frequently occurring recurrence form in the analysis of algorithms. It often occurs with the cn term replaced by such terms as c, or cn^2, or cn^3 etc. So, we would like to extend the result of Example 7.7 and obtain a general form for the solution of the recurrence:

$$t(n) = a * t(n/b) + g(n), \ n \geq b \text{ and } n \text{ a power of } b, \tag{7.15}$$

where a and b are known constants. We shall assume that $t(1)$ is also known. Clearly, (7.15) reduces to (7.14) when $t(1) = c$ and $g(n) = cn$. Using the substitution method, we obtain:

$$t(n) = a*t(n/b)+g(n)$$

$$= a[a*t(n/b^2)+g(n/b)]+g(n)$$

$$= a^2t(n/b^2)+ag(n/b)+g(n)$$

.

.

.

$$= a^kt(1)+\sum_{i=0}^{k-1}a^ig(n/b^i)$$

where $k = \log_b n$. This equation may be further simplified as below:

$$t(n) = a^kt(1)+\sum_{i=0}^{k-1}a^ig(n/b^i)$$

$$= a^kt(1)+\sum_{i=0}^{k-1}a^ig(b^{k-i})$$

$$= a^k[t(1)+\sum_{j=1}^{k}a^{-j}g(b^j)]$$

Since $a^k = a^{\log_b n} = n^{\log_b a}$, the expression for $t(n)$ becomes:

$$t(n) = n^{\log_b a}[t(1)+\sum_{j=1}^{k}a^{-j}g(b^j)]$$

$$= n^{\log_b a}[t(1)+\sum_{j=1}^{k}\{g(b^j)/(b^j)^{\log_b a}\}]$$

$$= n^{\log_b a}[t(1)+\sum_{j=1}^{k}h(b^j)]$$

where $h(n) = g(n)/n^{\log_b a}$. Our final form for $t(n)$ is:

$$t(n)=n^{\log_b a}[t(1)+f(n)] \tag{7.16}$$

where $f(n) = \sum_{j=1}^{k}h(b^j)$ and $h(n) = g(n)/n^{\log_b a}$. Table 7.1 tabulates the asymptotic value of $f(n)$ for various $h(n)$s. This table together with (7.16) allows one to easily obtain the asymptotic value of $t(n)$ for many of the recurrences one encounters when analyzing algorithms.

$h(n)$	$f(n)$
$O(n^r)$, $r < 0$	$O(1)$
$\Theta((\log n)^i)$, $i \geqslant 0$	$\Theta(((\log n)^{i+1})/(i+1))$
$\Omega(n^r)$, $r > 0$	$\Theta(h(n))$

Table 7.1 $f(n)$ values for various $h(n)$ values.

Let us consider some examples using this table. The recurrence for t_B^w when n is a power of 2 is:

$$t(n) = t(n/2) + c_2$$

and $t(1) = c_1$. Comparing with (7.15), we see that $a=1$, $b=2$, and $g(n) = c_2$. So, $\log_b(a) = 0$ and $h(n) = g(n)/n^{\log_b a} = c_2 = c_2(\log n)^0 = \Theta((\log n)^0)$. From Table 7.1, we obtain:

$$f(n) = \Theta(\log n).$$

So,

$$t(n) = n^{\log_b a}(c_1 + \Theta(\log n))$$

$$= \Theta(\log n) \text{ (as } t(n) > 0)$$

For the recurrence

$$t(n) = 7t(n/2) + 18n^2, \ n \geqslant 2 \text{ and } n \text{ a power of 2,}$$

we obtain: $a=7$, $b=2$, and $g(n) = 18n^2$. So, $\log_b a = \log_2 7 \simeq 2.81$ and $h(n) = 18n^2/n^{\log_2 7} = 18n^{2-\log_2 7} = O(n^r)$ where $r = 2 - \log_2 7 < 0$. So, $f(n) = O(1)$. The expression for $t(n)$ is:

$$t(n) = n^{\log_2^7}(t(1) + O(1))$$

$$= \Theta(n^{\log_2 7}) \text{ (as } t(n) > 0)$$

as $t(1)$ may be assumed to be constant.

As a final example, consider the recurrence:

$t(n) = 9t(n/3)+4n^6$, $n \geqslant 3$ and a power of 3.

Comparing with (7.15), w obtain $a=9$, $b=3$, and $g(n) = 4n^6$. So, $\log_b a = 2$ and $h(n) = 4n^6/n^2 = 4n^4 = \Omega(n^4)$. From Table 7.1, we see that $f(n) = \Theta(h(n)) = \Theta(n^4)$. So,

$t(n) = n^2(t(1)+\Theta(n^4))$

$= \Theta(n^6)$ (as $t(n) > 0$)

as $t(1)$ may be assumed constant.

7.3 INDUCTION

Induction is more of a verification method than a solution method. If we have an idea as to what the solution to a particular recurrence is then we can verify it by providing a proof by induction.

Example 7.4: Induction can be used to show that $t(n) = 3n+2$ is the solution to the recurrence:

$$t(n) = \begin{cases} 2 & n=0 \\ 3+t(n-1) & n>0 \end{cases}$$

For the induction base, we see that when $n=0$, $t(n) = 2$ and $3n+2 = 2$. Assume that $t(n) = 3n+2$ for some n, $n = m$. For the induction step, we shall show that $t(n) = 3n+2$ when $n = m+1$. From the recurrence for $t(n)$, we obtain $t(m+1) = 3+t(m)$. But from the induction hypothesis $t(m) = 3m+2$. So, $t(m+1) = 3+3m+2 = 3(m+1)+2$. $\square$

Example 7.5: Consider recurrence (7.10). We shall show that $R(n)<4cn$, $n\geqslant 1$.

Induction Base: For $n=1$, and 2, (7.10) yields: $R(1)\leqslant c< 4\ cn$ and $R(2)\leqslant 2.5c<4cn$.

Induction Hypothesis: Let m be an arbitrary natural number, $m\geqslant 3$. Assume that $R(n)<4cn$ for all n, $1\leqslant n<m$.

Induction Step: For $n = m$ and m even, (7.10) gives:

$$R(m) \leq cm + \frac{2}{m} \sum_{m/2}^{m-1} R(i)$$

$$< cm + \frac{8c}{m} \sum_{m/2}^{m-1} i \text{ (from the IH)}$$

$$< 4cm$$

When m is odd, (7.10) yields:

$$R(m) \leq cm + \frac{2}{m} \sum_{(m+1)/2}^{m-1} R(i)$$

$$< cm + \frac{8c}{m} \sum_{(m+1)/2}^{m-1} i$$

$$< 4cm$$

Since $R(n) < 4cn$, $R(n) = O(n)$. Hence, the average computing time of procedure *SEL* is $O(n)$. Since procedure *SEL* spends at least n units of time on each input of size n, $t_{SEL}^{a}(n) = \Omega(n)$. Combining these two, we get $t_{SEL}^{a}(n) = \Theta(n)$. $\square$

Example 7.6: Consider recurrence (7.9). Let $t(n)$ denote $t_{SEL}^{w}(n)$. We shall show that $t(n) = c_2 n(n+1)/2 + c_1 - c_2 = \Theta(n^2)$.

Induction Base: When $n=1$, (7.9) yields $t(n) = c_1$. Also, $c_2 n(n+1)/2 + c_1 - c_2 = c_1$.

Induction Hypothesis: Let m be an arbitrary natural number. Assume that $t(n) = c_2 n(n+1)/2 + c_1 - c_2$ when $n=m$.

Induction Step: When $n=m+1$, (7.9) yields:

$$t(m+1) = c_2(m+1) + t(m)$$

$$= c_2(m+1) + c_2 m(m+1)/2 + c_1 - c_2 \text{ (from the IH)}$$

$$= [2c_2(m+1) + c_2 m(m+1)]/2 + c_1 - c_2$$

$$= c_2(m+1)(m+2)/2 + c_1 - c_2. \quad \square$$

As mentioned earlier, the induction method cannot be used to find the solution to a recurrence relation. It can be used only to verify that a candidate solution is correct.

7.4 CHARACTERISTIC ROOTS

The recurrence relation for $f(n)$ is a *linear recurrence relation* iff it is of the form:

$$f(n) = \sum_{i=1}^{k} g_i(n) f(n-i) + g(n)$$

where the $g_i(n)$, $1 \leqslant i \leqslant k$, and $g(n)$ are functions of n but not of f. A linear recurrence relation is of *order k* iff it is of the form:

$$f(n) = \sum_{i=1}^{k} g_i(n) f(n-i) + g(n)$$

where k is a constant and $g_k(n)$ is not identically equal to zero. If $g_k(n) \equiv 0$ then the recurrence is of order less than k. A linear recurrence of order k is of *constant coefficients* iff there exist constants $a_1, a_2, ..., a_k$ such that $g_i(n) \equiv a_i$, $1 \leqslant i \leqslant k$. In this sub-section, we shall be concerned only with the solution of linear recurrence relations of order k that have constant coefficients. These recurrences are of the form:

$$f(n) = a_1 f(n-1) + a_2 f(n-2) + ... + a_k f(n-k) + g(n), n \geqslant k \quad (7.17)$$

where $a_k \neq 0$ and $g(n)$ is a function of n but not of f. (7.17) is a *homogeneous recurrence relation* iff $g(n) \equiv 0$. One may readily verify that for any set $f(0)$, $f(1)$, ..., $f(k-1)$, of initial values, the recurrence (7.17) uniquely determines $f(k)$, $f(k+1)$, ...

Many of the recurrence relations we have considered so far are (or can be transformed into equivalent) linear recurrence relations with constant coefficients. Using $t(n)$ to denote $t_M(n)$, the recurrence (7.3) takes the form:

$$t(n) = \begin{cases} c_1 & n=1 \\ 2t(n/2) + c_4 n & n \geqslant 2 \ \text{and a power of 2} \end{cases} \quad (7.18)$$

This isn't a linear recurrence relation of order k for any fixed k because of the occurrence of $t(n/2)$ on the right hand side. However, since n must be a power of 2, (7.18) may be rewritten as:

$$t(2^k) = \begin{cases} c_1 & k=0 \\ 2t(2^{k-1}) + c_4 2^k & k \geqslant 1 \end{cases} \quad (7.19)$$

Using $h(k)$ to denote $t(2^k)$, (7.19) becomes:

$$h(k) = \begin{cases} c_1 & k=0 \\ 2h(k-1)+c_4 2^k & k \geq 1 \end{cases} \qquad (7.20)$$

Recurrence (7.20) is readily seen to be a linear recurrence relation with constant coefficients. It is of order 1 and it is not homogeneous. Since $h(k) = t(2^k) = t(n)$ for n a power of 2, solving (7.20) is equivalent to solving (7.3).

Recurrence (7.5) is a linear recurrence relation. It is however not of order k for any fixed k. By performing some algebra on (7.5), we can transform it into an order 1 linear recurrence. We shall use $t(n)$ as an abbreviation for $t_Q^a(n)$. With this, (7.5) becomes (for $n > 1$):

$$t(n)=c_2 n+\frac{2}{n}\sum_{i=1}^{n-1} t(i) \qquad (7.21)$$

Multiplying (7.21) by n, we obtain:

$$nt(n)=c_2 n^2+2\sum_{1}^{n-1} t(i) \qquad (7.22)$$

Substituting $n-1$ for n in (7.22), we get:

$$(n-1)t(n-1)=c_2(n-1)^2+2\sum_{1}^{n-2} t(i) \qquad (7.23)$$

Subtracting (7.23) from (7.22) yields:

$$nt(n)-(n-1)t(n-1)=[2n-1]c_2+2t(n-1)$$

or

$$nt(n)=[2n-1]c_2+(n+1)t(n-1)$$

or

$$t(n) = \frac{n+1}{n}t(n-1) + (2 - \frac{1}{n})c_2 \qquad (7.24)$$

Recurrence (7.24) is not a linear recurrence with constant coefficients. Nonetheless, it can be solved fairly easily (see the exercises). Recurrence (7.8) can be transformed into an equivalent constant coefficient linear recurrence of order 1 in much the same way we transformed recurrence (7.3). (7.9) is already in the form of (7.17). The recurrence:

$$F(n) = F(n-1) + F(n-2), n \geq 2$$

defines the Fibonacci numbers when the initial values $F(0) = 0$ and $F(1) = 1$ are used. This is an order 2 homogeneous constant coefficient linear recurrence.

Linear recurrences of the form (7.17) are common occurrences in the analysis of algorithms. These recurrences can be solved by first obtaining a general solution for $f(n)$. This solution contains some unspecified constants and has the property that for any given set $f(0), f(1),...,f(k-1)$ of initial values, we can assign values to the unspecified constants such that the general solution defines the unique sequence $f(0), f(1),...$.

Consider the recurrence $f(n) = 5f(n-1) - 6f(n-2), n \geq 2$. Its general solution is $f(n) = c_1 2^n + c_2 3^n$ (we shall see later how to determine this). The unspecified constants are c_1 and c_2. If we are given that $f(0) = 0$ and $f(1) = 1$, then we substitute into $c_1 2^n + c_2 3^n$ to determine c_1 and c_2. Doing this, we get:

$$f(0) = c_1 + c_2 = 0$$

$$f(1) = 2c_1 + 3c_2 = 1$$

Solving for c_1 and c_2, we get $c_1 = -c_2 = -1$. So, $f(n) = 3^n - 2^n$, $n \geq 0$, is the solution to the recurrence $f(n) = 5f(n-1) - 6f(n-2)$, $n \geq 2$ when $f(0) = 0$ and $f(1) = 1$.   If we change the initial values to $f(0) = 0$ and $f(1) = 10$, then we get:

$$f(0) = c_1 + c_2 = 0$$

$$f(1) = 2c_1 + 3c_2 = 10$$

and so, $c_1 = -c_2 = -10$. Hence, $f(n) = 10(3^n - 2^n)$, $n \geq 0$.

The general solution to any recurrence of the form (7.17) can be represented as the sum of two solutions $f_h(n)$ and $f_p(n)$. $f_h(n)$ is the general solution to the homogeneous part of (7.17), i.e.

$$f_h(n) = a_1 f_h(n-1) + a_2 f_h(n-2) + ... + a_k f_h(n-k).$$

$f_p(n)$ is a particular solution for:

$$f_p(n) = a_1 f_p(n-1) + a_2 f_p(n-2) + ... + a_k f_p(n-k) + g(n).$$

While at first glance, it might appear sufficient to determine $f_p(n)$, it should be noted that $f_p(n) + f_h(n)$ is also a solution to (7.17). Since the methods used to determine $f_p(n)$ will give us an $f_p(n)$ form that does not explicity contain all the zeroes of $f(n)$ (i.e. all solutions to $f(n) - \sum_{i=1}^{k} a_i f(n-i) = 0$), it is necessary to determine $f_h(n)$ and add it to $f_p(n)$ to get the general solution to $f(n)$.

Solving For $f_h(n)$

To determine $f_h(n)$ we need to solve a recurrence of the form:

$$f_h(n) = a_1 f_h(n-1) + a_2 f_h(n-2) + ... + a_k f_h(n-k)$$

or

$$f_h(n) - a_1 f_h(n-1) - a_2 f_h(n-2) - ... - a_k f_h(n-k) = 0. \qquad (7.25)$$

We might suspect that (7.25) has a solution of the form $f_h(n) = Ax^n$. Substituting this into (7.25), we obtain:

$$A(x^n - a_1 x^{n-1} - a_2 x^{n-2} - \cdots - a_k x^{n-k}) = 0$$

We may assume that $A \neq 0$. So, we obtain:

$$x^{n-k}(x^k - \sum_{i=1}^{k} a_i x^{k-i}) = 0$$

The above equation has n roots. Because of the term x^{n-k}, $n-k$ of these roots are 0. The remaining k roots are roots of the equation:

$$x^k - a_1 x^{k-1} - a_2 x^{k-2} - ... - a_k = 0 \qquad (7.26)$$

(7.26) is called the *characteristic equation* of (7.25). From elementary polynomial root theory, we know that (7.26) has exactly k roots $r_1, r_2, ..., r_k$. Let r_1, r_2, ..., r_k be the k roots of this equation.

The roots of the characteristic equation

$$x^2-5x+6=0 \qquad (7.27)$$

are $r_1=2$ and $r_2=3$ (as 2^2-5*2+6=0 and 3^2-5*3+6=0). The characteristic equation:

$$x^3-8x^2+21x-18=0 \qquad (7.28)$$

has roots $r_1 = 2$, $r_2 = 3$, and $r_3 = 3$. As is evident, the roots of a characteristic equation need not be distinct. A root r_i is of *multiplicity j* iff $|\{a: r_a = r_i\}| = j$. The roots of (7.27) are distinct. So, they are all of multiplicity 1. 3 is a root of multiplicity 2 for (7.28) while 2 is a root of multiplicity 1. The distinct roots of (7.28) are 2 and 3. Theorem 7.1 tells us how to determine the general solution to a linear homogeneous recurrence relation of the form (7.25) from the roots of its characteristic equation.

Theorem 7.1: Let the distinct roots of the characteristic equation:

$$x^k-a_1x^{k-1}-a_2x^{k-2}-...-a_{k-1}x-a_k = 0$$

of the linear homogeneous recurrence relation:

$$f_h(n) = a_1f_h(n-1)+a_2f_h(n-2)+...+a_kf_h(n-k), a_k \neq 0$$

be $t_1, t_2, ..., t_s$ where $s \leqslant k$. There is a general solution $f_h(n)$ which is of the form:

$$f_h(n)=u_1(n)+u_2(n)+...+u_s(n)$$

where

$$u_i(n) = (c_{i_0}+c_{i_1}n+c_{i_2}n^2+...+c_{i_{w-1}}n^{w-1})t_i^n$$

and w is the multiplicity of the root t_i.

Proof: See the references for some texts where a proof of this theorem can be found. □

The characteristic equation for the recurrence $f(n)=5f(n-1)-6f(n-2), n \geqslant 2$ is $x^2-5x+6=0$. Recall that the roots of $ax^2+bx+c = 0$ are:

$$\frac{-b \pm \sqrt{b^2 - 4ac}}{2a}$$

So, the roots of our characteristic equation are $r_1 = 2$ and $r_2 = 3$. The distinct roots are $t_1 = 2$ and $t_2 = 3$. From Theorem 7.1, it follows that $f(n) = u_1(n) + u_2(n)$ where $u_1(n) = c_1 2^n$ and $u_2(n) = c_2 3^n$. So, $f(n) = c_1 2^n + c_2 3^n$.

(7.28) is the characteristic equation for the homogeneous recurrence:

$$f(n) = 8f(n-1) - 21f(n-2) + 18f(n-3)$$

Its distinct roots are $t_1 = 2$ and $t_2 = 3$. t_2 is a root of multiplicity 2. So, $u_1(n) = c_1 2^n$ and $u_2(n) = (c_2 + c_3 n) 3^n$. The general solution to this recurrence is therefore $f(n) = c_1 2^n + (c_2 + c_3 n) 3^n$.

The recurrence for the Fibonacci numbers is homogeneous and has the characteristic equation $x^2 - x - 1 = 0$. Its roots are $r_1 = (1 + \sqrt{5})/2$ and $r_2 = (1 - \sqrt{5})/2$. The roots are distinct. So, $u_1(n) = c_1((1 + \sqrt{5})/2)^n$ and $u_2(n) = c_2(1 - \sqrt{5})/2)^n$. Therefore,

$$F(n) = c_1 \left[\frac{1 + \sqrt{5}}{2} \right]^n + c_2 \left[\frac{1 - \sqrt{5}}{2} \right]^n$$

is a general solution to the Fibonacci recurrence. Using the initial values $F(0) = 0$ and $F(1) = 1$, we get $c_1 + c_2 = 0$ and $c_1(1 + \sqrt{5})/2 + c_2(1 - \sqrt{5})/2 = 1$. Solving for c_1 and c_2, we get $c_1 = -c_2 = 1/\sqrt{5}$. So the Fibonacci numbers satisfy the equality:

$$F(n) = \frac{1}{\sqrt{5}} \left[\frac{1 + \sqrt{5}}{2} \right]^n - \frac{1}{\sqrt{5}} \left[\frac{1 - \sqrt{5}}{2} \right]^n.$$

Theorem 7.1 gives us a straightforward way to determine a general solution for an order k linear homogeneous recurrence with constant coefficients. We need only determine the roots of its characteristic equation.

Solving For $f_p(n)$

There is no known general method to obtain the particular solution $f_p(n)$. The form of $f_p(n)$ depends very much on the form of $g(n)$. We shall consider only two cases. One where $g(n)$ is a polynomial in n and the other where $g(n)$ is an exponential function of n.

When $g(n) = 0$, the particular solution is $f_p(n) = 0$.

When $g(n) = \sum_{i=0}^{d} e_i n^i$, and $e_d \neq 0$, the particular solution is of the form:

$$f_p(n) = p_0 + p_1 n + p_2 n^2 + \cdots + p_{d+m} n^{d+m} \qquad (7.29)$$

where $m = 0$ if 1 is not a root of the characteristic equation corresponding to the homogeneous part of (7.17). If 1 is a root of this equation, then m equals the multiplicity of the root 1.

To determine $p_0, p_1, \ldots, p_{d+m}$, we merely substitute the right hand side of (7.29) into the recurrence for $f_p(\,)$; compare terms with like powers of n on the left and right hand side of the resulting equation and solve for $p_0, p_1, p_2, \ldots, p_{d+m}$.

As an example, consider the recurrence:

$$f(n) = 3f(n-1) + 6f(n-2) + 3n + 2 \qquad (7.30)$$

$g(n) = 3n + 2$. The characteristic equation is $x^2 - 3x - 6 = 0$. 1 is not one of its roots. So, the particular solution is of the form:

$f_p(n) = p_0 + p_1 n$.

Substituting into (7.30), we obtain:

$$p_0 + p_1 n = 3(p_0 + p_1 (n-1)) + 6(p_0 + p_1(n-2)) + 3n + 2$$

$$= 3p_0 + 3p_1 n - 3p_1 + 6p_0 + 6p_1 n - 12p_1 + 3n + 2$$

$$= (9p_0 - 15p_1 + 2) + (9p_1 + 3)n.$$

Comparing terms on the left and right hand sides, we see that:

$$p_0 = 9p_0 - 15p_1 + 2$$

and

$$p_1 = 9p_1 + 3.$$

So, $p_1 = -3/8$ and $p_0 = -61/64$. The particular solution for (7.30) is therefore:

$$f_p(n) = -\frac{61}{64} - \frac{3}{8}n$$

Consider the recurrence:

$$f(n) = 2f(n-1) - f(n-2) - 6 \tag{7.31}$$

The corresponding characteristic equation is $x^2 - 2x + 1 = 0$. Its roots are $r_1 = r_2 = 1$. So, $f_p(n)$ is of the form:

$$f_p(n) = p_0 + p_1 n + p_2 n^2$$

Substituting into (7.31), we obtain:

$$p_0 + p_1 n + p_2 n^2$$
$$= 2(p_0 + p_1 n - p_1 + p_2(n^2 - 2n + 1)) - p_0 - p_1 n + 2p_1 - p_2(n^2 - 4n + 4) - 6$$
$$= (2p_0 - 2p_1 + 2p_2 - p_0 + 2p_1 - 4p_2 - 6) + (2p_1 - 4p_2 - p_1 + 4p_2)n + p_2 n^2$$
$$= (p_0 - 2p_2 - 6) + p_1 n + p_2 n^2$$

Comparing terms, we get:

$$p_0 = p_0 - 2p_2 - 6$$

or

$$p_2 = -3.$$

So, $f_p(n) = p_0 + p_1 n - 3n^2$. $f_h(n) = (c_0 + c_1 n)(1)^n$. So, $f(n) = c_0 + c_1 n + p_0 + p_1 n - 3n^2 = c_2 + c_3 n - 3n^2$. c_2 and c_3 can be determined once the intial values $f(0)$ and $f(1)$ have been specified.

When $g(n)$ is of the form ca^n where c and a are constants, then the particular solution $f_p(n)$ is of the form:

$$f_p(n) = (p_0 + p_1 n + p_2 n^2 + \cdots + p_w n^w)a^n$$

where w is 0 if a is not a root of the characteristic equation corresponding to the homogeneous part of (7.17) and equals the multiplicity of a otherwise.

Consider the recurrence:

$$f(n)=3f(n-1)+2f(n-4)-6*2^n \qquad (7.32)$$

The corresponding homogeneous recurrence is:

$$f_h(n)=3f_h(n-1)+2f_h(n-4)$$

Its characteristic equation is:

$$x^4-3x^3-2=0$$

We may verify that 2 is not a root of this equation. So, the particular solution to (7.32) is of the form:

$$f_p(n)-p_02^n.$$

Substituting this into (7.32), we obtain:

$$p_02^n=3p_02^{n-1}+2p_02^{n-4}-6*2^n$$

Dividing out by 2^{n-4}, we obtain:

$$16p_0 = 24p_0+2p_0-96 = 26p_0-96$$

So, $p_0 = 96/10 = 9.6$. The particular solution to (7.32) is $f_p(n) = 9.6*2^n$
      The characteristic equation corresponding to the homogeneous part of the recurrence:

$$f(n)=5f(n-1)-6f(n-2)+4*3^n \qquad (7.33)$$

is $x^2-5x+6 = 0$. Its roots are $r_1 = 2$ and $r_2 = 3$. Since 3 is a root of multiplicity 1 of the characteristic equation, the particular solution is of the form:

$$f_p(n) = (p_0+p_1n)3^n.$$

Substituting into (7.33), we obtain:

$$p_03^n+p_1n3^n$$
$$= 5(p_0+p_1(n-1))3^{n-1}-6(p_0+p_1(n-2))3^{n-2}+4*3^n$$

Dividing by 3^{n-2}, we get:

$$9p_0+9p_1n = 15p_0+15p_1n-15p_1- 6p_0-6p_1n+12p_1+36$$
$$= (9p_0-3p_1+36)+9p_1n$$

Comparing terms, we obtain:

$$9p_1 = 9p_1$$

and

$$9p_0 = 9p_0-3p_1+36$$

These equations enable us to determine that $p_1 = 12$. The particular solution to (7.33) is:

$$f_p(n) = (p_0+12n)3^n$$

The homogeneous solution is:

$$f_h(n)=c_12^n+c_23^n$$

The general solution for $f(n)$ is therefore:

$$f(n) = f_h(n) + f_p(n)$$

$$= c_12^n+(c_2+p_0)3^n+12n3^n$$

$$= c_12^n+c_33^n+12n3^n$$

Given two initial values, $f(0)$ and $f(1)$, we can determine c_1 and c_3.

Obtaining The Complete Solution

We know that $f_h(n)+f_p(n)$ is a general solution to the recurrence:

$$f(n)=a_1f(n-1)+a_2f(n-2)+ \cdots +a_kf(n-k)+g(n),n \geqslant k. \quad (7.34)$$

By using the initial values $f(0)$, $f(1)$, ..., $f(k-1)$, we can solve for the k undetermined coefficients in $f_h(n) + f_p(n)$ to obtain the unique solution of (7.34) for which $f(0),\ldots,f(k-1)$ have the given values.

Summary

The characteristic roots method to solve the linear recurrence (7.34) consists of the following steps:

(1) Write down the characteristic equation:

$$x^k - \sum_{i=1}^{k}a_ix^{k-i} = 0$$

(2) Determine the distinct roots t_1, t_2, ..., t_s of the characteristic equation. Determine the multiplicity m_i of the root t_i, $1 \leqslant i \leqslant s$.

(3) Write down the form of $f_h(n)$. I.e.,

$$f_h(n) = u_1(n) + u_2(n) + \cdots + u_s(n)$$

where

$$u_i(n) = (c_{i_0} + c_{i_1}n + c_{i_2}n^2 + \cdots + c_{i_{w-1}}n^{w-1})t_i^n$$

and $w = m_i = $ multiplicity of the root t_i.

(4) Obtain the form of the particular solution $f_p(n)$.

  (a) If $g(n) = 0$ then $f_p(n) = 0$.
  (b) If $g(n) = \sum_{i=0}^{d} e_i n^i$ and $e_d \neq 0$, then $f_p(n)$ has the form:

$$f_p(n) = p_0 + p_1 n + p_2 n^2 + \cdots + p_{d+m}n^{d+m}$$

  where $m = 0$ if 1 is not a root of the characteristic equation. m is the multiplicity of 1 as a root of the characteristic equation otherwise.
  (c) If $g(n) = c * a^n$ then

$$f_p(n) = (p_0 + p_1 n + p_2 n^2 + \cdots + p_w n^w)a^n$$

  where w is zero if a is not a root of the characteristic equation. If a is a root of the characteristic equation, then w is the multiplicity of a.

(5) If $g(n) \neq 0$, then use the $f_p(n)$ obtained in (4) above to eliminate all occurrences of $f(n-i)$, $0 \leqslant i \leqslant k$ from (7.34). This is done by substituting the value of $f_p(n-i)$ for $f(n-i)$, $0 \leqslant i \leqslant k$ in (7.34). Following this substitution, a system of equations equating the coefficients of like powers of n is obtained. This system is solved to obtain the values of as many of the p_is as possible.

(6) Write down the form of the answer. I.e., $f(n) = f_h(n) + f_p(n)$. Solve for the remaining unknowns using the initial values $f(0)$, $f(1)$, ..., $f(k-1)$.

Theorem 7.6: The six step procedure outlined above always finds the unique solution to (7.34) with the given initial values.

Proof: See the text by Brualdi that is cited in the reference section. □

EXAMPLES

Example 7.7: The characteristic equation for the homogeneous recurrence:

$$t(n) = 6t(n-1) - 4t(n-2), \ n \geqslant 2$$

is

$$x^2 - 6x + 4 = 0$$

Its roots are $r_1 = 3 + \sqrt{5}$ and $r_2 = 3 - \sqrt{5}$. The roots are distinct and so $t(n) = c_1(3+\sqrt{5})^n + c_2(3-\sqrt{5})^n$. Suppose we are given that $t(0) = 0$ and $t(1) = 4\sqrt{5}$. Substituting $n = 0$ and 1 into $t(n)$, we get:

$$0 = c_1 + c_2$$

and

$$4\sqrt{5} = c_1(3+\sqrt{5}) + c_2(3-\sqrt{5})$$

The first equality yields $c_1 = -c_2$. The second then gives us $4\sqrt{5} = c_1(3+\sqrt{5}-3+\sqrt{5}) = 2\sqrt{5}c_1$ or $c_1 = 2$. The expression for $t(n)$ is therefore: $t(n) = 2(3+\sqrt{5})^n - 2(3-\sqrt{5})^n, \ n \geqslant 0$. $\square$

Example 7.8: In this example we shall obtain a closed form formula for the sum $s(n) = \sum_{i=0}^{n} i$. The recurrence for $s(n)$ is easily seen to be:

$$s(n) = s(n-1) + n, \ n \geqslant 1$$

Its characteristic equation is $x-1=0$. So, $s_h(n) = c_1(1)^n = c_1$. Since $g(n) = n$ and 1 is a root of multiplicity 1, the particular solution is of the form:

$$s_p(n) = p_0 + p_1 n + p_2 n^2$$

Substituting into the recurrence for $s(n)$, we obtain:

$$p_0 + p_1 n + p_2 n^2 = p_0 + p_1(n-1) + p_2(n-1)^2 + n$$

$$= p_0 + p_1 n - p_1 + p_2 n^2 - 2p_2 n + p_2 + n$$

$$= (p_0 - p_1 + p_2) + (p_1 - 2p_2 + 1)n + p_2 n^2$$

Equating the coefficients of like powers of n and solving the resulting equations, we get $p_1 = p_2$, and $2p_2 = 1$ or $p_2 = 1/2$. The particular solution is $s_p(n) = p_0 + n/2 + n^2/2$ The general solution becomes $s(n) = (c_1+p_0) + n/2 + n^2/2$. Since $s(0) = 0$, $c_1+p_0 = 0$. Hence, $s(n) = n(n+1)/2, \ n \geqslant 0$. $\square$

Example 7.9: Consider the recurrence:

$$f(n) = 5f(n-1) - 6f(n-2) + 3n^2, n \geqslant 2$$

and

$$f(0) = 2.5; f(1) = 4.5.$$

The characteristic equation for the homogeneous part is:

$$x^2 - 5x + 6 = 0$$

Its roots are $r_1 = 2$ and $r_2 = 3$. The general solution to the homogeneous part is therefore:

$$f_h(n) = c_1 2^n + c_2 3^n.$$

Since $g(n) = 3n^2$ and 1 is not a root of the characteristic equation, the particular solution has the form:

$$f_p(n) = p_0 + p_1 n + p_2 n^2$$

Substituting into the recurrence for $f(\ )$, we obtain:

$$p_0 + p_1 n + p_2 n^2$$

$$= 5(p_0 + p_1(n-1) + p_2(n-1)^2) - 6(p_0 + p_1(n-2) + p_2(n-2)^2) + 3n^2$$

$$= (7p_1 - p_0 - 19p_2) + (14p_2 - p_1)n + (3 - p_2)n^2$$

Comparing terms, we get:

$$p_2 = 3 - p_2$$

$$p_1 = 14p_2 - p_1$$

$$p_0 = 7p_1 - p_0 - 19p_2$$

Hence, $p_2 = 1.5$, $p_1 = 7p_2 = 10.5$, and $p_0 = 22.5$. So, the general solution for $f(n)$ is:

$$f(n) = c_1 2^n + c_2 3^n + 22.5 + 10.5n + 1.5n^2$$

Since $f(0)$ and $f(1)$ are known to be 2.5 and 4.5, respectively, we obtain:

$$2.5 = c_1 + c_2 + 22.5$$

and

$$4.5 = 2c_1 + 3c_2 + 34.5$$

Solving for c_1 and c_2, we get: $c_1 = -30$ and $c_2 = 10$. The solution to our recurrence is therefore:

$$f(n) = 22.5 + 10.5n + 1.5n^2 - 30*2^n + 10*3^n. \quad \square$$

Example 7.10: Let us solve the recurrence:

$$f(n) = 10f(n-1)-37f(n-2)+60f(n-3)-36f(n-4)+4, \ n \geqslant 4$$

and

$$f(0) = f(1) = f(2) = f(3) = 1.$$

The characteristic equation is:

$$x^4-10x^3+37x^2-60x+36 = 0$$

or

$$(x-2)^2(x-3)^2 = 0$$

The four roots are $r_1 = r_2 = 2$, and $r_3 = r_4 = 3$. Since each is a root of multiplicity 2, $u_1(n) = (c_1+c_2n)2^n$ and $u_2(n) = (c_3+c_4n)3^n$. The solution to the homogeneous part is:

$$f_h(n) = (c_1+c_2n)2^n + (c_3+c_4n)3^n$$

Since $g(n) = 4 = 4*n^0$ and 1 is not a root of the characteristic equation, $f_p(n)$ is of the form:

$$f_p(n) = p_0$$

Substituting into the recurrence for $f(n)$, we get:

$$p_0 = 10p_0-37p_0+60p_0-36p_0+4$$

or

$p_0 = 1$

The general solution for $f(n)$ is:

$$f(n) = (c_1+c_2n)2^n + (c_3+c_4n)3^n + 1$$

Substituting for $n = 0, 1, 2$, and 3, and using $f(0) = f(1) = f(2) = f(3) = 1$, we get:

$$0 = c_1+c_3 \tag{7.35a}$$

$$0 = 2c_1+2c_2+3c_3+3c_4 \tag{7.35b}$$

$$0 = 4c_1+8c_2+9c_3+18c_4 \tag{7.35c}$$

$$0 = 8c_1+24c_2+27c_3+81c_4 \tag{7.35d}$$

Solving for c_1, c_2, c_3, and c_4, we obtain $c_1 = c_2 = c_3 = c_4 = 0$. So, $f(n) = 1$, $n \geqslant 1$.

We may verify that $f(n) = 1$, $n \geqslant 0$ does indeed satisfy the given recurrence. We proceed by induction. For the induction base, we need to show that $f(n) = 1$, $0 \leqslant n \leqslant 3$. This is true by definition of $f()$. So, let m be an arbitrary natural number such that $m \geqslant 3$. Assume $f(n) = 1$, for $n \leqslant m$. When $n = m+1$, $f(m+1) = 10f(m) - 37f(m-1) + 60f(m-2) - 36f(m-3) + 4 = 10-37+60-36+4 = 1$.

Let us change the initial values to $f(0) = f(1) = f(2) = 1$, and $f(3) = 4$. Now, only equation (7.35d) changes. It becomes:

$$3=8c_1+24c_2+27c_3+81c_4 \tag{7.35e}$$

Solving (7.35 a to c) and (7.35e) for c_1, c_2, c_3, and c_4, we obtain $c_1 = 6$, $c_2 = 1.5$, $c_3 = -6$, and $c_4 = 1$. So,

$$f(n) = (6+1.5n)2^n + (n-6)3^n + 1, \; n \geqslant 0.$$

Once again, one may verify the correctness of this formula using induction on n. □

Solving Other Recurrences

Certain non-linear recurrences as well as linear ones with non constant coefficients may also be solved using the method of this section. In all cases, we need to first perform a suitable transformation on the given recurrence so as to obtain a linear recurrence with constant coefficients. For example, recurrences of the form:

$$f^c(n) = \sum_{i=1}^{k} a_i f^c(n-i) + g(n), \, n \geq k$$

may be solved by first substituting $f^c(n) = q(n)$ to obtain the recurrence:

$$q(n) = \sum_{i=1}^{k} a_i q(n-i) + g(n), \, n \geq k$$

This can be solved for $q(n)$ as described earlier. From $q(n)$ we may obtain $f(n)$ by noting that $f(n) = (q(n))^{1/c}$.

Recurrences of the form:

$$nf(n) = \sum_{i=1}^{k} (n-i) a_i f(n-i) + g(n), \, n \geq k$$

may be solved by substituting $q(n) = nf(n)$ to obtain:

$$q(n) = \sum_{i=1}^{k} a_i q(n-i) + g(n), \, n \geq k$$

Since $f(n) = q(n)/n$, $f(n)$ is determined once $q(n)$ is.

7.5 GENERATING FUNCTIONS

A generating function $G(z)$ is an infinite power series

$$G(z) = \sum_{i \geq 0} c_i z^i \tag{7.36}$$

We shall say that the generating function $G(z)$ corresponds to the function f: $N \rightarrow R$ iff $c_i = f(i), \, i \geq 0$.

Example 7.11: $G(z) = \sum_{i \geq 0} 2z^i$ generates the function $f(n) = 2, \, n \geq 0$; $G(z) = \sum_{i \geq 0} iz^i$ generates the function $f(n) = n, \, n \geq 0$; $G(z) = \sum_{i \geq 8} 2z^i$ generates the function:

$$f(n) = \begin{cases} 0 & 0 \leqslant n \leqslant 7 \\ 2 & n \geqslant 8 \end{cases} \quad \square$$

A generating function may be specified in two forms. One of these is called the *power series* form. This is the form given in equation (7.36). The other form is called the *closed form*. In this form there are no occurrences of the symbol Σ.

Example 7.12: The power series form for the generating function for $f(n) = 1$, $n \geqslant 0$ is $G(z) = \sum_{i \geqslant 0} z^i$. So, $zG(z) = \sum_{i \geqslant 1} z^i$. Subtracting, we obtain: $G(z)$

$- zG(z) = \sum_{i \geqslant 0} z^i - \sum_{i \geqslant 1} z^i = 1$. So, $G(z) = \dfrac{1}{1-z}$. The closed form for the

power series $\sum_{i \geqslant 0} z^i$ is therefore $\dfrac{1}{1-z}$.

Note that $\dfrac{1}{1-z} = \sum_{i \geqslant 0} z^i$ only for those values of z for which the series $\sum_{i \geqslant 0} z^i$ converges. The values of z for which this series converges are not relevant to our discussion here. $\square$

Example 7.13: Let n be an integer and i a natural number. The *binomial coefficient* $\binom{n}{i}$ is defined to be:

$$\binom{n}{i} = \frac{n(n-1)(n-2)...(n-i+1)}{i(i-1)(i-2)...(1)}$$

So, $\binom{3}{2} = \dfrac{3*2}{2*1} = 3$; $\binom{4}{2} = \dfrac{4*3}{2*1} = 6$; $\binom{-3}{2} = \dfrac{(-3)(-4)}{2*1} = 6$.

The *binomial theorem* states that:

$$(1+z)^n = \sum_{i=0}^{n} \binom{n}{i} z^i, \; n \geqslant 0$$

A more general form of the binomial theorem is:

$$(1+z)^n = \sum_{i=0}^{m} \binom{n}{i} z^i \tag{7.37}$$

where $m = n$ if $n \geqslant 0$ and $m = \infty$ otherwise.

(7.37) leads us to some important closed forms. When $n = -2$, we obtain:

$$\frac{1}{(1+z)^2} = \sum_{i \geqslant 0} \binom{-2}{i} z^i$$

But,

$$\binom{-2}{i} = \frac{(-2)(-3)...(-i-1)}{i(i-1)...1} = (-1)^i (i+1)$$

So,

$$\frac{1}{(1+z)^2} = \sum_{i \geqslant 0} (-1)^i (i+1) z^i \qquad (7.38)$$

Substituting $-z$ for z in (7.38), we obtain:

$$\frac{1}{(1-z)^2} = \sum_{i \geqslant 0} (i+1) z^i$$

Hence, $\dfrac{1}{(1-z)^2}$ is the closed form for $\sum_{i \geqslant 0} (i+1) z^i$. (7.37) may be used to obtain the power series form for $\dfrac{1}{(1-z)^n}$, $n \geqslant 1$. $\square$

As we shall soon see, generating functions can be used to solve recurrence relations. First, let us look at the calculus of generating functions.

Generating Function Operations

Addition and Subtraction: If $G_1(z) = \sum_{i \geqslant 0} c_i z^i$ and $G_2(z) = \sum_{i \geqslant 0} d_i z^i$ are the generating functions for f_1 and f_2, then the generating function for $f_1 + f_2$ is:

$$G_3(z) = \sum_{i \geqslant 0} (c_i + d_i) z^i$$

and that for $f_1 - f_2$ is:

$$G_4(z) = \sum_{i \geqslant 0} (c_i - d_i) z^i.$$

These two equalities follow directly from the definition of a generating function.

Multiplication: If $G_1(z) = \sum\limits_{i > 0} c_i z^i$ is the generating function for f, then $G_2(z)$
$= aG_1(z) = \sum\limits_{i \geqslant 0} (ac_i) z^i$ is the generating function for $a * f$ (a is a constant).

Since $z^k G_1(z) = \sum\limits_{i \geqslant 0} c_i z^{k+i}$, it is the generating function for a function g
such that $g(j) = 0$, $0 \leqslant j < k$ and $g(j) = f(j - k)$, $j \geqslant k$. So, multiplying a generating function by z^k corresponds to shifting the function it generates by k.

Example 7.14: In Example 7.13, we showed that $\dfrac{1}{(1-z)^2} = \sum\limits_{i \geqslant 0} (i+1) z^i$.

Multiplying both sides by z, we obtain $\dfrac{z}{(1-z)^2} = \sum\limits_{i \geqslant 0} (i+1) z^{i+1} = \sum\limits_{i \geqslant 0} i z^i$. So,

$\dfrac{z}{(1-z)^2}$ is the closed form for the generating function for $f(i) = i$, $i \geqslant 0$. $\square$

The product $G_1(z) * G_2(z)$ of the two generating functions $G_1(z) = \sum\limits_{i \geqslant 0} c_i z^i$ and $G_2(z) = \sum\limits_{i \geqslant 0} d_i z^i$ is a third generating function $G_3(z) = \sum\limits_{i \geqslant 0} e_i z^i$. One may verify that e_i is given by:

$$e_i = \sum_{j=0}^{i} c_j d_{i-j} \tag{7.39}$$

Note that $*$ is commutative (i.e., $G_1(z) * G_2(z) = G_2(z) * G_1(z)$).

An examination of (7.39) indicates that the product of generating functions might be useful in computing sums. In particular, if $G_2(z) = \sum\limits_{i \geqslant 0} z^i$
$= \dfrac{1}{1-z}$ (Example 7.12) (i.e., $d_i = 1$, $i \geqslant 0$), then (7.39) becomes

$$e_i = \sum_{j=0}^{i} c_j \tag{7.40}$$

Example 7.15: Let us try to find the closed form for the sum $s(n) = \sum\limits_{i=0}^{n} i$.

From Example 7.14, we know that $\dfrac{z}{(1-z)^2}$ is the closed form for the generating function for $f(i) = i$, $i \geq 0$. Also, from Example 7.12, we know that $\dfrac{1}{1-z}$ generates $f(i) = 1$, $i \geq 0$. So, $\dfrac{z}{(1-z)^2} * \dfrac{1}{1-z} = \dfrac{z}{(1-z)^3}$ is the closed form for $(\sum\limits_{i \geq 0} iz^i)(\sum\limits_{i \geq 0} z^i)$. Let the power series form of $\dfrac{z}{(1-z)^3}$ be $\sum\limits_{i \geq 0} e_i z^i$. From (7.40), it follows that $e_n = \sum\limits_{i=0}^{n} i = s(n)$, $n \geq 0$. Let us proceed to determine e_n. Using the binomial theorem (7.37), we obtain:

$$(1-z)^{-3} = \sum_{i \geq 0} \binom{-3}{i}(-1)^i z^i$$

The coefficient of z^{n-1} in the expansion of $(1-z)^{-3}$ is therefore:

$$\binom{-3}{n-1}(-1)^{n-1} = \frac{(-3)(-4)...(-3-n+2)}{(n-1)(n-2)...(1)}(-1)^{n-1}$$
$$= \frac{(n+1)n(n-1)...(3)}{(n-1)(n-2)...(1)}$$
$$= \frac{n(n+1)}{2}$$

So, the coefficient e_n of z^n in the power series form of $\dfrac{z}{(1-z)^3}$ is $n(n+1)/2$ $= s(n)$, $n \geq 0$. $\square$

Differentiation: Differentiating (7.36) with respect to z gives:

$$\frac{d}{dz}G(z) = \sum_{i \geq 0} ic_i z^{i-1}$$

or

$$z\frac{d}{dz}G(z) = \sum_{i \geq 0} (ic_i)z^i \qquad (7.41)$$

Example 7.16: In Example 7.13, the binomial theorem was used to obtain the closed form for $\sum\limits_{i \geq 0}(i+1)z^i$. This closed form can also be obtained using differentiation. From Example 7.12, we know that $\dfrac{1}{1-z} = \sum\limits_{i \geq 0} z^i$. From (7.41), it follows that:

$$\frac{d}{dz}\frac{1}{1-z} = \sum_{i \geq 0} iz^{i-1}$$

or

$$\frac{1}{(1-z)^2} = \sum_{i \geq 0} (i+1) z^i \quad \square$$

Integration: Integrating (7.36), we get

$$\int_0^z G(u) \, du = \sum_{j \geq 1} c_{j-1} z^j / j \tag{7.42}$$

Example 7.17: The closed form of the generating function for $f(n) = 1/n$, $n \geq 1$ can be obtained by integrating the generating function for $f(n) = 1$. From Example 7.12, we obtain:

$$\frac{1}{1-u} = \sum_{i \geq 0} u^i$$

Therefore

$$\int_0^z \frac{1}{1-u} \, du = \sum_{i \geq 0} \int_0^z u^i du$$

$$= \sum_{i \geq 0} \frac{1}{i+1} z^{i+1}$$

$$= \sum_{i > 0} \frac{1}{i} z^i$$

But,

$$\int_0^z \frac{1}{1-u} \, du = -\ln(1-z)$$

So, the generating function for $f(n) = 1/n$, $n \geq 1$ and $f(0) = 0$, is $-\ln(1-z)$. $\square$

The five operations: addition, subtraction, multiplication, differentiation, and integration prove useful in obtaining generating functions for $f:N \rightarrow R$.

Table 7.2 lists some of the more important generating functions in both power series and closed forms.

	Closed Form	Power Series
1	$(1-az)^{-1}$	$\sum_{i \geq 0} a^i z^i$
2	$(1-az)^{-2}$	$\sum_{i \geq 0} (i+1) a^i z^i$
3	$(1+az)^n$	$\sum_{i=0}^{m} \binom{n}{i} a^i z^i$ $m = n$ if $n \geq 0$ $m = \infty$ otherwise
4	$ln(1+az)$	$\sum_{i \geq 1} \frac{(-1)^{i+1}}{i} a^i z^i$
5	$-ln(1-az)$	$\sum_{i \geq 1} \frac{1}{i} a^i z^i$
6	e^{az}	$\sum_{i \geq 0} \frac{1}{i!} a^i z^i$

Table 7.2 Some power series.

Solving Recurrence Equations

The generating function method of solving recurrences is best illustrated by an example. Consider the recurrence:

$$F(n) = 2F(n-1) + 7, \; n \geq 1; \; F(0) = 0$$

The steps to follow in solving any recurrence using the generating function method are:

(1) Let $G(z) = \sum_{i \geq 0} a_i z^i$ be the generating function for $F()$. So, $a_i = F(i), \; i \geq 0$.

(2) Replace all occurrences of $F()$ in the given recurrence by the corresponding a_i. Doing this on the example recurrence yields:

$$a_n = 2a_{n-1} + 7, \, n \geqslant 1$$

(3) Multiply both sides of the resulting equation by z^n and sum up both sides for all n for which the equation is valid. For the example, we obtain:

$$\sum_{n \geqslant 1} a_n z^n = 2 \sum_{n \geqslant 1} a_{n-1} z^n + \sum_{n \geqslant 1} 7 z^n$$

(4) Replace all infinite sums involving the a_is by equivalent expressions involving only $G(z)$, z, and a finite number of the a_is. For a degree k recurrence only a_0, a_1, ..., a_{k-1} will remain. The example yields:

$$G(z) - a_0 = 2zG(z) + \sum_{n \geqslant 1} 7 z^n$$

(5) Substitute the known values of a_0, a_1, ...,a_{k-1} (recall that $F(i) = a_i$, $0 \leqslant i < k$. Our example reduces to:

$$G(z) = 2zG(z) + \sum_{n \geqslant 1} 7 z^n$$

(6) Solve the resulting equation for $G(z)$. The example equation is easily solved for $G(z)$ by collecting the $G(z)$ terms on the left and then dividing by the coefficient of $G(z)$. We get:

$$G(z) = \sum_{n \geqslant 1} 7 z^n * \frac{1}{1-2z}$$

(7) Determine the coefficient of z^n in the power series expansion of the expression obtained for $G(z)$ in step 6. This coefficient is $a_n = F(n)$. For our example, we get:

$$G(z) = \sum_{n \geqslant 1} 7 z^n * \frac{1}{1-2z}$$

$$= \sum_{n \geqslant 1} 7 z^n * \sum_{i \geqslant 0} 2^i z^i$$

The coefficient of z^n in the above series product is:

$$\sum_{i=1}^{n} 7*2^{n-i} = 7(2^n - 1)$$

So, $F(n) = 7(2^n - 1)$, $n \geqslant 0$.

The next several examples illustrate the technique further.

Example 7.18: Let us reconsider the recurrence for the Fibonacci numbers:

$$F(n) = F(n-1) + F(n-2), n \geqslant 2$$

and

$$F(0) = 0, F(1) = 1.$$

Let $G(z) = \sum_{i \geqslant 0} c_i z^i$ be the generating function for F. From the definition of a generating function, it follows that $F(j) = c_j$, $j \geqslant 0$. So, $F(n) = c_n$, $F(n-1) = c_{n-1}$, and $F(n-2) = c_{n-2}$. From the recurrence relation for F, we see that:

$$c_n = c_{n-1} + c_{n-2}, n \geqslant 2$$

Multiplying both sides by z^n and summing from $n=2$ to ∞, we get:

$$\sum_{n \geqslant 2} c_n z^n = \sum_{n \geqslant 2} c_{n-1} z^n + \sum_{n \geqslant 2} c_{n-2} z^n \qquad (7.43)$$

Observe that the sum cannot be performed from $n = 0$ to ∞ as the recurrence $F(n) = F(n-1) + F(n-2)$ is valid only for $n \geqslant 2$. (7.43) may be rewritten as:

$$G(z) - c_1 z - c_0 = z \sum_{n \geqslant 2} c_{n-1} z^{n-1} + z^2 \sum_{n \geqslant 2} c_{n-2} z^{n-2}$$

$$= z \sum_{i \geqslant 1} c_i z^i + z^2 \sum_{i \geqslant 0} c_i z^i$$

$$= zG(z) - c_0 z + z^2 G(z)$$

Collecting terms and substituting $c_0 = F(0) = 0$ and $c_1 = F(1) = 1$, we get:

$$G(z) = \frac{z}{1 - z - z^2}$$

$$= \frac{z}{(1 - az)(1 - bz)} , a = \frac{1 + \sqrt{5}}{2} \text{ and } b = \frac{1 - \sqrt{5}}{2}$$

$$= \frac{1}{\sqrt{5}} \left[\frac{1}{1 - az} - \frac{1}{1 - bz} \right]$$

From Table 7.2, we see that the power series expansion of $(1-az)^{-1}$ is $\sum\limits_{i\geqslant 0}(az)^i$. Using this, we obtain:

$$G(z) = \frac{1}{\sqrt{5}}\left[\sum_{i\geqslant 0}a^i z^i - \sum_{i\geqslant 0}b^i z^i\right]$$

$$= \sum_{i\geqslant 0}\frac{1}{\sqrt{5}}\left[a^i-b^i\right]z^i$$

Hence, $F(n) = c_n = \dfrac{1}{\sqrt{5}}\left[\left[\dfrac{1+\sqrt{5}}{2}\right]^n - \left[\dfrac{1-\sqrt{5}}{2}\right]^n\right]$, $n\geqslant 0$. □

Example 7.19: Consider the recurrence:

$$t(n)=\begin{cases} 0 & n=0 \\ at(n-1)+bn & n\geqslant 1 \end{cases}$$

Let $G(z) = \sum\limits_{i\geqslant 0}c_i z^i$ be the generating function for $t(n)$. So, $t(n) = c_n$, $n\geqslant 0$. From the recurrence, it follows that:

$$c_n = ac_{n-1} + bn, \; n\geqslant 1$$

Multiplying both sides by z^n and summing from $n=1$ to ∞ yields:

$$\sum_{n>1}c_n z^n = \sum_{n\geqslant 1}ac_{n-1}z^n + \sum_{n>1}bnz^n$$

or

$$G(z)-c_0 = az\sum_{n>1}c_{n-1}z^{n-1} + \sum_{n>1}bnz^n$$

$$= azG(z)+\sum_{n>1}bnz^n$$

Substituting $c_0 = 0$ and collecting terms, we get:

$$G(z) = \left(\sum_{n>1}bnz^n\right)/(1-az)$$

$$= \left(\sum_{n>1}bnz^n\right)\left(\sum_{i>0}a^i z^i\right)$$

Using the formula for the product of two generating functions, we obtain:

$$c_n = b \sum_{i=1}^{n} i a^{n-i}$$

$$= b a^n \sum_{i=1}^{n} \frac{i}{a^i}$$

Hence, $t(n) = b a^n \sum_{i=1}^{n} \frac{i}{a^i}, \ n \geqslant 0. \ \square$

Example 7.20: In the previous example, we determined that $c_n = b a^n \sum_{i=0}^{n} \frac{i}{a^i}$. A closed form for c_n can be obtained from a closed form for $d_n = \sum_{i=0}^{n} \frac{i}{a^i}$, $n \geqslant 0$. First, let us find the generating function for $f(i) = i/a^i$. We know that $(1-z)^{-1} = \sum_{i \geqslant 0} z^i$. So, $(1-z/a)^{-1} = \sum_{i \geqslant 0} (z/a)^i$. Differentiating with respect to z, we obtain:

$$\frac{d}{dz} \frac{1}{1-z/a} = \sum_{i \geqslant 0} \frac{d}{dz} (z/a)^i = \sum_{i \geqslant 0} \frac{i}{a^i} z^{i-1}$$

or

$$\frac{1}{a} \frac{1}{(1-z/a)^2} = \sum_{i \geqslant 0} \frac{i}{a^i} z^{i-1}$$

Multiplying both sides by z, we get:

$$\frac{z}{a(1-z/a)^2} = \sum_{i \geqslant 0} \frac{i}{a^i} z^i$$

The generating function for $\sum_{i=0}^{n} i/a^i$ can now be obtained by multiplying by $1/(1-z)$ (see Equation (7.40)). So,

$$\frac{z}{a(1-z/a)^2(1-z)} = \sum_{n \geqslant 0} \left[\sum_{i=0}^{n} \frac{i}{a^i} \right] z^n$$

$$= \sum_{n \geqslant 0} d_n z^n$$

We now need to find the form of the coefficient of z^n in the expansion of

$$\frac{z}{a(1-z/a)^2(1-z)}$$

Expanding this, we get:

$$\frac{z}{a(1-z/a)^2(1-z)} = \frac{z}{a}\sum_{i \geq 0}(z/a)^i\sum_{i \geq 0}(z/a)^i\sum_{i \geq 0}z^i$$

So,

$$d_n = \frac{1}{a}\sum_{i=0}^{n-1}\left[\frac{1}{a^i} \sum_{j=0}^{n-1-i}\frac{1}{a^j}\right]$$

$$= \frac{1}{a}\sum_{i=0}^{n-1}\left[\frac{1}{a^i} \frac{(1/a)^{n-i}-1}{1/a-1}\right], \ a \neq 1$$

$$= \frac{1}{a(1/a-1)}\sum_{i=0}^{n-1}\left[\frac{1}{a^n} - \frac{1}{a^i}\right], \ a \neq 1$$

$$= \frac{1}{1-a}\left[\frac{n}{a^n} - \frac{(1/a)^n-1}{(1/a)-1}\right], \ a \neq 1$$

$$= \frac{n}{(1-a)a^n} - \frac{a(1/a^n-1)}{(1-a)^2}, \ a \neq 1$$

When $a = 1$, $d_n = \sum_{i=0}^{n} i = n(n+1)/2$. Observe that the recurrence for d_n is:

$$d_n = d_{n-1} + \frac{n}{a^n}, \ n \geq 1$$

Since the general form of the particular solution is not known when $g(n) = n/a^n$, it would be difficult to obtain the solution for d_n using the characteristic roots method. □

Example 7.21: An alternate approach to obtain the power series form of $G(z) = (\sum_{n \geq 1} bnz^n)/(1-az)$ (see Example 7.19) is:

$$G(z) = (\sum_{n \geq 1} bnz^n)/(1-az)$$

$$= (b\sum_{n \geq 0} nz^n)/(1-az).$$

$$= \frac{bz}{(1-z)^2(1-az)}$$

When $a \neq 1$, we obtain:

$$G(z) = \frac{Az+B}{(1-z)^2} + \frac{C}{1-az}$$

Solving for A, B, and C, we obtain:

$$A = \frac{b}{(1-a)^2}$$

$$B = -\frac{ab}{(1-a)^2}$$

$$C = \frac{ab}{(1-a)^2}$$

So,

$$G(z) = \frac{Az}{(1-z)^2} + \frac{B}{(1-z)^2} + \frac{C}{1-az}$$

$$= A\sum_{i>0} iz^i + B\sum_{i>0}(i+1)z^i + C\sum_{i>0}a^iz^i$$

The coefficient of z^n is therefore:

$$t(n) = An + B(n+1) + Ca^n$$

$$= \frac{b}{(1-a)^2}n - \frac{ab}{(1-a)^2}(n+1) + \frac{ab}{(1-a)^2}a^n$$

$$= \frac{bn}{1-a} - \frac{ab(1-a^n)}{(1-a)^2}, a \neq 1, \ n \geq 0$$

When $a = 1$,

$$G(z) = \frac{bz}{(1-z)^3}$$

$$= \frac{b}{2}\sum_{i>0} i(i+1)z^i \ \text{(Example 7.15)}$$

So, $f(n) = bn(n+1)/2, \ n \geq 0, \ a=1.$ $\square$

Example 7.22: Consider the recurrence:

$$f(n) = 5f(n-1) - 6f(n-2) + 2n, \ n \geq 2$$

and

$$f(0) = f(1) = 0$$

Let $G(z) = \sum\limits_{i \geq 0} c_i z^i$ be the generating function for f. So, $f(n) = c_n$, $f(n-1) = c_{n-1}$ and $f(n-2) = c_{n-2}$. Therefore:

$$c_n = 5c_{n-1} - 6c_{n-2} + 2n, \, n \geq 2$$

or

$$c_n z^n = 5c_{n-1}z^n - 6c_{n-2}z^n + 2nz^n, \, n \geq 2$$

Summing up for n from 2 to ∞ yields:

$$\sum_{n>2} c_n z^n = 5z \sum_{n>2} c_{n-1}z^{n-1} - 6z^2 \sum_{n \geq 2} c_{n-2}z^{n-2} + \sum_{n>2} 2nz^n$$

or

$$G(z) - c_1 z - c_0 = 5z(G(z) - c_0) - 6z^2 G(z) + \sum_{n>2} 2nz^n$$

Substituting $c_1 = c_0 = 0$, we get:

$$G(z)(1 - 5z + 6z^2) = \sum_{n>2} 2nz^n$$

or

$$G(z) = \frac{\sum\limits_{n>2} 2nz^n}{(1-3z)(1-2z)}$$

$$= \sum_{j>2} 2jz^j \left[\frac{3}{1-3z} - \frac{2}{1-2z} \right]$$

$$= \sum_{j>2} 2jz^j [3 \sum_{i \geq 0} 3^i z^i - 2 \sum_{i \geq 0} 2^i z^i]$$

The coefficient c_n of z^n is now seen to be:

$$c_n = \sum_{j=2}^{n} 6j3^{n-j} - \sum_{j=2}^{n} 4j2^{n-j}$$

$$= 6*3^n \sum_{j=2}^{n} (j/3^j) - 4*2^n \sum_{j=2}^{n} j/2^j$$

From Example 7.20, we know that:

$$\sum_{2}^{n} \frac{j}{3^j} = -\frac{1}{2}\frac{n}{3^n} - \frac{3(3^{-n}-1)}{4} - \frac{1}{3}$$

and

$$\sum_{j=2}^{n} \frac{j}{2^j} = -\frac{n}{2^n} - 2(2^{-n}-1) - \frac{1}{2}$$

So,

$$c_n = -3n - 4\cdot5 + 4\cdot5*3^n - 2*3^n + 4n + 8 - 8*2^n + 2*2^n$$

$$= n + 3\cdot5 + 2\cdot5*3^n - 6*2^n$$

So, $f(n) = n + 3\cdot5 + 2\cdot5*3^n - 6*2^n$, $n \geqslant 0$. □

REFERENCES AND SELECTED READINGS

Table 7.1 is due to J. Bentley, D. Haken, and J. Saxe. They presented it in their solution of recurrence (7.15). Their work appears in the report:

> *A general method for solving divide-and-conquer recurrences*, by J. Bentley, D. Haken, and J. Saxe, *SIGACT News*, 12(3), 1980, pp. 36-44.

A proof of Theorem 7.1 may be found in

> *Introduction to combinatorial mathematics*, by C.L. Liu, McGraw Hill Book Co., New York 1968.

and

> *Introductory combinatorics*, by R. Brualdi, Elsevier North-Holland Inc., New York, 1977.

Both of the aforementioned texts contain solutions to several recurrence relations. They are good references for the characteristic roots method and for generating functions.

EXERCISES

1. Prove that equation (7.10) is correct.

2. Express the worst case time complexity of procedure INSORT (Algorithm 5.19) as a recurrence relation.

3. Obtain a recurrence relation for the worst case computing time of Algorithm 7.1.

```
procedure AL (X,n)
    //n is a power of 2//
    integer i,j,n
    declare X(n,n),A(n/2,n/2),B(n/2,n/2), C(n/2,n/2)
    for i ← 1 to n/2 do
      for j ← 1 to n/2 do
        A (i,j) ← X (i,j)
        B (i,j) ← X (i,j+n/2)
        C (i,j) ← X (i+n/2,j)
      end for
    end for
    call AL (A,n/2)
    call AL (B,n/2)
    call AL (C,n/2)
end AL
```

Algorithm 7.1

4. Solve the following recurrences using the substitution method. Do not use Table 7.1 or the result of Example 7.3. In each case, obtain a $g(n)$ such that $t(n) = \Theta(g(n))$ or obtain $g_1(n)$ and $g_2(n)$ such that $t(n) = \Omega(g_1(n))$ and $t(n)=O(g_2(n))$. Verify the correctness of your answers using induction on n. In each case, assume $t(1) = 1$.

(a) $t(n) = 6t(n/3)+2n^2$, n a power of 3
(b) $t(n) = 3t(n/4)+3n$, n a power of 4
(c) $t(n) = t(n-2)+2n$, $n > 1$, $t(0) = 1$
(d) $t(n) = t(n-1)+2n^2$, $n > 1$
(e) $t(n) = 3t(n/2)+2$, n a power of 2

5. Use Table 7.1 to solve the following recurrences. In each case, assume $t(1) = 1$. Obtain a $g(n)$ such that $t(n) = \Theta(g(n))$.

(a) $t(n) = 10t(n/3)+11n, n \geqslant 3$ and a power of 3
(b) $t(n) = 10t(n/3)+11n^5$, $n \geqslant 3$ and a power of 3
(c) $t(n) = 27t(n/3)+11n^3$, $n \geqslant 3$ and a power of 3
(d) $t(n) = 64t(n/4)+10n^3\log^2 n$, $n \geqslant 4$ and a power of 4
(e) $t(n) = 9t(n/2)+n^2 2^n$, $n \geqslant 2$ and a power of 2
(f) $t(n) = 3t(n/8)+n^2 2^n \log n$, $n \geqslant 8$ and a power of 8
(g) $t(n) = 128t(n/2)+6n$, $n \geqslant 2$ and a power of 2
(h) $t(n) = 128t(n/2)+6n^8$, $n \geqslant 2$ and a power of 2
(i) $t(n) = 128t(n/2)+2^n/\sqrt{n}$, $n \geqslant 2$ and a power of 2
(j) $t(n) = 128t(n/2)+\log^3 n$, $n \geqslant 2$ and a power of 2

6. Solve the following recurrences using the characteristic roots method.

(a) $t(n) = 6t(n-1)-12t(n-2)+8t(n-3)$, $n \geqslant 3$, $t(0) = t(1) = t(2) = 4$
(b) $t(n) = 5t(n-1)-8t(n-2)+4t(n-3)$, $n > 3$, $t(0) = t(1) = 0$, $t(2) = 1$
(c) $t(n) = 2t(n-1)+7n$, $n \geqslant 1$, $t(0)=0$
(d) $t(n) = t(n-1)+n$, $n \geqslant 1$, $t(0)=0$
(e) $t(n) = 4t(n-1)+2^n$, $n \geqslant 1$, $t(0)=1$
(f) $t(n) = 7t(n-1)-12t(n-2)+2n$, $t(0) = 0$, $t(1) = 1$
(g) $t(n) = 7t(n-1)-12t(n-2)+6*2^n$, $t(0) = 0, t(1) = 1$
(h) $t(n) = 4t(n-1)-2t(n-2)+2n^2+3n$, $t(0) = 0$, $t(1) = 1$
(i) $t(n) = 2t(n-1)-t(n-2)+4$, $t(0) = 0, t(1) = 2$
(j) $t(n) = 2t(n-1)-t(n-2)+4n+2$, $t(0) = 0$, $t(1) = 2$
(k) $t(n) = 2t(n-1)-t(n-2)+2^n$, $t(0) = 0$, $t(1) = 1$
(l) $t(n) = 4t(n-1)-5t(n-2)+2t(n-3)+4$, $t(0)=0$, $t(1) = t(2) = 1$
(m) $t(n) = 8t(n-1)-20t(n-2)+16t(n-3)+2^n$, $t(0:2) = !$
(n) $t(n) = t(n-2)+2n$, $t(0) = 0$, $t(1) = 1$

7. (a) Transform the recurrence:

$$t(n) = \begin{cases} 1 & n=4 \\ 2t(n/4)+2\log_2 n-1 & n>4 \text{ and a power of } 4 \end{cases}$$

into an order one linear recurrence with constant coefficients. This recurrence should be such that the solution for $t(n)$ is readily available from the solution for this recurrence.

(b) Solve the recurrence obtained in (a) using the characteristic roots method.

(c) From the solution to (b) obtain the solution for $t(n)$.

8. Do Exercise 7 for the following recurrences:

(a) $t(n) = 10t(n/3)+11n,\ t(1) = 1$
(b) $t(n) = 4t(n/2)-4t(n/4)+\log_2 n,\ t(1) = t(4) = 1$
(c) $t(n) = 5t(n/3)-6t(n/9)+2\log_3 n,\ t(1) = 0,\ t(9) = 1$

9. Solve the following recurrences using the characteristic roots method:
   (a) $nF(n) = (n-1)F(n-1) + (n-2)F(n-2) + 2,\ n \geqslant 2;\ F(0) = F(1) = 0$ (substitute $q(n) = nF(n)$).
   (b) $nF^2(n) = 4(n-1)F^2(n-1) + 4F^2(n-2) + n^2,\ n \geqslant 2;\ F(0) = F(1) = 0$ (substitute $q(n) = nF^2(n)$).
   (c) $\sqrt{n}\,F^3(n) = 4\sqrt{n-2}F^3(n-2),\ n \geqslant 2;\ F(0) = F(1) = 1.$
   (d) $2^{F(n)} = 3*2^{F(n-1)} + 2^{F(n-2)} + 6*2^n,\ n \geqslant 2;\ F(0) = F(1) = 0.$

10. Use the generating function method to determine a closed form expression for each of the following sums:

   (a) $\displaystyle\sum_{i=0}^{n} i^2$

   (b) $\displaystyle\sum_{i=1}^{n} \frac{1}{i}$

   (c) $\displaystyle\sum_{i=0}^{n} \frac{i}{2^i}$

   (d) $\displaystyle\sum_{i=0}^{n} \frac{i^2}{a^i}$

   (e) $\displaystyle\sum_{i=0}^{n} i2^i$

   (f) $\displaystyle\sum_{i=0}^{n} i^2 2^i$

11. Use the generating function method to solve the recurrences of Exercise 6.

12. Prove the validity of Equation (7.39).

13. Let $G(z)$ be the generating function for $f(n)$.

   (a) Show that $1/2(G(z)+G(-z))$ is the generating function for $g(n)=f(n)$ if n is even and $g(n)=0$ if n is odd.

   (b) Suppose that $(z^4+z^2-1)/(z^8-z^6+z^4+2)$ is the generating function for $f(n)$. What is the generating function for $h(n)=f(2n)$, $n \geqslant 0$?

   (c) If $G(z)$ is the generating function for $f(n)$. What is the generating function for $g(n)=f(n)$ if n is odd and $g(n)=0$ if n is even.

   (d) How may the generating function for $h(n) = f(2n+1)$, $n \geqslant 0$ be obtained from that for $f(n)$?

CHAPTER 8

COMBINATORICS AND DISCRETE PROBABILITY

8.1 THE BINOMIAL COEFFICIENT

The binomial coefficient $\binom{r}{i}$ often arises when the generating function method is used to solve a recurrence relation. In Section 7.5 we used binomial coefficients in Examples 7.13 and 7.15. In algorithms that work explicitly with subsets of sets, the binomial coefficient may appear directly in the expression for the algorithm's complexity. This section is concerned with developing some facility in manipulating expressions involving the binomial coefficient.

In the most general form of the binomial coefficient $\binom{r}{i}$, r is a real number and i an integer. The value of $\binom{r}{i}$ is defined to be

$$\binom{r}{i} = \begin{cases} 0 & i<0 \\ 1 & i=0 \\ \dfrac{r(r-1)(r-2)...(r-i+1)}{i(i-1)(i-2)...1} & i>0 \end{cases} \qquad (8.1)$$

From (8.1), it follows that $\binom{2.5}{2} = (2.5*1.5)/(2*1) = 1.875$; $\binom{2.5}{-2} = 0$; $\binom{2.5}{0} = 1$; $\binom{-2}{3} = (-2)(-3)(-4)/(3*2*1) = -4$.

When r and i are natural numbers such that $i \leqslant r$, (8.1) becomes:

$$\binom{r}{i} = \frac{r!}{i!(r-i)!} \quad , \quad r \text{ and } i \text{ natural numbers and } i \leqslant r \quad (8.2)$$

where $i! = i(i-1)(i-2)...1$ when $i \geqslant 1$ and $i! = 1$ when $i = 0$.

A useful formula for obtaining an approximation to $n!$ is:

$$n! = \sqrt{2\pi n}\,(n/e)^n(1 + O(1/n))$$
$$\simeq \sqrt{2\pi n}\,(n/e)^n \tag{8.3}$$

where $e = 2.7182...$ is the base of the natural logarithm. (8.3) is known as Stirling's approximation for $n!$.

When r and i are natural numbers and $i > r$, one of the terms in the numerator of (8.1) becomes zero and we get:

$$\binom{r}{i} = 0, \; r \text{ and } i \text{ are natural numbers and } i > r \tag{8.4}$$

As noted earlier, binomial coefficients arise in the binomial theorem. The most general form of this theorem is:

$$(x+y)^r = \sum_{i \geq 0} \binom{r}{i} x^i y^{r-i} \tag{8.5}$$

When r is a natural number, (8.5) reduces to the more familiar form:

$$(x+y)^r = \sum_{i=0}^{r} \binom{r}{i} x^i y^{r-i}, \; r \text{ a natural number} \tag{8.6}$$

as $\binom{r}{i} = 0$ for $i > r$. Substituting $y = 1$ into (8.5) we get the form of the binomial theorem given in (7.37). Substituting $x = y = 1$ and $r = n$ into (8.6), we get the important identity:

$$2^n = \sum_{i=0}^{n} \binom{n}{i}, \; n \text{ a natural number} \tag{8.7}$$

Equation (8.6) can be used to obtain other useful identities involving the binomial coefficient. Setting $x = 2$, $y = 1$, and $r = n$, we obtain:

$$3^n = \sum_{i=0}^{n} 2^i \binom{n}{i}, \; n \text{ a natural number} \tag{8.8}$$

In general, we see that:

$$(k+1)^n = \sum_{i=0}^{n} k^i \binom{n}{i}, \quad n \text{ a natural number} \tag{8.9}$$

By setting $x = -1$, $y = 1$, and $r = n$, we obtain:

$$0 = \sum_{i=0}^{n} (-1)^i \binom{n}{i}, \quad n \text{ a natural number} \tag{8.10}$$

Some other useful identities may be obtained by setting $y = 1$ and $r = n$ in (8.6) and then differentiating both sides with respect to x. Doing this yields:

$$n(1+x)^{n-1} = \sum_{i=1}^{n} i \binom{n}{i} x^{i-1} \tag{8.11}$$

Setting $x = 1$ gives us:

$$n2^{n-1} = \sum_{i=1}^{n} i \binom{n}{i}, \quad n \text{ a natural number} \tag{8.12}$$

Setting $x = 2$ yields:

$$n3^{n-1} = \sum_{i=1}^{n} i \binom{n}{i} 2^{i-1}$$

$$= \frac{1}{2} \sum_{i=1}^{n} i \binom{n}{i} 2^i, \quad n \text{ a natural number} \tag{8.13}$$

Two other useful formulas for the sum of binomial coefficients can be obtained from the identity:

$$\binom{r}{i} = \binom{r-1}{i} + \binom{r-1}{i-1}, \quad i \text{ integer} \tag{8.14}$$

These formulas are:

$$\sum_{i=0}^{n} \binom{r+i}{i} = \binom{r+n+1}{n}, \quad n \text{ a natural number} \tag{8.15}$$

and

$$\sum_{i=0}^{n}\binom{i}{m} = \binom{n+1}{m+1}, \quad m \text{ and } n \text{ are natural numbers.} \qquad (8.16)$$

We leave the proofs of (8.14) through (8.16) as an exercise. The exercises develop several other identities involving binomial coefficients. Table 8.1 contains some of the more useful identities involving the binomial coefficient.

I1 $\binom{n}{i} = \binom{n}{n-i}$, $n \in N$ and $\in I$

I2 $\binom{r}{i} = \frac{r}{i}\binom{r-1}{i-1}$, $i \in I - \{0\}$

I3 $\sum_{i=0}^{n}\binom{n}{i} = 2^n$, $n \in N$

I4 $\sum_{i=0}^{n} i\binom{n}{i} = n2^{n-1}$, $n \in N$

I5 $\sum_{i=0}^{n} i^2\binom{n}{i} = n(n+1)2^{n-2}$, $n \in N$

I6 $\sum_{i=0}^{n} k^i\binom{n}{i} = (k+1)^n$, $n \in N$

I7 $\sum_{i=0}^{n} (-1)^i\binom{n}{i} = 0$, $n \in N$

I8 $\sum_{i=0}^{n}\binom{r+i}{i} = \binom{r+n-1}{n}$, $n \in N$

I9 $\sum_{i=0}^{n}\binom{i}{m} = \binom{n+1}{m+1}$, $n \in N$

I10 $\sum_{i \geqslant 0}\binom{r}{i}\binom{t}{n-i} = \binom{r+t}{n}$, $n \in I$

Table 8.1 Some identities.

8.2 PERMUTATIONS AND COMBINATIONS

Let $S=\{x_1, x_2, ..., x_n\}$ be a set. $Y = y_1y_2...y_k$ is a *selection sequence* from S iff $y_i \in S$, $1 \leqslant i \leqslant k$. k is the *length* of the selection sequence. If $S = \{a,b,c\}$, then *aba*, *aaa*, *abb*, *bbb*, *abc*, etc. are selection sequences of length three; *abca*, *abbc*, *bbcc*, etc. are selection sequences of length 4; etc. Two selection sequences $Y = y_1y_2...y_k$ and $Z = z_1z_2...z_r$ are *different* iff $k \neq r$ or $y_i \neq z_i$ for some i. *ab* and *abc* are different selection sequences. So also are *abc* and *bbc*; *a* and *c*; *aab* and *bca*.

Since y_1 can be any of $x_1, x_2, ..., x_n$ and since the x_is are distinct (as S is a set), the number of different selection sequences of length 1 is n. The number of different selection sequences of length 2 is n^2 as there are n choices for y_1 and n choices for y_2. In general, there are n^k different selection sequences of size k. This leads to the following proposition:

Proposition 8.1: If y_i is selected from n_i distinct items, $1 \leqslant i \leqslant k$, then there are $n_1 \times n_2 \times ... \times n_k$ different selection sequences $y_1y_2...y_k$. $\square$

Example 8.1: An identifier is defined to be a letter from the alphabet $\{A,B, ..., Z\}$ followed by one or more characters from the set $\{A,B, ..., Z, 0, 1, ...9\}$. From Proposition 8.1, it follows that there are 26 different identifiers of length 1; 26×36 of length 2; $26 \times 36 \times 36$ of length 3; and 26×36^i of length $i+1$, $i \geqslant 0$. $\square$

Example 8.2: The Computer Science department at some university advertised three openings for assistant professors; one in each of the following specialty areas: software, systems, and computer architecture. 6 applications were received for the first opening; 8 for the second; and 2 for the third. No one applied for more than one of the positions. The number of different outcomes of the recruiting process is $7 \times 9 \times 3 = 189$. Note that $6 \times 8 \times 2$ is not correct as it is quite likely that a position may not get filled. So, there are 7 outcomes for the software position (i.e., it stays open or one of the six applicants fill it) [Historical remark: These three positions are still unfilled]. $\square$

Example 8.3: A green ball and a red ball are to be placed into two boxes marked 1 and 2. Each box is of infinite capacity. Every placement of the balls is a selection sequence from the set $\{1,2\}$. The selection sequence 21 corresponds to placing the green ball into box 2 and the red one into box 1. The sequence 11 corresponds to placing both balls into box 1. From Proposition 8.1, we see that there are $2*2 = 4$ different ways to place the two balls into the two boxes. These four ways are 11, 12, 21, and 22. Note that if both balls are of the same color, then the sequences 12 and 21 are indistinguishable and there are only three different ways to place 2 identical balls into two marked boxes. The number of ways to place 3 distinguishable balls into 5 marked boxes is $5*5*5 = 125$. In general, there are m^n ways in which n distinguishable balls can be placed into m distinguishable boxes when the boxes have a capacity that is at least n. □

Let $S = \{x_1, x_2, ..., x_n\}$ be a set. A $k-permutation$ from S is a selection sequence $Y = y_1 y_2 ... y_k$ such that $y_i \neq y_j$, $i \neq j$. So, all the y_is are distinct. When $k = n$, a k-permutation is simply called a permutation of A. The number of different k-permutations, $k \leqslant n$, is denoted by $P(n,k)$. It is readily seen that $P(n, 0) = 1$ and $P(n, 1) = n$. $P(n, 2) = n(n-1)$ as there are n choices for y_1 and only $n-1$ for y_2 (y_2 must be different from y_1); $P(n, 3) = n(n-1)(n-2)$ as there are n choices for y_1, $n-1$ for y_2, and $n-2$ for y_3 (y_3 is different from both y_1 and y_2).

Proposition 8.2: $P(n,k) = n(n-1) \ ... \ (n-k+1)$, $1 \leqslant k \leqslant n$. $P(n,n) = n(n-1) \ ... \ 1 = n!$. By convention, $0! = 1$. So, $P(n,k) = \dfrac{n!}{(n-k)!}$. □

Example 8.4: Suppose that we have invited 8 guests to dinner and that 8 places have been set at the table. The number of different ways to seat these 8 guests is $P(8, 8) = 8!$ If only 6 of the 8 invitees actually come, then there are $P(8,6) = 8!/2!$ ways to seat the 6 guests who come. Every seating arrangement is a 6-permutation from the set of 8 seats. □

Example 8.5: A certain Computer Science department has 6 faculty offices but only 3 faculty. Every assignment of offices to faculty is a 3-permutation from the set of 6 offices. Hence, there are $P(6,3) = 6*5*4 = 120$ different ways in which the offices may be assigned. □

Example 8.6: Suppose that k differently colored balls are to be placed in n boxes marked 1, 2, ..., n, $k \leqslant n$. If each box has a capacity of 1, every placement of the k balls corresponds to a k-permutation from the set $\{1, 2, ..., n\}$. For example, the 3-permutation 263 corresponds to placing the first ball into box 2; the second into box 6; and the third into box 3. The number of different placements of the k balls is $P(n,k)$. When $k=n=2$, there are two placements (12 and 21). The placements 11 and 22 of Example 8.3 are invalid as a box can now hold at most 1 ball. There are $P(5,3) = 60$ ways to place 3 different balls into 5 marked boxes when each box has a capacity of 1. □

Let $Y = y_1 y_2 ... y_k$ be a permutation from the set $S = \{x_1, ..., x_n\}$. The set $\{y_1, ..., y_k\}$ is called a $k-combination$ from S. Let $C(n,k)$ denote the number of different k-combinations from S. Observe that there are $k!$ permutations of $\{y_1,...,y_k\}$. Hence, the set of $P(n,k)$ k-permutations from S may be partitioned into $P(n,k)/k!$ equivalance classes; each equivalence class consisting of k! permutations each of which is a permutation of the other. Let $S = \{a,b,c\}$. Its permutations of size 2 are ab, ba, ac, ca, bc, and cb. These may be partitioned into the equivalence classes $\{ab,ba\}$, $\{ac,ca\}$, and $\{bc,cb\}$. Each of the $P(n,k)/k!$ equivalence classes defines a different k-combination from S. For example, the class $\{ab,ba\}$ defines the combination $\{a,b\}$. Hence,

$$C(n,k) = P(n,k)/k!$$

$$= \frac{n!}{k!(n-k)!}$$

$$= \binom{n}{k}$$

Proposition 8.3: The number of k-combinations of the set $S = \{x_1, x_2, ..., x_n\}$ is $C(n,k) = \binom{n}{k}$. Observe that $C(n, 0) = 1$. □

Example 8.7: A bridge hand consists of 13 cards. Since there are 52 cards in a deck, the number of different bridge hands is $\binom{52}{13} \simeq 1.27*10^{11}$. □

Example 8.8: Let S be a set of size n. The number of different subsets of S that have size k is $C(n,k)$. So, the total number of different subsets of S is $\sum_{k=0}^{n} C(n,k) = \sum_{k=0}^{n} \binom{n}{k} = 2^n$, Eq. (8.7). □

Example 8.9: Let $|S| = n$. We wish to select a subset T of S for which $f(T)$ is minimum over all choices for T. Assume that the time needed to compute $f(Q) = |Q|^2$. One way to find T is to generate all subsets of S and compute $f()$ for each. The one with least $f()$ is T. As there are $C(n,k)$ subsets of size k, it takes $\Theta(k^2\binom{n}{k})$ time to compute $f()$ for all of them. The time complexity of the resulting algorithm is therefore $\Theta(\sum_{k=0}^{n} k^2\binom{n}{k}) = \Theta(n(n+1)2^{n-2})$ (I5, Table 8.1). □.

Example 8.10: k identical balls are to be placed in n different boxes, $k \leqslant n$. Assume that the capacity of each box is 1. One readily sees that every placement of the balls corresponds to a k-combination of the boxes (i.e., the k boxes that are not empty). Hence, there are $C(n,k)$ different ways in which k identical balls may be placed into n boxes when the box capacity is 1. So, there are $C(5, 3) = 10$ ways to place 3 balls into 5 boxes. □

Example 8.11: Mr Jones is planning a vacation to Europe. After much thought, Mr Jones decides that there are 10 cities that interest him. Unfortunately, because of budgetary and time constraints, Mr Jones can visit only 6 of these cities. The number of ways in which these 6 cities may be selected is $C(10, 6) = 210$. An *itinerary* is the order in which the cities are visited. The number of different itineraries is $P(10, 6) = 151,200$.

Suppose that Mr Jones has decided that of the 10 cities in his list, Paris and Zurich must be visited. Now, there are only $C(8, 4) = 70$ ways in which the remaining 4 cities may be selected. The number of possible itineraries is $70*6! = 50,400$.

If Mr Jones decides that at least one of Paris and Zurich must be visited, then the number of itineraries may be obtained as follows. Let A be the number of ways to select 6 cities such that Paris is always included. Let B be the corresponding number when Zurich is always included. Let C be the number of ways to select 6 cities subject to the restriction that Paris and Zurich are always selected. We see that $A = B = C(9, 5) = 126$ and $C = C(8, 4) = 70$. From the inclusion - exclusion theorem (Theorem 3.3), it follows that there are $A + B - C = 182$ ways to select 6 cities such that each selection includes at least one of Paris and Zurich. The number of different itineraries is therefore $182*6! = 131,040$. $\square$

8.3 THE MULTINOMIAL COEFFICIENT

Suppose that $M = \{x_1, x_2, ..., x_n\}$ is a multiset. Let $a_1, a_2, ..., a_q, q \leqslant n$, be the distinct elements of M. Let n_i be the number of times a_i occurs in M, $1 \leqslant i \leqslant q$. Clearly, $\sum\limits_{i=1}^{q} n_i = n$. For convenience, we shall write M as $\{n_1.a_1, n_2.a_2, n_3.a_3, ..., n_q.a_q\}$. $Y = y_1y_2...y_k$ is a *k-permutation* of the multiset M iff at most n_i of the y_is is a_i, $1 \leqslant i \leqslant q$. So, if $M = \{4.a, 1.b, 2.c\}$, then *aaaa, aabc, aacc,* etc. are 4-permutations of M. Note that if $n_i = \infty$, $1 \leqslant i \leqslant q$, then every selection sequence of $\{a_1, ..., a_1\}$ of size k is a *k*-permutation of $M = \{n_1.a_1, n_2.a_2, ..., n_q.a_q\}$ and vice-versa. So, from Proposition 8.1, we obtain:

Proposition 8.4: If $M = \{\infty.a_1, \infty.a_2, ..., \infty.a_q\}$, then M has exactly q^k *k*-permutations. $\square$

When all the n_is are finite, the number of permutations (i.e., n-permutations where $n = \sum\limits_{i=1}^{q} n_i$) may be determined as follows. Let $y_1y_2...y_n$

be an n-permutation. The number of ways to select the n_1 positions for the a_1s is $C(n,n_1)$. This leaves us with $n-n_1$ possible positions for the a_2s. The number of ways to select the n_2 positions for the a_2s is $C(n-n_1,n_2)$. Continuing this line of reasoning (and using Proposition 8.1), we see that the total number of permutations of M is:

$$C(n,n_1)*C(n-n_1,n_2)*C(n-n_1-n_2,n_3) * ... * C(n_q,n_q)$$

$$= \binom{n}{n_1}\binom{n-n_1}{n_2}\binom{n-n_1-n_2}{n_3} \cdots \binom{n_q}{n_q}$$

$$= \frac{n!}{n_1!(n-n_1)!} \frac{(n-n_1)!}{n_2!(n-n_1-n_2)!} \frac{(n-n_1-n_2)!}{n_3!(n-n_1-n_2-n_3)!} \cdots \frac{n_q!}{n_q!0!}$$

$$= \frac{n!}{n_1!\, n_2! \cdots n_q!}$$

$\dfrac{n!}{n_1!\, n_2! \cdots n_q!}$ is called the *multinomial coefficient* and is written as $\begin{pmatrix} n \\ n_1\, n_2 \cdots n_q \end{pmatrix}$. Note that the binomial coefficient $\binom{n}{k}$ equals the multinomial coefficient $\begin{pmatrix} n \\ k\ \ n-k \end{pmatrix}$ when n and k are natural numbers.

Proposition 8.5: The number of permutations of the multiset $m = \{n_1.a_1, n_2.a_2, ..., n_q.a_q\}$ when $n = n_1+n_2+...+n_q$ is given by the multinomial coefficient $\begin{pmatrix} n \\ n_1\, n_2 \cdots n_q \end{pmatrix}$. □

Example 8.12: The number of different ways in which 2 red balls, 3 blue balls, and 4 green ones can be laid out in a line is $\begin{pmatrix} 9 \\ 2\ 3\ 4 \end{pmatrix} = 1260$. □

Corresponding to the binomial theorem (8.5), there is the *multinomial theorem*:

$$(x_1+x_2+...+x_q)^n = \sum \begin{pmatrix} n \\ n_1\, n_2 \cdots n_q \end{pmatrix} x_1^{n_1} x_2^{n_2} \cdots x_q^{n_q} \qquad (8.17)$$

where n is a natural number and the sum is taken over all sequences n_1, n_2, ..., n_q such that the n_is are natural numbers and $\sum_{i=1}^{q} n_i = n$.

The multinomial theorem leads us to the interesting question: How many ways can we assign natural numbers to n_1, n_2, ..., n_q such that $n_1 + n_2 + ... + n_q = n$? Let this number be $S(n,q)$. When $q = 2$ and $n = 4$, the possible combinations are $(0,4)$, $(1,3)$, $(2,2)$, $(3,1)$, and $(4,0)$. So, $S(4,2) = 5$. The number $S(n,q)$ gives us the number of ways in which n identical objects can be assigned to q persons. So, for example, there are $S(n,q)$ ways to allocate n dollars to q projects.

Any assignment of n identical objects to q persons can be viewed pictorially as below:

$$AA\,|\,AAA\,||\,AA\,|\,A\,|||$$

This figure illustrates an assignment of 8 objects (labeled A) to 7 persons (each denoted by $|$). The number of A's to the left of the first $|$ gives n_1. The number between consecutive $|$s give the remaining n_is. So, the above figure represents the assignment $n_1 = 2$, $n_2 = 3$, $n_3 = 0$, $n_4 = 2$, $n_5 = 1$, and $n_6 = n_7 = 0$. The assignment $n_1 = 0$, $n_2 = 1$, $n_3 = 4$, $n_4 = 0$, $n_5 = 1$, $n_6 = 0$, and $n_7 = 2$ is given by the figure:

$$|\,A\,|\,AAAA\,||\,A\,||\,AA\,|$$

One may verify that every sequence of As and $|$s that ends in a $|$ and contains exactly n As and q $|$s represents a possible assignment of n objects to q persons. Since one of the $|$s is required to be at the end, the number of different sequences one can get is the number of permutations of the multiset $\{n.A, q{-}1.|\}$. From Proposition 8.5, it follows that this number is the multinomial coefficient $\begin{pmatrix} n+q-1 \\ n \quad q-1 \end{pmatrix}$ which equals the binomial coefficient $\begin{pmatrix} n+q-1 \\ n \end{pmatrix}$. Hence, $S(n,q) = C(n+q-1, n) = \begin{pmatrix} n+q-1 \\ n \end{pmatrix}$.

Example 8.13: The number of ways to assign k identical balls to n different boxes with infinite capacity is $S(n,k)$. So, 2 identical balls may be placed in 2 boxes in $S(2,2) = C(3,2) = 3$ different ways (Example 8.3). There are $S(5,3) = C(7,5) = 21$ ways to place 3 identical balls into 5 boxes. □

A *k-combination* of the multiset M is a sub-multiset Q of M such that $|Q| = k$. For example, if $M = \{a, a, a, b, c, d, d, e\}$, then $\{a, a, d, d, e\}$ is a 5-combination of M. The number of k-combinations of the multiset $M = \{n_1.a_1, n_2.a_2, ..., n_q.a_q\}$ where $n_i \geqslant k$, $1 \leqslant i \leqslant q$, is also $S(k,q) = \begin{pmatrix} k+q-1 \\ k \end{pmatrix}$. To see this, simply observe that every sequence $m_1, m_2, ..., m_q$ of natural numbers such that $m_i \leqslant n_i$, $1 \leqslant i \leqslant q$, and $m_1 + m_2 + ... + m_q = k$ defines a k-combination.

Proposition 8.6: Let $M = \{n_1.a_1, n_2.a_2, ..., n_q.a_q\}$ be a multiset. The number of k-combinations of M is $\binom{k+q-1}{k}$ when $n_i \geqslant k$, $1 \leqslant i \leqslant q$. $\square$

8.4 DISCRETE PROBABILITY

8.4.1 Events, Trials, and Probability

Consider the following experiments and possible outcomes:

(1) When a coin is tossed, there are two possible outcomes: it lands heads up or tails up. We shall rule out the possibility that the coin lands on its end in a crack or against a wall.

(2) When a dice is thrown, there are six possible outcomes: 1, 2, 3, 4, 5, 6.

(3) There are 52 possible outcomes when a card is drawn from a complete deck. If two cards are drawn, there are $52*51 = 2652$ possible outcomes if the order in which cards are drawn is important. If this order is not important, then there are only $C(52,2) = 1326$ different possible outcomes.

(4) When a red and a blue ball are placed into two boxes, each of capacity at least 2, there are 4 possible outcomes.

We shall use the term *event* to denote the outcome of an experiment. For experiment (1) above, there are two possible events: the coin lands heads up and the coin lands tails up. A *compound event* is an event that can be decomposed into simpler events. For example, if we draw one card from a deck, the event that a 4 is drawn occurs when the four of spades or the four of hearts, or the four of diamonds, or the four of clubs is drawn. Hence, the event that a four is drawn can be decomposed into four simpler events. We shall say that a compound event is the union of the simpler events into which it can be decomposed.

An event that cannot be decomposed further is called a *simple event*. The event that a four of spades is drawn is a simple event. When a coin is tossed, the event that it lands heads up is a simple event. Simple events are also called *sample points* and the union of all sample points is called the *sample space*. A sample space is said to *discrete* iff it is either finite or countably infinite. In this section, we shall be concerned only with discrete sample spaces.

Example 8.14: When two colored balls ($r = $ *red*, $b = $ *blue*) are placed into 3 boxes (1,2,3) of infinite capacity, there are 9 possible simple events (Figure 8.1). Each of these 9 simple events is a sample point and the sample space consists of exactly these 9 points. The event that at least one box contains more than one ball is a compound event composed of the sample points 1, 2, and 3. The event that each box contains at most one ball is composed of the sample points 4 through 9. □

| event | 1 | 2 | 3 | 4 | 5 | 6 | 7 | 8 | 9 |
box									
1	rb	-	-	r	b	r	b	-	-
2	-	rb	-	b	r	-	-	r	b
3	-	-	rb	-	-	b	r	b	r

Figure 8.1 Simple events for Example 8.14.

Example 8.15: If a coin is tossed 3 times, the simple events are *HHH, HHT, HTH, HTT, THH, THT, TTH,* and *TTT* (*H* denotes heads up and *T* denotes tails up). The compound event: all three tosses have the same outcome, consists of the sample points *HHH* and *TTT*. □

Let $S = \{e_1, e_2, ...\}$ be a discrete sample space. Let $P(e_i)$ denote the probability that event e_i occurs when the experiment is conducted. Intuitively, the probability of an event is its likelihood of occurrence. An event with a probability of zero has no chance of occurring while an event with a probability of one will occur with certainty. It is easy to see that:

(1) $0 \leqslant P(e_i) \leqslant 1$, for every event e_i

and

(2) $\sum_i P(e_i) = 1$

Example 8.16: If an unbiased coin is tossed, the probability of getting a head equals that of getting a tail. So, $P(H) = P(T) = 0.5$. If this coin is tossed 3 times, then there are 8 sample points (Example 8.15). Each of these can occur with equal likelihood. So, $P(e_i) = 1/8$ for each of these 8 points. □

Let $S = \{e_1, ..., e_k\}$ be a finite sample space. If a sample point is drawn from this space at *random*, then the sample point drawn is e_i with probability $1/k$, $1 \leqslant i \leqslant k$. The probability that any compound event occurs is the sum of the probabilities of the simple events it is composed of.

Example 8.17: Consider placing 2 colored balls into 3 marked boxes. There are 9 sample points corresponding to this experiment (Example 8.14). With each of these sample points, we may associate a probability of 1/9. The probability that a random placement of the balls results in two empty boxes is $P(1) + P(2) + P(3) = 1/3$. The probability that a random placement results in having the red ball in box 1 is $P(1) + P(4) + P(6) = 1/3$. □

Example 8.18: Suppose that an unbiased coin is tossed 3 times (see Examples 8.15 and 8.16). The probability that all three tosses have the same outcome, is $P(HHH) + P(TTT) = 1/4$. The probability that exactly two of the tosses have the same outcome is $P(HHT) + P(HTH) + P(HTT) + P(THT) + P(THH) + P(TTH) = 3/4$. □

Example 8.19: Assume that 4 cards are drawn at random from a deck of 52 cards. The number of sample points for this experiment is $C(52,4) = 270{,}725$. Since the cards are drawn at random, each sample point has the same probability. The probability of drawing 4 aces is $1/270{,}725$. The probability that all 4 cards have the same face value is $13/270{,}725$ as there are exactly 13 sample points corresponding to this compound event. There are $C(13,4) = 65$ sample points consisting only of spades. Hence, the probability that all four of the cards drawn are spades is $65/270{,}725$. The probability that all four cards are of the same suit is four times this, i.e., $260/270{,}725$. □

From the inclusion - exclusion theorem (Theorem 3.3), we see that if A and B are two compound events, then:

$$P(A \cup B) = P(A) + P(B) - P(A \cap B) \qquad (8.18)$$

If $A \cap B = \varnothing$ then the events A and B are *mutually exclusive*. For example, if a coin is tossed three times, the two events (1) all three tosses have the same outcome and (2) exactly two tosses have the same outcome (Example 8.18) are mutually exclusive. From (8.18), we see that the probability that at least two tosses have the same outcome is P(exactly three outcomes the same) + P(exactly two outcomes the same) = 1. The event A: the first toss has outcome heads, consists of the four sample points HHH, HHT, HTH, and HTT. The probability of this event is 1/2. The event B: the third toss results in a tail, consists of the sample points TTT, HTT, HHT, and THT. This event has probability 1/2. From (8.18), we see that:

$$P(A \cup B) = P(A) + P(B) - P(A \cap B)$$
$$= 1 - P(\{HHT, HTT\})$$
$$= 1 - 1/4$$
$$= 3/4$$

So, the probability that the first toss is a head or the third is a tail is 3/4.

From the inclusion - exclusion theorem, Boole's inequality:

$$P(A_1 \cup A_2 \cup \cdots \cup A_k) \leqslant P(A_1) + P(A_2) + \cdots + P(A_k)$$

is readily obtained.

8.4.2 Conditional Probability

Let A and B be two events such that $P(B) > 0$. The probability of A given that B has occurred is called the *conditional probability* of A given B. It is denoted $P(A|B)$. It is not too difficult to justify:

$$P(A|B) = \frac{P(A \cap B)}{P(B)} \qquad (8.19)$$

When all sample points in the sample space have equal probability, then (8.19) becomes:

$$P(A|B) = |A \cap B| / |B|$$

where $|B|$ is the number of sample points composing B.

Example 8.20: A coin is tossed three times. Assume that the first toss resulted in a head. The probability of getting at least two heads given that the first toss resulted in a head is $P(\{HHH, HHT, HTH\}) / P(\{HHH, HHT, HTH, HTT\}) = 3/4$. □

Example 8.21: Assume that 60% of the population is male and that 80% of the males are over 5ft tall. Further, assume that only 50% of the females are over 5ft tall. If a person is selected at random from the population, then the probability that this person is male is 0.6. The probability that the person is a male over 5ft is $0.6*0.8 = 0.48$. If we are given that the person selected is male, then the probability that the selected person is over 5ft tall is 0.8. The probability that the person selected is a female under 5ft is $0.4*0.5 = 0.2$ while the probability that the person selected is under 5ft given that this person is female is 0.5. □

Example 8.22: A certain money bag contains 5 coins. Two of these are fair (i.e., unbiased coins with a head on one side and a tail on the other) and the remaining three are double headed (i.e., both sides have a head on them). Let $F1$, $F2$ denote the fair coins, and let $Z1$, $Z2$, $Z3$ denote the three double headed coins.

If three coins are drawn at random from this money bag, the set of coins drawn could be any one of: $\{F1, \ F2, \ Z1\}$; $\{F1, \ F2, \ Z2\}$; $\{F1, \ F2, \ Z3\}$; $\{F1, \ Z1, \ Z2\}$; $\{F1, \ Z1, \ Z3\}$; $\{F1, \ Z2, \ Z3\}$; $\{F2, \ Z1, \ Z2\}$; $\{F2, \ Z1, \ Z3\}$; $\{F2, \ Z2, \ Z3\}$; and $\{Z1, \ Z2, \ Z3\}$. So, there are 10 possible simple events for the drawing of three coins. Since, the coins are drawn at random, the probability of each of these events is 1/10.

The event "three double headed coins are drawn" has probability 1/10 as there is only one sample point (or simple event) that corresponds to this event. The event "two fair coins are drawn" has probability 3/10 as 3 sample points correspond to this event.

Now, suppose that the three coins selected are dropped onto a table. Let n_d denote the number of double headed coins selected. The probability that all three coins fall heads up on the table is:

$$P(3 \text{ heads up}) = P(n_d = 3) + P(n_d = 2 \text{ and the fair coin lands heads up}) +$$
$$P(n_d = 1 \text{ and both fair coins land heads up})$$
$$= 1/10 + (6/10)*(1/2) + (3/10)*(1/4)$$
$$= 19/40$$

In general, if we have p fair coins and q double headed coins and r coins are drawn, the number of sample points with exactly k fair coins is $\binom{p}{k}\binom{q}{r-k}$ (observe that $\binom{p}{k}$ is the number of ways to choose k fair coins while $\binom{q}{r-k}$ is the number of ways to select $r-k$ double headed coins).

If we are given that the 3 coins on the table are heads up, the probability that these three coins are double headed is:

$$P(n_d = 3 \mid 3 \text{ heads up}) = \frac{P(n_d = 3 \text{ and 3 heads up})}{P(3 \text{ heads up})}$$

$$= \frac{P(n_d = 3)}{P(3 \text{ heads up})}$$

$$= \frac{1/10}{19/40}$$

$$= 4/19$$

The probability that $n_d = 2$ given that the 3 coins on the table are heads up is:

$$P(n_d = 2 \mid 3 \text{ heads up}) = \frac{P(n_d = 2 \text{ and } 3 \text{ heads up})}{P(3 \text{ heads up})}$$

$$= \frac{P(n_d = 2 \text{ and the fair coin is heads up})}{P(3 \text{ heads up})}$$

$$= \frac{6/20}{19/40}$$

$$= 12/19 \quad \square$$

Two events A and B are said to be *independent* (or stochastically or statistically independent) iff $P(A \mid B) = P(A)$. From (8.19), we see that this requirement is equivalent to the requirement that $P(A \cap B) = P(A)*P(B)$. This in turn implies that $P(A \mid B) = P(A)$ iff $P(B \mid A) = P(B)$.

Example 8.23: If a card is drawn at random from a deck, then the events "it is a club" and "it is a 4" are independent. In fact, we see that P(it is a club) = 1/4 , P(it is a 4) = 1/13, and P(it is a 4 of clubs) = 1/52 = (1/4)*(1/13). $\square$

Example 8.24: Suppose that on a multiple choice test there are 2 choices for question 1, 4 for question 2, and 5 for question 3. If a student selects the answers at random, then the probability of selecting the right answer is 1/2 for question 1, 1/4 for question 2, and 1/5 for question 3. The probability that all three answers are correct is 1/(2*4*5) = 1/40. (If you want to do well on a multiple choice test, study for it.) $\square$

8.4.3 Bernoulli Trials

Any experiment may be repeated several times. For example, a coin may be repeatedly tossed. A *trial* is the performance of an experiment once. Two trials are independent iff the outcome of the second is in no way influenced by the first trial. A trial is a *Bernoulli trial* iff it has only two possible outcomes. The two possible outcomes of a Bernoulli trial may be called *success* (S) and *failure* (F).

The tossing of a coin is a Bernoulli trial as there are only two possible outcomes (H or T). The drawing of a card from a deck can have 52 different outcomes. So, this is not a Bernoulli trial. If we are interested only in whether the card drawn is a spade or not, there are only two outcomes possible. The trial is a success if a spade is drawn and a failure otherwise. There are 5 possible outcomes of taking the discrete structures course: A, B, C, D, N (we exclude the outcome: insanity). If we are interested only in whether one passes (A, B, C, or D) or fails (N), then we have a Bernoulli trial.

Let p be the probability that the outcome of a Bernoulli trial is a success. Let $q = 1-p$ be the probability of a failure. If n *independent Bernoulli trials* are made, the probability of success on the kth trial is p, $1 \leqslant k \leqslant n$.

Example 8.25: If an unbiased coin is tossed, the probability of a head, $P(H)$, is 1/2. Assume that the coin does in fact land heads up. If it is tossed again, the probability of getting a head the second time is still 1/2. I.e., the results of successive throws of a coin are independent. The probability of getting two heads in two throws is $1/2*1/2 = 1/4$. The probability of n heads in n throws is $1/2^n$. $\square$

We are often interested in the number of successes in n trials. For instance, we would like to know how many times a coin landed heads up in n tosses; or how many times a spade is drawn in n draws; etc.

The outcome of n independent Bernoulli trials may be represented by an n-vector $(T_1, ..., T_n)$ where $T_i = S$ or F, $1 \leqslant i \leqslant n$. If $(T_1, ..., T_n)$ has j Ss, then its probability is $p^j q^{n-j}$. The number of vectors with j Ss is $C(n,j)$. Hence, the probability, $B(n,p,j)$, of j successes in n independent Bernoulli trials is:

$$B(n,p,j) = \binom{n}{j} p^j q^{n-j} \tag{8.20}$$

As one would expect, $\sum_{j=0}^{n} B(n,p,j) = \sum_{j=0}^{n} \binom{n}{j} p^j q^{n-j} = (p+q)^n = 1$ (Eq. (8.6)). $B(n,p,j)$ is also called the *binomial distribution with parameter p*.

Example 8.26: If an unbiased coin is tossed n times, the probability of getting a head exactly j times is $B(n, 0.5, j) = \binom{n}{j}(0.5)^n$. The probability of getting a head exactly once is 1/2 when one or two tosses are made; it is 3/8 when three tosses are made; and 1/4 when four tosses are made. The probability of getting at most k heads is $\sum_{j=0}^{k} B(n,.5,j)$. So, the probability of getting at most one head is 1 when the number of tosses is 1. This probability is 3/4 when the number of tosses is 2; 1/2 when 3 tosses are made; and 5/16 when 4 tosses are made. $\square$

Example 8.27: Let the probability of success in a Bernoulli trial be 0.1. What is the minimum number of trials that must be performed so that the probability of at least one success is no less than 0.99? This number may be determined as follows. If n trials are performed, the probability of at least one success is $1 - P(\text{zero successes}) = 1 - \binom{n}{0}q^n = 1 - q^n$. So, we want that $1-q^n \geqslant 0.99$, or $1-(0.9)^n \geqslant 0.99$, or $(0.9)^n \leqslant 0.01$. This inequality is satisfied whenever $n \geqslant 44$. Hence, if at least 44 trials are performed, there will be at least one success with probability at least 0.99 $(1-q^{44} = 0.99031)$. $\square$

8.4.4 Random Variables

A *random variable* is a function with domain equal to the sample space of some experiment. The range of this function is some subset of the real numbers. For instance, the sample space corresponding to three tosses of a coin is $\{HHH, HHT, HTH, HTT, THH, THT, TTH, TTT\}$. The function X such that $X(e)$ equals the number of heads in the simple event e ($X(HHH)=3$, $X(TTT)=0$, etc.) is an example of a random variable.

Let X be a random variable with range $\{r_1, r_2,...\}$. The union of the events e in the sample space S for which $X(e) = r_i$ defines the event $X=r_i$. Let $f(r_i) = P(X=r_i)$ be the probability of the event $X=r_i$. f is called the *distribution* (or probability distribution) of the random variable X. The *expected* or *mean* value of X (denoted $E(X)$ or $\bar{X}$) is defined to be:

$$E(X)=\bar{X}=\sum_i r_i f(r_i) \tag{8.21}$$

provided that $\sum_i |r_i f(r_i)|$ converges. $E(X)$ is undefined otherwise.

Example 8.28: Let Q be a set of size n. The sample space S defined by the set of all subsets of Q is of size 2^n. With each point in S, we may associate a probability $1/2^n$. On the sample space S, we may define a random variable X with value equal to the size of the subset. So, $X(\{1\}) = 1$, and $X(\{4,6,8,9\}) = 4$. The random variable X may take on the values 0, 1, 2, ..., n. As there are $\binom{n}{k}$ subsets in S of size k, $P(X=k) = \binom{n}{k}/2^n$, $0 \leqslant k \leqslant n$. Hence, the expected value $\bar{X}$ of X is $(\sum_{k=0}^{n} k\binom{n}{k})/2^n = n2^{n-1}/2^n = n/2$ (14, Table 8.1). In other words, if a subset of S is selected at random, its size is expected to be $n/2$. $\square$

Example 8.29: Suppose that n Bernoulli trials are performed. Let S_n be the random variable with value equal to the number of successes. The distribution of S_n is given by $B(n,p,j)$. The expected value of S_n is:

$$E(S_n) = \bar{S}_n = \sum_j jB(n,p,j)$$

$$= \sum_{j=0}^{n} j\binom{n}{j}p^j q^{n-j}$$

$$= \sum_{j=1}^{n} n\binom{n-1}{j-1}p^j q^{n-j}$$

$$= np\sum_{j=1}^{n} \binom{n-1}{j-1}p^{j-1}q^{n-j}$$

$$= np\sum_{i=0}^{n-1} \binom{n-1}{i}p^i q^{n-1-i}$$

$$= np(p+q)^{n-1}$$

$$= np$$

The expected number of successes in n Bernoulli trials is np. □

Let X_i, $1 \leq i \leq n$ be n random variables. Assume that $\bar{X}_i$ is defined for all i. From (8.21), it follows that $E(\sum_i X_i)$ is defined and is given by:

$$E(\sum_i X_i) = \sum_i E(X_i) \tag{8.22}$$

Example 8.30: (8.22) may be used to obtain $\bar{S}_n$ (Example 8.29) in a relatively easy way. Consider the random variable X_i such that X_i equals one if the i th Bernoulli trial is a success and is 0 otherwise. Since, $P(X_i=1) = p$ and $P(X_i=0) = q$, $E(X_i) = 1.p + 0.q = p$. As $S_n = \sum_{i=1}^{n} X_i$, from (8.22) we obtain:

$$\bar{S}_n = \sum_{i=1}^{n} \bar{X}_i = np \quad \square$$

$E(X^k)$ is called the *kth moment* of X. From (8.21), we see that:

$$E(X^k) = \sum_i r_i^k f(r_i) \tag{8.23}$$

provided that $\sum_i |r_i^k f(r_i)|$ converges.

The *variance* of X is defined to be:

$$var(X) = E((X - \bar{X})^2) = E(X^2) - \bar{X}^2 \tag{8.24}$$

$\sqrt{var(X)}$ (nonnegative root is used) is the *standard deviation* of X.

Example 8.31: Let S_n be as in Example 8.29. The variance of S_n is:

$$var(S_n) = E(S_n^2) - \bar{S}_n^2$$

$$= \sum_{j=0}^{n} j^2 \binom{n}{j} p^j q^{n-j} - n^2 p^2$$

$$= np \sum_{j=1}^{n} j \binom{n-1}{j-1} p^{j-1} q^{n-j} - n^2 p^2 \quad (\text{Eg. (8.29)})$$

$$= np \sum_{i=0}^{n-1} (i+1) \binom{n-1}{i} p^i q^{n-1-i} - n^2 p^2$$

$$= np \sum_{i=0}^{n-1} i \binom{n-1}{i} p^i q^{n-1-i} + np \sum_{i=0}^{n-1} \binom{n-1}{i} p^i q^{n-1-i} - n^2 p^2$$

$$= np \sum_{i=0}^{n-1} i \binom{n-1}{i} p^i q^{n-1-i} + np - n^2 p^2$$

$$= np(n-1)p + np - n^2 p^2 \quad (\text{Eg. (8.29)})$$

$$= n^2 p^2 - np^2 + np - n^2 p^2$$

$$= np(1-p)$$

$$= npq \quad \square$$

Let X and Y be two random variables defined on the same sample space $S = \{e_1, e_2, \cdots\}$. Let f and g, respectively, be the probability distributions of X and Y. Assume that the values X and Y can, respectively, take on are $r_1, r_2, \cdots$ and $s_1, s_2, \ldots$. The union of the events in S for which $X = r_i$ and $Y = s_j$ defines the event $(X = r_i$ and $Y = s_j)$. The probability of this event is denoted $P(X = r_i$ and $Y = s_j)$ and is called the *joint probability distribution* of X and Y.

Let $h(r_i, s_j) = P(X = r_i$ and $Y = s_j)$. We readily see that:

$$h(r_i, s_j) = P(X = r_i | Y = s_j) * g(s_j) \tag{8.25}$$

$$= P(Y = s_j | X = r_i) * f(r_i)$$

The random variables X and Y are *independent* iff $h(r_i,s_j) = f(r_i)g(s_j)$ for every i and j. From this and (8.25), we see that X and Y are independent iff $P(Y=s_j|X=r_i) = g(s_j)$ and $P(X=r_i|Y=s_j) = f(r_i)$.   Hence,   the definition of independence for random variables is consistent with that for events.

The expectation, $E(XY)$, of $X*Y$ is given by:

$$E(XY) = \sum_{i,j} r_i s_j h(r_i,s_j)$$

provided that $\sum_{i,j} |r_i s_j h(r_i,s_j)|$ converges. Since,

$$|r_i s_j| = \left| \frac{(r_i+s_j)^2 - r_i^2 - s_j^2}{2} \right|,$$

$E(XY)$ is defined whenever $E(X^2)$ and $E(Y^2)$ are.

The *covariance*, $cov(X,Y)$, of X and Y is defined to be:

$$cov(X,Y) = E((X-\bar{X})(Y-\bar{Y})) \tag{8.26}$$

If we assume that $E(X^2)$ and $E(Y^2)$ are defined, then we obtain

$$E((X-\bar{X})(Y-\bar{Y})) = E(XY-X\bar{Y}-\bar{X}Y+\bar{X}\bar{Y})$$
$$= E(XY)-\bar{Y}E(X)-\bar{X}E(Y)+\bar{X}\bar{Y}$$
$$= E(XY)-\bar{X}\bar{Y}$$

Hence, (8.26) is equivalent to:

$$cov(X,Y) = E(XY)-\bar{X}\bar{Y} \tag{8.27}$$

We note that if X and Y are mutually independent then $cov(X,Y) = 0$. The reverse is not true (see the exercises).

Let X_1, X_2, ..., X_n be n random variables. Assume that $var(X_i)$ is defined for all i. One may show (see the exercises) that:

$$var\left(\sum_{i=1}^{k} X_i\right) = \sum_{i=1}^{k} var(X_i) + 2\sum_{\substack{i,j \\ i<j}} cov(X_i,X_j) \tag{8.28}$$

Example 8.32: The variance of S_n (Example 8.31) is easily obtained using (8.28). Let X_i, $1 \leqslant i \leqslant n$ be the random variables defined in Example 8.30. Since the n Bernoulli trials are independent, X_1, X_2, ..., X_n are independent random variables. Hence, $cov(X_i, X_j) = 0$, $i \neq j$. From (8.28), we obtain:

$$
\begin{aligned}
var(S_n) &= \sum_{i=1}^{n} var(X_i) \\
&= n \ var(X_1) \\
&= n(E(X_1^2) - \overline{X}_1^2) \\
&= n(1.p + 0.q - p^2) \\
&= np(1-p) \\
&= npq \quad \square
\end{aligned}
$$

8.4.5 Chebychev's Inequality

Let X be a random variable with mean $\overline{X}$. What is the probability that the value of X is within ϵ, $\epsilon > 0$, of $\overline{X}$? In this section, we shall provide an upper bound on this probability.

Theorem 8.1: [Chebychev's Inequality] Let X be a random variable with mean $\overline{X}$ and variance $var(X)$. Let $\epsilon > 0$ be a constant.

$$
P(|X - \overline{X}| \geqslant \epsilon) \leqslant \frac{var(X)}{\epsilon^2}
$$

Proof: Let $R = \{r_1, r_2, ..., \}$ be the range of X. Let $Y = \{y | y \in R$ and $|y - \overline{X}| \geqslant \epsilon\}$. We obtain:

$$
P(|X - \overline{X}| \geqslant \epsilon) = \sum_{y \in Y} P(y)
$$

$$
\leqslant \sum_{y \in Y} \left[\frac{y - \overline{X}}{\epsilon}\right]^2 P(y) \ (\text{ as } \left|\frac{y - \overline{X}}{\epsilon}\right|^2 \geqslant 1 \text{ for } every \ y \in Y)
$$

$$
\leqslant \sum_{i} \left[\frac{r_i - \overline{X}}{\epsilon}\right]^2 P(r_i)
$$

$$
= \frac{var(X)}{\epsilon^2} \quad \square
$$

From Chebychev's inequality, we see that if X has a "small" variance, then there is a "high" probability that it will have a value "near" its mean.

Example 8.33: Let S_n be the number of successes in n independent Bernoulli trials. $\overline{S}_n$ = np and var(S_n) = npq (Examples 8.30 and 8.32). S_n/n is the average number of successes in n trials and $E(S_n/n) = p$ and $var(S_n/n) = pq/n$. From Chebychev's inequality, it follows that:

$$P(|S_n/n - p| \geqslant \epsilon) \leqslant \frac{pq}{n\epsilon^2}$$
$$\rightarrow 0 \text{ as } n \rightarrow \infty$$

In words, the probability that S_n/n deviates from its mean approaches 0 as n gets larger and larger. This agrees with our intuitive understanding of the mean of a random variable. □

Theorem 8.2: [One-sided Chebychev's Inequality] Let X be as in Theorem 8.1. Let $\epsilon > 0$ be a constant.

(1) $P(X \geqslant \overline{X} + \epsilon) \leqslant \dfrac{var(X)}{var(X) + \epsilon^2}$

(2) $P(X \leqslant \overline{X} - \epsilon) \leqslant \dfrac{var(X)}{var(X) + \epsilon^2}$

Proof: Left as an exercise. □

8.4.6 Law Of Large Numbers

Theorem 8.3: [Weak law of large numbers] Let $X_1, ..., X_n$ be n independent random variables.   Let $\overline{X}_i$ and var(X_i), respectively, be the mean and variance of X_i, $1 \leqslant i \leqslant n$. Let $Q_n = \sum\limits_{i=1}^{n} X_i$ and let $\epsilon > 0$ be a constant.

$$P\left(\left|\frac{Q_n - \overline{Q}_n}{n}\right| \geqslant \epsilon\right) \leqslant \frac{1}{\epsilon^2 n^2} \sum_{i=1}^{n} var(X_i)$$

Proof: From Chebychev's inequality (Theorem 8.1), we obtain:

$$P\left(\left|\frac{Q_n - \overline{Q}_n}{n}\right| \geqslant \epsilon\right) \leqslant \frac{var(Q_n/n)}{\epsilon^2}$$

$$= \frac{1}{\epsilon^2 n^2} var(Q_n)$$

$$= \frac{1}{\epsilon^2 n^2} \sum_{i=1}^{n} var(X_i)$$

The last equality comes from (8.28) and the fact that the X_is are independent. $\square$

From Theorem 8.3, we see that when the X_is have the same distribution,

$$P\left(\left| \frac{Q_n - \bar{Q}_n}{n} \right| \geqslant \epsilon \right) \leqslant \frac{var(X_1)}{\epsilon^2 n} \tag{8.29}$$

Example 8.34: Let $X_i = 1$ if the i th Bernoulli trial is a success and 0 otherwise. Let $S_n = \sum_{i=1}^{n} X_i$. In Example 8.32, we showed that $var(X_i) = pq$. If the n Bernoulli trials are independent, then from (8.29) we obtain:

$$P\left(\left| \frac{S_n}{n} - p \right| \geqslant \epsilon \right) \leqslant \frac{pq}{\epsilon^2 n}, \epsilon > 0$$

$$\rightarrow 0 \text{ as } n \rightarrow \infty$$

This is the same result as obtained in Example 8.33. $\square$

Let X_i, $1 \leqslant i \leqslant m$ be m independent random variables with finite mean and variance. Let n, $1 \leqslant n \leqslant m$ be a natural number and let $\epsilon_i > 0$, $n \leqslant i \leqslant m$ be constants. The weak law of large numbers asserts that if n is sufficiently large, then $\frac{1}{n} Q_n$ will be close (within ϵ) to $\frac{1}{n} \bar{Q}_n$ with probability "close to" 1. It however does not assert that $\frac{1}{n+i} Q_{n+i}$ will also be close (within ϵ_{n+i}), to $\frac{1}{n+i} \bar{Q}_{n+i}$ for every i, $1 \leqslant i \leqslant m-n$ with probability "close to" 1. This stronger assertion is made by the strong law of large numbers.

Theorem 8.4: [Strong law of large numbers] Let X_i, $1 \leqslant i \leqslant m$ be m independent random variables with finite mean and variance. Let n, $1 \leqslant n \leqslant m$ be a natural number and let $\epsilon_i > 0$, $n \leqslant i \leqslant m$ be constants.

$$P\left(\left|\frac{Q_{n+i}-\bar{Q}_{n+i}}{n+i}\right| \geqslant \epsilon \text{ for at least one } i,\ 0 \leqslant i \leqslant m-n\right)$$

$$\leqslant \frac{1}{n^2 \epsilon_n^2} \sum_{i=1}^{n} var\,(X_i) + \sum_{i=n+1}^{m} \frac{var\,(X_i)}{i^2 \epsilon_i^2}$$

Proof: See the text by Thomasian cited at the end of this chapter. □
For random variables with the same distribution and $\epsilon = \epsilon_n = \epsilon_{n+1} = \dots = \epsilon_m$, Theorem 8.4 reduces to:

$$P\left(\left|\frac{Q_{n+i}-\bar{Q}_{n+i}}{n}\right| \geqslant \epsilon \text{ for at least one } i,\ 0 \leqslant i \leqslant m-n\right) \qquad (8.30)$$

$$\leqslant \frac{var\,(X_1)}{\epsilon^2}\left(\frac{1}{n} + \sum_{j=n+1}^{m} \frac{1}{j^2}\right)$$

$$\leqslant \frac{2\,var\,(X_1)}{n\epsilon^2}$$

Example 8.35: Let X_i be 1 if the i th Bernoulli trial is a success and 0 otherwise, $1 \leqslant i \leqslant m$. Let n, $1 \leqslant n \leqslant m$ be a natural number. Let $S_j = \sum_{i=1}^{j} X_i$. The probability that the fraction of successes, S_j/j, is between $p - \epsilon$ and $p + \epsilon$ for every j, $n \leqslant j \leqslant m$ is:

$$P\,|S_j/j-p|<\epsilon, n \leqslant j \leqslant m)$$

$$= 1 - P(|S_j/j-p| \geqslant \epsilon, \text{ for at least one } j,\ n \leqslant j \leqslant m)$$

$$\geqslant 1 - \frac{2\,var\,(X_1)}{n\epsilon^2}$$

$$= 1 - \frac{2pq}{n\epsilon^2}$$

If $p = 0.4$ and $\epsilon = 0.01$, then the probability that the fraction of successes is 0.4 ± 0.01 for every n, $n \geqslant 10,000$ is at least 0.52. □

REFERENCES AND SELECTED READINGS

Some excellent references for this chapter are:

Introduction to combinatorial mathematics, by C.L. Liu, McGraw-Hill Book Co., New York 1968.

Introductory combinatorics, by R. Brualdi, Elsevier North-Holland Inc., New York, 1977.

The art of computer programming: fundamental algorithms, by D.E.Knuth, Vol. 1, 2nd Edition, Addison-Wesley, 1973.

An introduction to probability theory and its applications, by W. Feller, Vol. 1, John Wiley and Sons, 1968.

and

The structure of probability theory with applications, by A. Thomasian, McGraw-Hill Book Co., NY, 1969.

EXERCISES

1. Compute the following:

   (a) $\binom{9}{3}$

   (b) $\binom{10}{2}$

   (c) $\binom{10}{-1}$

   (d) $\binom{3}{0}$

   (e) $\binom{-5}{2}$

   (f) $\binom{3.6}{2}$

   (g) $\binom{-3.6}{3}$

   (h) $\binom{4}{8}$

2. Determine the error in Stirling's approximation (8.3) for $n!$ when n = 5, 10, 15, and 20.

3. Use (8.14) to obtain a table that gives the value of $\binom{r}{i}$ for $r \in N$, $i \in N$, and $0 \leqslant r \leqslant 10$, $0 \leqslant i \leqslant 10$.

4. Show that $\binom{n}{\lfloor n/2 \rfloor} \geqslant \binom{n}{i}$, $0 \leqslant i \leqslant n$. n is a natural number.

5. By differentiating (8.11) with respect to x, show that $\sum_{i=1}^{n} i^2 \binom{n}{i} = n(n+1)2^{n-2}$, n a natural number

6. Obtain a formula for $\sum_{i=1}^{n} i^2 \binom{n}{i} 2^i$, (Hint: differentiate (8.11)).

7. Obtain a formula for $\sum_{i=1}^{n} i^3 \binom{n}{i}$, n a natural number.

8. Show that $\sum_{i \text{ odd}} \binom{n}{i} = \sum_{i \text{ even}} \binom{n}{i}$ when n is an odd natural number. Is this equality true when n is even?

9. Prove the identities (8.14), (8.15), and (8.16).

10. Prove identity I10 of Table 8.1.

11. Use induction on n to show that $F(n) = \sum_{i=0}^{m} \binom{n-i}{i}$ where $m = \lfloor n/2 \rfloor$. $F(n)$ is the nth Fibonacci number.

12. Show that:

$$\binom{r}{n}\binom{n}{i} = \binom{r}{i}\binom{r-i}{n-i}$$

for all real r and all $k \in I$ and all $m \in I$.

13. Suppose that algorithm A generates every subset of the set S, $|S| = n$, exactly once; and that for each subset of size i, it spends $\Theta(i^3)$ time. What can you say about the time complexity of A?

14. Show that the number of ways to allocate n dollars to q projects such that each project gets at least one dollar is $\binom{n-1}{q-1}$ whenever $n \geqslant q$.

15. A box contains 5 red balls, 3 green balls, and 4 blue balls. How many 3-combinations are there?

16. How many even numbers are there in the range 0 to $10^{10}-1$ that contain only distinct decimal digits? (Example 4, 42, 24, 1236, but not 111238).

17. [Stirling Numbers] Stirling numbers of the first kind are denoted $\left[\begin{matrix}n\\j\end{matrix}\right]$ while those of the second kind are denoted $\left\{\begin{matrix}n\\j\end{matrix}\right\}$. These numbers are useful for conversion between binomial coefficients and powers. Stirling numbers are defined as below:

$$n!\left[\begin{matrix}x\\n\end{matrix}\right] = x(x-1)(x-2)...(x-n+1) \tag{8.31}$$

$$= \sum_{i=0}^{n}(-1)^{n-i}\left[\begin{matrix}n\\i\end{matrix}\right]x^i$$

and

$$x^n = \sum_{i=0}^{n}\left\{\begin{matrix}n\\i\end{matrix}\right\}\left(\begin{matrix}x\\i\end{matrix}\right)i! \tag{8.32}$$

We see that

$$3!\left[\begin{matrix}x\\3\end{matrix}\right] = x(x-1)(x-2)$$

$$= x^3 - 3x^2 + 2x$$

Comparing with (8.31), we obtain: $\left[\begin{matrix}3\\0\end{matrix}\right] = 0;\ \left[\begin{matrix}3\\1\end{matrix}\right] = 2;\ \left[\begin{matrix}3\\2\end{matrix}\right] = 3;$ and $\left[\begin{matrix}3\\3\end{matrix}\right] = 1.$

Also,

$$x^2 = 2\left\{\begin{matrix}x\\2\end{matrix}\right\} + \left\{\begin{matrix}x\\1\end{matrix}\right\}$$

Comparing with (8.32), we obtain: $\left\{\begin{matrix}2\\2\end{matrix}\right\} = \left\{\begin{matrix}2\\1\end{matrix}\right\} = 1;$ and $\left\{\begin{matrix}2\\0\end{matrix}\right\} = 0.$ We define $\left[{n \atop j} \right] = \left\{ {n \atop j} \right\} = 0$ for $j > n.$

Use (8.31) and (8.32) to show the following:

(a) $\left[{n \atop n} \right] = \left\{ {n \atop n} \right\} = 1,\ n \geqslant 0$

(b) $\left[{n \atop 0} \right] = \left\{ {n \atop 0} \right\} = 0,\ n > 0$

(c) $\left\{ {n \atop 1} \right\} = 1,\ n > 0$

(d) $\left[{n \atop 1} \right] = (n-1)!,\ n > 0$

(e) $\left[{n \atop j} \right] = (n-1)\left[{n-1 \atop j} \right] + \left[{n-1 \atop j-1} \right],\ n > 0$

(f) $\left\{ {n \atop j} \right\} = j \left\{ {n-1 \atop j} \right\} + \left\{ {n-1 \atop j-1} \right\},\ n > 0$

18. [Stirling Numbers]
    (a) Use the results of Exercise 17 to obtain a table of values for $\left[{n \atop j} \right]$
        and $\left\{ {n \atop j} \right\}$ for $0 \leqslant n \leqslant 10,\ 0 \leqslant j \leqslant n.$
    (b) Use this table to obtain the expansions of:
$$\text{(i) } 4!\binom{x}{4}$$
$$\text{(ii) } 10!\binom{x}{10}$$
$$\text{(iii) } 7!\binom{x}{7}$$

19. [Stirling Numbers] Show that:
    (a) $\left[{n \atop n-1} \right] = \left\{ {n \atop n-1} \right\} = \binom{n}{2},\ n > 0$
    (b) $\left\{ {n \atop 2} \right\} = 2^{n-1} - 1,\ n > 0$

20. A coin is tossed k times.

(a) How many points are there in the sample space?

(b) What are the simple events when $k=4$?

(c) Which points is the event "the coin lands heads up exactly twice" composed of?

(d) Which points is the event "the coin lands heads up at least twice" composed of?

(e) Are the events of (c) and (d) independent?

(f) Assume that the coin is unbiased. What is the probability of event (c)? What is the probability of event (d)? What is the conditional probability of (c) given (d)? What is the conditional probability of (d) given (c)?

21. (a) Let X and Y be two independent random variables for which $E(X)$ and $E(Y)$ are defined. Show that $E(XY) = E(X)E(Y)$.

(b) Give an example of two random variables X and Y that are not independent and for which $\text{cov}(X,Y) = 0$.

22. Prove Eq. (8.28).

23. Suppose that a money bag contains 10 coins; five of which are fair and five double headed. Assume that 5 coins are drawn at random from this bag. These 5 coins are then dropped onto a table and it is observed that all five land heads up. Let n be a random variable with value equal to the number of double headed coins drawn. What is the distribution of n?

24. (a) Show that $B(n,p,j) = \dfrac{n-j+1}{j} \dfrac{p}{q} B(n,p,j-1), j>0.$

(b) Use the result of (a) to obtain $B(10,0.6,j), 0 \leqslant j \leqslant 10.$

(c) From (a) conclude that $B(n,p,j) < B(n,p,j-1)$ for $j<(n+1)p$ and $B(n,p,j)>B(n,p,j-1)$ for $j>(n+1)p.$

(d) Show that if $(n+1)p = m$ is an integer, then $B(n,p,m) = B(n,p,m-1).$

25. [Poisson distribution] The Poisson distribution $R(\lambda,j)$ with parameter λ and j a natural number is defined to be $e^{-\lambda}\lambda^j/j!.$

(a) Let $\lambda=0.5$. Obtain the values $R(\lambda,j), 0 \leqslant j \leqslant 5.$

(b) Show that $\sum\limits_{j=0}^{\infty} R(\lambda,j) = 1$ whenever $\lambda \geqslant 0.$

(c) Obtain the mean of $R(\lambda,j).$

(d) Obtain the variance of $R(\lambda,j).$

26. Show that

$$\lim_{n \to \infty} B(n, \lambda/n, j) = R(\lambda, j)$$

where B is the binomial distribution (8.20) and R is the Poisson distribution (Exercise 25).

27. Let $B(n,p,j)$ be the binomial distribution. Show that $pq \leqslant 1/4$ and hence, $P(|S_n/n - p| \geqslant \epsilon) \leqslant \dfrac{1}{4n\epsilon^2}$ (see Example 8.33).

28. Prove Theorem 8.2.

CHAPTER 9

GRAPHS

9.1 THE BASICS

9.1.1 Introduction

A *graph* $G = (V, E)$ is an ordered pair of finite sets V and E. The elements of V are called *vertices* (vertices are also called *nodes* and *points*). The elements of E are called *edges* (edges are also called *arcs* and *lines*). Each edge in E joins two distinct vertices in V. Figure 9.1 gives some examples of graphs. Vertices are represented by circles and edges by lines. In the graphs of Figures 9.1(a) and (b), the vertices are not labeled. The vertices in the remaining graphs are labeled 1, 2, 3, ... In a labeled graph, the vertex labels are distinct. Henceforth, unless otherwise stated, by a graph we shall mean a labeled graph.

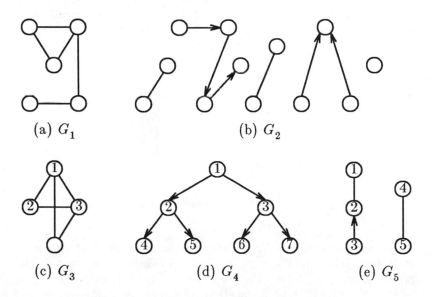

Figure 9.1 Some example graphs.

Some of the edges in Figure 9.1 have arrow heads on them while others do not. An edge with an arrow head is called a *directed* edge while one without an arrow head is called an *undirected* edge. An undirected edge that joins vertices i and j is denoted (i,j). The edges (i,j) and (j, i) are the same. A directed edge from vertex i to vertex j (i.e., the arrow head points towards j) is denoted $<i,j>$. $<i,j>$ and $<j,i>$ denote two different edges. These two edges differ in their orientation.

Using set notation, the graphs of Figures 9.1(c), (d), and (e) may be specified as $G_3 = (V_3, E_3)$, $G_4 = (V_4, E_4)$, $G_5 = (V_5, E_5)$. The sets V_3, V_4, V_5, E_3, E_4, and E_5 are defined below:

$$V_3 = \{1,2,3,4\}; \qquad E_3 = \{(1,2),(1,3),(2,3),(1,4),\ (3,4)\}$$
$$V_4 = \{1,2,3,4,5,6,7\}; E_4 = \{<1,2>,<1,3>,<3,6>,$$
$$<3,7>,<2,4>,<2,5>\}$$
$$V_5 = \{1,2,3,4,5\}; \qquad E_5 = \{(1,2),<3,2>,(5,4)\}$$

If all the edges in the graph are undirected, then the graph is an *undirected* graph. The graphs of Figures 9.1(a) and (c) are undirected graphs. If all the edges are directed, then the graph is a *directed* graph. The graph G_4 of Figure 9.1(d) is a directed graph. A directed graph is also referred to as a *digraph*. When a graph contains both directed and undirected edges, it is customary to regard each undirected edge (i,j) as equivalent to the two directed edges $<i,j>$ and $<j,i>$. So, a graph containing both directed and undirected edges is also a directed graph. The graph of Figure 9.1(e) will generally be drawn as in Figure 9.2.

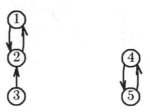

Figure 9.2 Equivalent graph for Figure 9.1(e).

While the notion of a graph was introduced in terms of sets, it is customary to refer to the diagrams of Figure 9.1 and 9.2 as graphs. The set representations (i.e., $G_3 = (V_3, E_3)$, $V_3 = \{1, 2, 3, 4\}$, $E_3 = \{(1, 2), (1, 3), (2, 3), (1, 4), (3, 4)\}$, etc.) are an alternate way of specifying graphs. In fact, for unlabeled graphs a set representation can be arrived at only by imposing some labelling on the vertices.

By definition, a graph does not contain multiple copies of the same edge. For an undirected graph this means that there can be at most one edge between any pair of vertices. In the case of a directed graph there can be at most one edge from vertex i to vertex j and one from j to i. Also, we require that a graph contain no *self edges* (i.e., no edges of the form (i,i) or $<i,i>$). A self edge is also called a *loop*. Thus the diagrams of Figure 9.3 are not graphs.

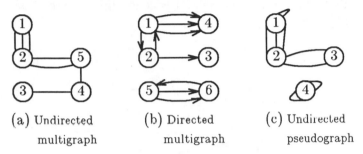

(a) Undirected   (b) Directed   (c) Undirected
  multigraph   multigraph   pseudograph

Figure 9.3 Some multigraphs and pseudographs.

A *multigraph* is an ordered pair (V,E) where V is a finite set of vertices and E is a finite multiset of edges. E does not contain any self edges. The diagrams of Figures 9.3(a) and (b) are multigraphs. Figure 9.3(c) is not a multigraph. If E is permitted to both be a multiset and contain self edges, then the resulting ordered pair (V,E) defines a *pseudograph*. Figure 9.3(c) is a pseudograph.

The first recorded use of graphs dates back to 1736 when Euler used them to solve the now classical Koenigsberg bridge problem. In the town of Koenigsberg (in Eastern Prussia) the river Pregal flows around the island Kneiphof and then divides into two. There are, therefore, four land areas bordering this river (see Figure 9.4(a)). These land areas are connected by means of seven bridges a, b, ..., g. The land areas are labeled A, B, C, and D. The Koenigsberg bridge problem is to determine whether starting from some land area it is possible to walk across all the bridges exactly once returning to the starting land area. One possible walk would be to start at land area C and walk across bridge d to island A ; from A walk across bridge a to land area B; from here take bridge b back to island A and then take bridge e to D; from D take bridge g to C. We have returned to the starting land area but we haven't walked across bridges c and f.

Euler answered the Koenigsberg bridge problem in the negative. The people of Koenigsberg will not be able to walk across each bridge exactly once and return to the starting point. He solved the problem by representing the land areas as vertices and the bridges as edges. This gave him the multigraph of Figure 9.4(b). Defining the degree of a vertex to be the number of edges incident on it, Euler showed that a multigraph has a walk starting at some vertex, going through each edge exactly once and returning to the start vertex iff every vertex of the multigragh is of even degree. Vertex C (amongst others) of Figure 9.4 (b) is of odd degree. So, the multigraph of this figure has no such walk. Hence the answer to the Koenigsberg bridge problem is: no.

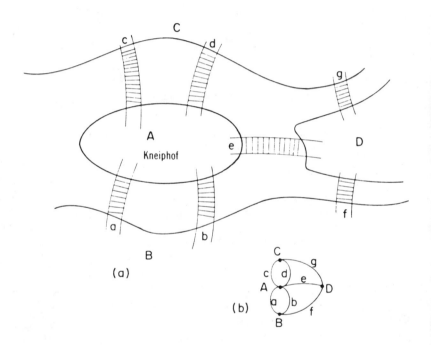

Figure 9.4 The Koenigsberg bridges and Euler's multigraph.

Of all mathematical structures, graphs (and their extensions: multigraphs and pseudographs) are probably the most widely used. In fact, we have already used pseudographs in our earlier chapters. Every binary relation on a finite set can be represented as a directed pseudograph (or more precisely as a directed graph with self edges). Every function with a finite range and a finite domain can be represented as a directed graph. When the range and domain are not disjoint, it is necessary to relabel the vertices so that all labels are distinct.

The graph corresponding to a function has the interesting property that the vertex set can be partitioned into two disjoint subsets R and D (R corresponds to the set of range vertices and D to the set of domain vertices) such that all edges in the graph join a vertex in D to one in R. There are no edges that join two vertices in R or two vertices in D. A graph whose vertex set can be partitioned in this way is called a *bipartite graph*. All graphs that correspond to functions are bipartite. The graph of Figure 9.1 (c) is not bipartite. Graph G_4 is bipartite. To see this, observe that if V_4 is partitioned into $A = \{1, 4, 5, 6, 7\}$ and $B = V - A = \{2, 3\}$, then no edges join two vertices in A or two in B. Figure 9.5 shows two other bipartite graphs.

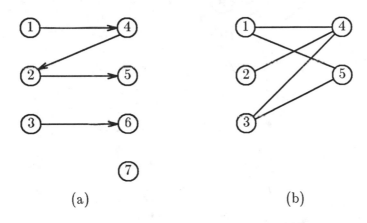

(a)   (b)

Figure 9.5 Two bipartite graphs.

Bipartite graphs are useful in representing several interesting situations. Suppose we have 4 interpreters and 7 languages. We can draw a bipartite graph in which the vertices represent interpreters and languages. There is a directed edge from an interpreter vertex i to a language vertex j iff interpreter i can interpret language j. Figure 9.6(a) shows a possible bipartite graph for this situation. Figure 9.6(b) shows a bipartite graph in which the vertices represent men and women. The graph is undirected and there is an edge between a man vertex i and a woman vertex j iff i and j like each other.

Graphs have found application in a variety of situations. They are used in the analysis of electrical networks, the study of the molecular structure of chemical compounds (particularly hydrocarbons), the representation of airline routes, the representation of communication networks, planning projects, genetic studies, in statistical mechanics, in social sciences, etc.

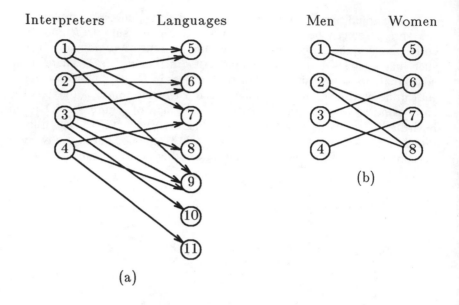

Interpreters   Languages   Men   Women

(a)

(b)

Figure 9.6 Bipartite graphs.

9.1.2 Terminology

Undirected Graphs

Let $G = (V,E)$ be an undirected graph. If $|V| = 0$, then we say that the graph G is *empty* or that G is the *empty* graph. Note that $|E|$ is necessarily zero for the empty graph. If $|V| = 1$, then $|E| = 0$ as G can contain no self edges. Assume that $|V| > 1$. Let i and j be two distinct vertices of the graph (i.e., $i \in V$ and $j \in V$, and $i \neq j$). Vertices i and j are *adjacent* in G iff the edge (i,j) is in E (note that since (i,j) and (j,i) represent the same edge, this statement should be taken to mean i and j are adjacent iff either $(i,j) \in E$ or $(j,i) \in E$). The edge (i,j) is *incident* on the vertices i and j. In the graph of Figure 9.6(b) the vertices 1 and 5 are adjacent. So also are the vertices 1 and 6; 2 and 7; 3 and 6; 2 and 8; 3 and 8; and 4 and 7. There are no other pairs of adjacent vertices in this graph. The edge $(1, 5)$ is incident on the vertices 1 and 5; the edge $(2, 7)$ is incident on the vertices 2 and 7; etc.

The *degree* d_i of vertex i is the number of edges incident on vertex i. For the graph of Figure 9.6(b), $d_1 = 2$; $d_5 = 1$; $d_4 = 1$; $d_8 = 2$; etc. For the graph of Figure 9.7(a), the degree d_1 of vertex 1 is 0. Since each edge in an undirected graph is incident on exactly two vertices, the sum of the degrees of the vertices equals two times the number of edges. If $|V| = n$ and $|E| = e$, then $\sum_{i=1}^{n} d_i = 2*e$. Since $d_i \leqslant n-1$, it follows that $e \leqslant n(n-1)/2$. An n vertex undirected graph that contains exactly $n(n-1)/2$ edges is a *complete graph* on n vertices. Figure 9.7 gives the complete graphs for the cases $n = 1, 2, 3, 4,$ and 5. The complete graph on n vertices is denoted K_n.

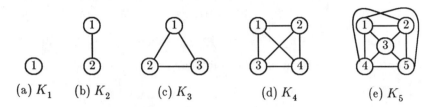

(a) K_1 (b) K_2 (c) K_3 (d) K_4 (e) K_5

Figure 9.7 Complete graphs.

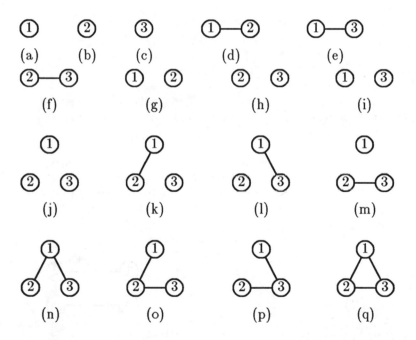

Figure 9.8 The subgraphs of K_3.

The undirected graph $G_1 = (V_1, E_1)$ is a *subgraph* of the graph $G = (V, E)$ iff $V_1 \subseteq V$ and $E_1 \subseteq E$. The graph of Figure 9.7(a) is a subgraph of the graph of Figure 9.7(b); the graph of Figure 9.7(b) is a subgraph of the graph of Figure 9.7(c) which is in turn a subgraph of the graph of Figure 9.7(d).

K_4 is a subgraph of K_5. From the definition of a complete graph, it follows that every n vertex undirected graph is a subgraph of K_n. Figure 9.8 shows all the subgraphs (except the empty subgraph) of K_3. Note that a graph is a subgraph of itself.

Of all the n vertex undirected graphs, K_n has the maximum number of subgraphs. Since there are $i(i-1)/2$ possible different edges in an i vertex graph, there are $2^{i(i-1)/2}$ different i vertex graphs. The number of ways to select i vertices out of n is $\binom{n}{i}$. So, K_n has exactly $\binom{n}{i}*2^{i(i-1)/2}$ subgraphs containing exactly i vertices. The number of subgraphs of K_n is therefore $\sum_{i=0}^{n}\binom{n}{i}*2^{i(i-1)/2}$ (this count includes the empty subgraph).

An examination of the graph of Figure 9.8 reveals that several of the graphs of this figure are similar. For example, graphs (a), (b), and (c) are identical except that the vertex labels differ. Similarly, graphs (d), (e), and (f) are identical except for vertex labels. Two graphs $G_1 = (V_1,E_1)$ and $G_2 = (V_2,E_2)$ are *isomorphic* iff V_1 and V_2 can be put into a one-to-one correspondence f (i.e., for all i in V_1, i corresponds to $f(i)$ in V_2) such that (i,j) is an edge in G_1 iff $(f(i),\ f(j))$ is an edge in G_2. So, graphs (a), (b), and (c) of Figure 9.8 are isomorphic. The graphs (d), (e), and (f) are also isomorphic, and so also are the graphs (g), (h), and (i); (k), (l), and (m); and (n), (o), and (p).

One may readily verify that isomorphism is an equivalence relation and that two graphs can be isomorphic only if they have the same number of vertices and the same number of edges. The two graphs of Figure 9.9 have the same number of vertices and the same number of edges. They are, however, not isomorphic.

(a)   (b)

Figure 9.9 Two non-isomorphic graphs.

A sequence of vertices $P = i_1, i_2, ..., i_k$ is an i_1 to i_k *path* in the graph $G = (V,E)$ iff the edge (i_j,i_{j+1}) is in E for every j, $1 \leqslant j < k$. Vertex u is on this path iff $u = i_j$ for some j. Edge (u,v) *is on the path* P iff there exists a j

such that $u = i_j$ and $v = i_{j+1}$; or $u = i_j$ and $v = i_{j-1}$. The path P *uses* the edge (i,j) iff this edge is on the path. If $u = i_j$ and $v = i_{j+1}$ for some j, then P uses edge (u,v) *in the direction u to v*. 1, 2, 3, 4 is a path from 1 to 4 in the graph of Figure 9.9(b). {1, 2, 3, 4} is the set of vertices on this path. {(1, 2), (2, 3), (3, 4)} is the set of edges on this path. (1,2) is used in the direction 1 to 2. 1, 3, 4 is a 1 to 4 path in the graph of Figure 9.9(a) but is not a path in the graph of Figure 9.9(b). The *length* of a path is the number of edges on it. The path 1, 2, 3, 2, 1, 3, 4 in Figure 9.9(a) has a length of 6.

The path P is a *simple path* iff all vertices (except possibly the first and the last) and all edges on it are distinct. The paths 1, 2, 3, 2, 1, 3, 4 and 1, 2, 1 are not simple paths. The paths 1, 2, 3, 1 and 1, 2, 3, 4 (see Figure 9.9(a)) are simple paths. Observe that G has a u to v path iff G has a simple u to v path. A *cycle* is a simple path in which the first and last vertices are the same. 1, 2, 3, 1 is a cycle in the graph of Figure 9.9(a) while 1, 2, 3, 4, 1 is a cycle in the graph of Figure 9.9(b). The path 1, 2, 1 is not a cycle as it is not a simple path. Observe that all cycles have length more than 2.

An undirected graph $G = (V,E)$ is *connected* iff for every pair of vertices i and j in V, $i \neq j$, there is an i to j path in G. The graphs of Figures 9.9(a) and (b) are connected. The graphs of Figures 9.8(a) to (f) and (n) to (q) are also connected. Those of Figures 9.8(g) to (m) are not connected. A subgraph $G_1 = (V_1,E_1)$ of the graph G is a *connected component* iff G_1 is a connected graph and G contains no connected subgraph that properly contains G_1 (i.e., there is no connected subgraph $G_2 = (V_2,E_2)$ of G such that $V_1 \subseteq V_2$ and $E_1 \subset E_2$). Figure 9.10(a) shows a graph that is not connected. It has three connected components (abbreviated components). These are shown in Figures 9.10(b), (c), and (d). While the graph of Figure 9.10(a) contains many other connected subgraphs, each of these is a subgraph of one of the graphs of Figures 9.10(b), (c), and (d). Hence, the graph of Figure 9.10(a) has no other components.

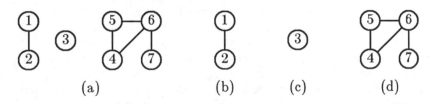

(a)   (b)   (c)   (d)

Figure 9.10 A graph and its components.

When one is designing a communication network for n cities, it is required that it be possible to communicate between every pair of cities in the network. It is possible to communicate between two cities i and j iff there is a sequence $i, i_1, i_2, ..., i_k, j$ of cities such that there is a direct link in the network between i and i_1, i_j and i_{j+1}, $1 \leqslant j < k$, and between i_k and j. The communication from i to j utilizes the path $i, i_1, ..., i_k, j$ while that from j to i utilizes the path $j, i_k, ..., i_1, i$. It is easy to see that it is possible to communicate between every pair of cities iff the communication network can be represented as a connected graph (the vertices in the graph represent cities and the edges represent direct communication links).

Since each direct link involves some capital expense, it is desirable to know the minimum number of edges needed to obtain a connected n vertex graph. Theorem 9.3 shows that every connected n vertex graph contains at least $n-1$ edges, $n \geqslant 1$. The proof of this theorem makes use of Theorems 9.1 and 9.2.

Theorem 9.1: Let $G = (V,E)$ be a connected graph with $|V| > 1$. G contains either a vertex of degree 1 or a cycle (or both).

Proof: If G contains a vertex of degree 1, then nothing is to be proved. So, assume G does not contain a vertex of degree 1. Now, we must show that G contains a cycle. Our proof of this is constructive. Since G is connected and has no vertices of degree 1, every vertex in G has a degree at least 2. A cycle of G can be constructed in the following way.

Start at any vertex (say i_1) in G. Use one of the edges incident on i_1 to get to another vertex i_2. Since the degree of i_2 is at least 2, there is an edge (i_2, i_3) incident on i_2 such that $i_1 \neq i_3$. Use one of the edges incident on i_2 with this property to build the simple path i_1, i_2, i_3. Since the degree of i_3 is at least 2, G contains an edge (i_3, i_4) such that $i_4 \neq i_3$. If $i_4 = i_1$, then we have obtained the cycle i_1, i_2, i_3, i_1. If $i_4 \neq i_1$, then i_1, i_2, i_3, i_4 is a simple path in G. Using the same reasoning as before, we conclude that G contains an edge (i_4, i_5) such that $i_5 \neq i_3$. If i_5 is equal to either i_1 or i_2, then the existence of a cycle has been demonstrated. If $i_5 \notin \{i_1, i_2\}$ then i_1, i_2, i_3, i_4, i_5 is a simple path in G.

By repeating this path extension step several times, we will either construct a cycle or a simple path $i_1, i_2, ..., i_n$ such that $i_n \notin \{i_1, i_2, ..., i_{n-1}\}$. In the latter case, there is an edge (i_n, i_{n+1}) such that $i_{n+1} \neq i_{n-1}$. As G has only n vertices, $i_{n+1} \in \{i_1, i_2, ..., i_{n-2}\}$. If $i_{n+1} = i_k$, then $i_k, i_{k+1}, ..., i_n, i_{n+1}$ is a cycle in G. $\square$

Theorem 9.2: Let $G = (V,E)$ be a connected graph that contains at least one cycle. Let $(i,j) \in E$ be an edge that is on at least one cycle of G. The graph $H = (V, E-\{(i,j)\})$ is also connected.

Proof: First, let us look at an example. Figure 9.11(a) shows a graph that is both connected and contains some cycles. 1, 2, 3, 1 is one of the cycles in this graph. When any one of the edges (1,2), (2,3), and (3,1) is removed from the graph, the remaining graph is still connected (see Figures 9.11(b), (c), and (d)).

We shall show that for every pair (u,v) of distinct vertices in V, there exists a u to v path in G that does not use the edge (i,j). Hence, this path is also present in $H = (V, E-\{(i,j)\}$ and so H is connected.

Let u and v be two arbitrary but distinct vertices in V. Since G is connected, G contains a u to v path, P. Since, (i,j) is on a cycle of G, there exists a simple path C of the form:

$$C = i,j,i_1,i_2, \ldots , i_k,i$$

A u to v path that does not use the edge (i,j) may be obtained from P as follows:

(1) Replace each use of the edge (i,j) in the direction i to j by the path i, i_k, ..., i_2, i_1, j.

(2) Replace each use of the edge (i,j) in the direction j to i by the path j, i_1, i_2, ..., i_k, i..

Hence, G contains a u to v path that does not use the edge (i,j). This path is therefore a u to v path in H. So, H is connected. ☐

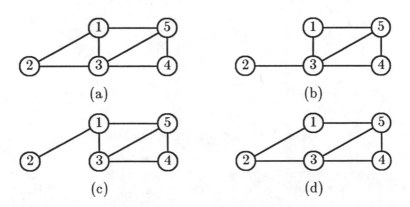

Figure 9.11 Example for Theorem 9.2.

Theorem 9.3: (i) For every n there exists a connected undirected graph containing exactly $n-1$ edges, $n \geqslant 1$.

(ii) Every n vertex connected undirected graph contains at least $n-1$ edges.

Proof: (i) The proof of this part is constructive. For any n, $n \geqslant 1$, consider the graph $G_n = (\{1, 2, ..., n\}, \{(1, 2), (2, 3), ..., (n-1,n)\})$. Note that when $n = 1$, $E = \emptyset$. A diagram for G_n is shown in Figure 9.12. G_n is readily seen to be connected. Also, $|E| = n-1$.

(ii) The proof of this part of the theorem is by contradiction. Suppose that this part of the theorem is not true. Then there exist n vertex graphs (for some values of n) that are connected and contain fewer than $n-1$ edges. Let m be the least n for which such graphs exist. Let $G = (V,E)$ be an m vertex connected graph containing the fewest number of edges amongst all m vertex connected graphs. By assumption, $|E| < m-1$. We also know that $m > 1$ as there are no graphs with fewer than 0 edges.   From the minimality of the edge set E and Theorem 9.2, it follows that G contains no cycles. Now, from Theorem 9.1 it follows that G contains a vertex u of degree 1. Let $(u,v) \in E$ be the sole edge in E that is incident on vertex u. It is readily seen that the graph $H = (V-\{u\}, E-\{(u, v)\})$ is connected. So, H is an $m-1$ vertex graph that is connected and contains fewer than $m-2$ edges. This contradicts the assumption that m is the least value of n for which there exist connected graphs with fewer than $n-1$ edges. Hence, there is no n for which part (ii) of the theorem is not true. $\square$

Figure 9.12 A connected graph with $e = n-1$.

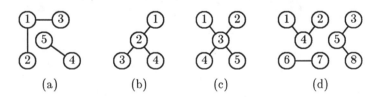

(a)   (b)   (c)   (d)

Figure 9.13 Some acyclic graphs (forests).

An *acyclic* graph is an undirected graph that contains no cycles. Figure 9.13 shows some acyclic graphs. A graph is a *tree* iff it is both connected and acyclic. The graph of Figure 9.12 is a tree. The graphs of Figures 9.13(b) and (c) are trees but those of Figures 9.11 and 9.13(a) and

(d) are not. Each connected component of Figures 9.13(a) and (d) is however a tree. An acyclic graph is therefore also called a *forest*. The number of components in an acyclic graph is the number of trees in that forest.

Theorem 9.4 provides some equivalent definitions of a tree.

Theorem 9.4: Let $G = (V,E)$ be an undirected graph. The following statements are equivalent:
(i) G is a tree;
(ii) For every pair of distinct vertices u and v in V there is a unique simple path from u to v in G.
(iii) G is acyclic and $|E| = |V| - 1$.
(iv) G is connected and $|E| = |V| - 1$.

Proof: The equivalence of these four statements can be established by showing that (i) $\implies$ (ii); (ii) $\implies$ (iii); (iii) $\implies$ (iv); and (iv) $\implies$ (i). Once this has been done, the equivalence of the four statements will follow from the transitivity of $\implies$. The details of the proof are left as an exercise. $\square$

Directed Graphs

Most of the terminology developed for undirected graphs extends in an obvious way to digraphs. The directed edge $<i,j>$ is *incident from i* and *incident to j*. If the edge $<i,j>$ is in the graph, then vertex *i* is *adjacent to* vertex *j* and vertex *j* is *adjacent from i*. In the digraph of Figure 9.14, the edge $<1,2>$ is incident to the vertex 2 and incident from the vertex 1. The edges $<1,2>$, $<5,2>$, and $<3,2>$ are incident to vertex 2. The edge $<2,3>$ is incident from vertex 2. The vertices 2 and 3 are adjacent from 1. 1, 5, and 3 are adjacent to 2. Vertex 3 is also adjacent from 2.

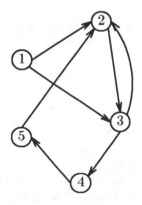

Figure 9.14 A digraph.

The *in-degree*, d_i^{in}, of vertex i is the number of edges incident to i (i.e., the number of edges coming into this vertex). The *out-degree*, d_i^{out}, of vertex i is the number of edges incident from this vertex (i.e., the number of edges leaving vertex i). For the digraph of Figure 9.14, $d_1^{in} = 0$, $d_1^{out} = 2$, $d_2^{in} = 3$, $d_2^{out} = 1$, $d_3^{in} = 2$, $d_3^{out} = 2$, etc. Since each edge is incident to exactly one vertex and incident from exactly one vertex, it follows that $\sum_{i=1}^{n} d_i^{in} = \sum_{i=1}^{n} d_i^{out} = e$ for every n vertex e edge digraph.

A *complete* digraph on n vertices contains exactly $n(n-1)$ directed edges. Figure 9.15 gives the complete digraphs for $n = 1, 2, 3,$ and 4.

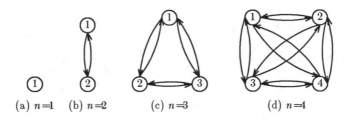

(a) $n=1$ (b) $n=2$ (c) $n=3$ (d) $n=4$

Figure 9.15 Complete digraphs.

The digraph $G_1 = (V_1, E_1)$ is a *subgraph* of the graph $G = (V,E)$ iff $V_1 \subseteq V$ and $E_1 \subseteq E$. Two digraphs G_1 and G_2 are *isomorphic* iff V_1 and V_2 can be put into a one-to-one correspondence f such that $<i,j> \in E_1$ iff $<f(i),f(j)> \in E_2$. The digraphs of Figures 9.16(a) and (b) are isomorphic while those of Figures 9.16(a) and (c) are not. To see that Figures 9.16(a) and (b) are isomorphic, just consider the one-to-one correspondence: $f(1:4) = (3,4,1,2)$. One may verify that isomorphism of digraphs is an equivalence relation.

Let $G = (V,E)$ be a digraph. The sequence of vertices $P = i_1, i_2, ..., i_k$ is a *directed path* from i_1 to i_k iff $<i_j, i_{j+1}> \in E$, $1 \leq j < k$. P is a *simple directed path* iff all vertices (except possibly the first and last) are distinct. P is a *directed cycle* iff it is a simple directed path and $i_1 = i_k$. In the digraph of Figure 9.14, 1, 2, 3, 4, 5, 2, 3, 2, 3, 4 is a directed path but not a simple directed path. 2, 3, 2 and 2, 3, 4, 5, 2 are simple directed paths that are also directed cycles. 1, 2, 3, 4 is a simple directed path that is not a directed cycle.

The sequence of vertices $P = i_1, i_2, ..., i_k$ is a *semipath* in the digraph $G = (V,E)$ iff either $<i_j, i_{j+1}>$ or $<i_{j+1}, i_j>$ (or both) is in E, $1 \leq j < k$. 2, 5, 4, 3, 1, 2, 3, 1 is a semipath in the digraph of Figure 9.14. Simple semipaths and semicycles are defined similarly. 1, 4, 3 is a simple semipath in the digraph of Figure 9.16(b). 1, 4, 3, 2, 1 is a simple semipath that is also a semicycle in this graph. The *length* of a directed path or simple directed path or semipath, etc. is the number of edges on it. The semipath 2, 5, 4, 3, 1, 2, 3, 1 is of length 7.

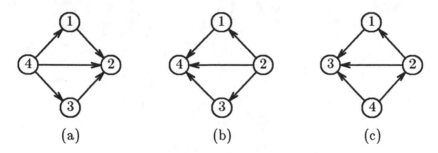

Figure 9.16 Digraphs.

A digraph is *strongly connected* iff it contains a directed path from i to j and from j to i for every pair of distinct vertices i and j. The digraph of Figure 9.16(a) is not strongly connected as it contains no directed path from 2 to 1. A theorem analogous to Theorem 9.3 can be established for the case of strongly connected digraphs.

Theorem 9.5: (i) For every n, $n \geqslant 2$, there exists a strongly connected digraph that contains exactly n edges.

(ii) Every n vertex strongly connected digraph contains at least n edges, $n \geqslant 2$.

Proof: (i) This can be proved by construction. For any n, $n \geqslant 2$, consider the digraph $G_n = (\{1, 2, ..., n\}, \{<1, 2>, <2, 3>, <3, 4>, ..., <n-1, n>, <n, 1>\})$. G_n is simply an n vertex directed cycle. G_n is readily seen to be strongly connected. Also, $|E| = n$.

(ii) Let $G = (V,E)$ be any n vertex strongly connected digraph. Since G contains a directed path from i to j and from j to i for every pair of vertices i and j, $d_i^{in} \geqslant 1$ and $d_i^{out} \geqslant 1$, $1 \leqslant i \leqslant n$. Hence, $\sum_{i=1}^{n} d_i^{in} \geqslant n$. If $|E| = e$, then $\sum_{i=1}^{n} d_i^{in} = e$. So, $e \geqslant n$. $\square$

A strongly connected subgraph $G_1 = (V_1, E_1)$ of the digraph $G = (V,E)$ is a *strongly connected component* of G iff G contains no strongly connected subgraph $G_2 = (V_2, E_2)$ such that $V_1 \subseteq V_2$ and $E_1 \subset E_2$. The strongly connected components of the digraph of Figure 9.14 are shown in Figure 9.17.

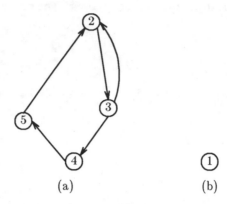

(a)   (b)

Figure 9.17 Strongly connected components of Figure 9.14.

A digraph $G = (V,E)$ is *unilaterally connected* iff for every pair of vertices i and j in V, G contains a directed path from either i to j or from j to i (or both). The digraph of Figure 9.14 is unilaterally connected. The digraph of Figure 9.16(a) is not unilaterally connected as it does not contain a directed path from either 1 to 3 or 3 to 1. *Unilaterally connected components* may be defined in a manner analogous to the definition of strongly connected and connected components. Figure 9.18 shows the unilaterally connected components of the graph of Figure 9.16(a).

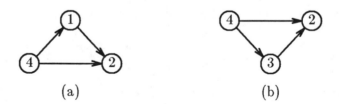

(a)   (b)

Figure 9.18 Unilaterally connected components of Figure 9.16(a).

$G = (V,E)$ is a *weakly connected* digraph iff G contains an i to j semipath for every pair of vertices i and j. The digraph of Figure 9.16(a) is weakly connected while that of Figure 9.19(a) is not. The weakly connected components of the digraph of Figure 9.19(a) are given in Figures 9.19(b), (c), and (d). Observe that a digraph G is weakly connected iff the undirected graph that results from the deletion of the arrow heads from the edges of G and the subsequent collapsing of pairs of identical undirected edges into a single undirected edge is connected. Figure 9.20(a) shows the undirected multigraph that results when the arrow heads are removed from the edges of Figure 9.19(b). Figure 9.20(b) shows the undirected graph that results when all pairs of identical edges are collapsed into single edges.

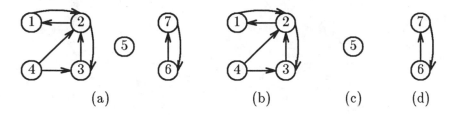

(a)   (b)   (c)   (d)

Figure 9.19 A digraph and its weakly connected components.

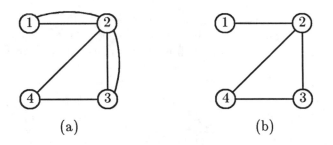

(a)   (b)

Figure 9.20.

From the definitions, it follows that every strongly connected digraph is also unilaterally connected. Every unilaterally connected digraph is also weakly connected.

A *directed acyclic graph* (dag) is a digraph that contains no directed cycles. No strongly connected digraph with more than one vertex is a dag. The digraphs of Figure 9.18 are dags while those of Figures 9.17 and 9.19(a) are not.

Finally, corresponding to the notion of a tree, there is the notion of a directed tree. A *directed tree* is a weakly connected digraph that contains no semicycles. Hence, the removal of the arrow heads from the edges of a directed tree results in a tree. The digraphs of Figure 9.21 are directed trees. A *rooted tree* is a directed tree in which exactly one vertex has an in-degree of 0 and the remaining vertices have an in-degree of 1. The vertex with in-degree 0 is called the *root* of the tree. The directed tree of Figure 9.21 (c) is a rooted tree. Its root is vertex 2. The directed trees of Figures 9.21 (a) and (b) are not rooted trees. From Theorem 9.4 and the definition of a directed tree, it immediately follows that an n vertex directed tree contains exactly $n-1$ edges.

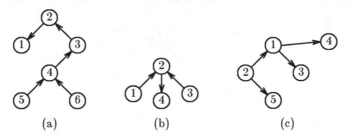

(a)   (b)   (c)

Figure 9.21 Some directed trees.

Since every rooted tree must contain a root, it follows that there is no rooted tree with zero vertices. It is often convenient to be able to talk about an empty rooted tree. So, the definition is usually extended to include the empty tree.

9.1.3 Graph Representation

While the diagrammatic representation of a graph is well suited for paper, it is not suitable as a representation scheme within a computer. The set representation scheme can easily be used within a computer. For any graph $G = (V,E)$ we simply need two arrays $V(1:n)$ and $E(1:e,1:2)$ where $n = |V|$ and $e = |E|$. The array V is used to hold the vertex labels (if these labels are simply $1,2,...,n$, then this array may be dispensed with). The edges of G are stored in the array E. For example, the graph of Figure 9.18 (a) could be represented as $V(1:3) = (1,4,2)$ and $E(1,1:2) = (4,1)$, $E(2,1:2) = (4,2)$, and $E(3,1:2) = (1,2)$. This array representation isn't a very convenient representation for algorithms to work with. For instance, one must examine all e edges in order to determine the vertices adjacent from any given vertex.

There are several ways to represent a graph within a computer so as to permit easy graph manipulation by algorithms. The most frequently used representation schemes are adjacency based: adjacency matrices, packed adjacency lists, and linked adjacency lists. Corresponding to these representation schemes, there are the less frequently used incidence based schemes: incidence matrices, packed incidence lists, and linked incidence lists.

Adjacency Schemes

The *adjacency matrix* of an n vertex graph $G = (V,E)$ is an $n \times n$ matrix $A(1{:}n, 1{:}n)$. We shall assume that $V = \{1, 2, ..., n\}$. If G is an undirected graph, then the elements of A are defined as:

$$A(i,j) = \begin{cases} 1 & \text{if } (i,j) \in E \text{ or } (j,i) \in E \\ 0 & \text{otherwise} \end{cases} \qquad (9.1)$$

If G is a digraph, then the elements of A are defined as:

$$A(i,j) = \begin{cases} 1 & \text{if } <i,j> \in E \\ 0 & \text{otherwise} \end{cases} \qquad (9.2)$$

The adjacency matrices for the graphs of Figures 9.11(a), 9.13(a), 9.14, and 9.16(a) are given in Figures 9.22(a) to (d) respectively.

The validity of the following statements is an immediate consequence of (9.1) and (9.2):

(1) $A(i,i) = 0, 1 \leqslant i \leqslant n$ for all n vertex graphs.
(2) The adjacency matrix of an undirected graph is symmetric. I.e., $A(i,j) = A(j,i), 1 \leqslant i \leqslant n, 1 \leqslant j \leqslant n$.
(3) For an n vertex undirected graph, $\sum_{j=1}^{n} A(i,j) = \sum_{j=1}^{n} A(j,i) = d_i$ (recall that d_i is the degree of vertex i).
(4) For an n vertex digraph, $\sum_{j=1}^{n} A(i,j) = d_i^{out}, 1 \leqslant i \leqslant n$ and $\sum_{i=1}^{n} A(i,j) = d_j^{in}, 1 \leqslant j \leqslant n$.

Since every entry in A is either 0 or 1, it takes n^2 bits to store the adjacency matrix of an n vertex graph in a computer. The space requirement can be reduced to $n^2 - n$ by not explicitly storing the diagonal of A. All diagonal entries are known to be zero. For undirected graphs the adjacency matrix is symmetric. So, only the elements above (or below) the diagonal need to be stored explicitly. Hence only $(n^2-n)/2$ bits are needed. When adjacency matrices are used, $\theta(n)$ time is needed to determine the set of vertices adjacent to or from any given vertex. $\theta(n^2)$ time is needed to determine the

$$
\begin{array}{c}
\begin{array}{ccccc} 1 & 2 & 3 & 4 & 5 \end{array} \\
\begin{array}{c} 1 \\ 2 \\ 3 \\ 4 \\ 5 \end{array}
\begin{bmatrix}
0 & 1 & 1 & 0 & 1 \\
1 & 0 & 1 & 0 & 0 \\
1 & 1 & 0 & 1 & 1 \\
0 & 0 & 1 & 0 & 1 \\
1 & 0 & 1 & 1 & 0
\end{bmatrix}
\end{array}
\qquad
\begin{array}{c}
\begin{array}{ccccc} 1 & 2 & 3 & 4 & 5 \end{array} \\
\begin{array}{c} 1 \\ 2 \\ 3 \\ 4 \\ 5 \end{array}
\begin{bmatrix}
0 & 1 & 1 & 0 & 0 \\
1 & 0 & 0 & 0 & 0 \\
1 & 0 & 0 & 0 & 0 \\
0 & 0 & 0 & 0 & 1 \\
0 & 0 & 0 & 1 & 0
\end{bmatrix}
\end{array}
$$

(a)   (b)

$$
\begin{array}{c}
\begin{array}{ccccc} 1 & 2 & 3 & 4 & 5 \end{array} \\
\begin{array}{c} 1 \\ 2 \\ 3 \\ 4 \\ 5 \end{array}
\begin{bmatrix}
0 & 1 & 1 & 0 & 0 \\
0 & 0 & 1 & 0 & 0 \\
0 & 1 & 0 & 1 & 0 \\
0 & 0 & 0 & 0 & 1 \\
0 & 1 & 0 & 0 & 0
\end{bmatrix}
\end{array}
\qquad
\begin{array}{c}
\begin{array}{cccc} 1 & 2 & 3 & 4 \end{array} \\
\begin{array}{c} 1 \\ 2 \\ 3 \\ 4 \end{array}
\begin{bmatrix}
0 & 1 & 0 & 0 \\
0 & 0 & 0 & 0 \\
0 & 1 & 0 & 0 \\
1 & 1 & 1 & 0
\end{bmatrix}
\end{array}
$$

(c)   (d)

Figure 9.22 Adjacency matrices.

number of edges in the graph. However, a new edge can be added or an old one deleted in $\theta(1)$ time.

In the *packed adjacency list* representation of a graph $G = (V,E)$ with $|V| = n$ and $|E| = e$, we use two one dimensional arrays $H(1{:}n{+}1)$ and $L(1{:}x)$ where $x = e$ if G is a digraph and $x = 2e$ if G is an undirected graph. First, all vertices adjacent from vertex 1 are put into L; then all vertices adjacent from 2 are put into L; then all vertices adjacent from 3 are put into L; and so on. (If i and j are adjacent vertices in an undirected graph, then i is adjacent from j and j is adjacent from i). H is set-up so that the vertices adjacent from vertex i are in positions $L(H(i))$, $L(H(i){+}1)$, ..., $L(H(i{+}1){-}1)$ if $H(i) < H(i{+}1)$. If $H(i) \geqslant H(i{+}1)$ then there are no vertices adjacent from i. We say that $L(H(i))$, $L(H(i){+}1)$, ..., $L(H(i{+}1){-}1)$ is the packed adjacency list for vertex i. The order in which vertices appear in this list is not important. Figures 9.23(a) to (d) give the packed adjacency lists corresponding to the graphs of Figures 9.11(a), 9.13(a), 9.14, and 9.16(a) respectively.

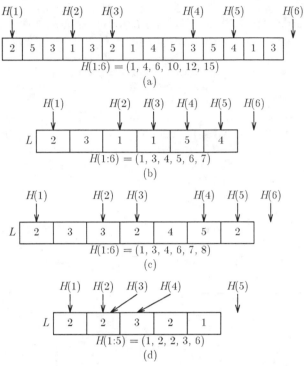

Figure 9.23 Packed adjacency lists.

For an undirected graph, the values of H are in the range 1 to $2e+1$. As this range represents only $2e+1$ distinct values, each $H(i)$ need be at most $\lceil \log(2e+1) \rceil$ bits long. The L entries are each in the range 1 to n. Hence, each of these need be at most $\lceil \log n \rceil$ bits long. The total number of bits needed to store the packed adjacency lists of an undirected n vertex e edge graph is therefore at most $(n + 1)\lceil \log(2e+1) \rceil + 2e\lceil \log n \rceil = O((n + e)\log n)$. When e is much less than n^2, the space needed by packed adjacency lists is less than that needed by an adjacency matrix. If G is an undirected graph, the degree of vertex i is simply $H(i+1) - H(i)$ and the number of edges in G is $H(n+1) - 1$. So, it is easier to determine these quantities when adjacency lists are being used than when adjacency matrices are used. The addition of a new edge or the deletion of an old one, however, requires $O(n+e)$ time (see the exercises).

In packed adjacency lists, the vertices adjacent from any given vertex are stored sequentially in L. In the case of linked adjacency lists, there is a pointer (called *LINK*) associated with each vertex in an adjacency list. This pointer tells us where the next element in the list is stored. If there is no next element, then the pointer is zero. The linked adjacency lists of a graph use three arrays: $H(1:n)$, $V(1:x)$, and $LINK(1:x)$. $x = 2e$ for an undirected graph and $x = e$ for a directed graph. $H(i)$ is the head node for the adjacency list of vertex i, $1 \leqslant i \leqslant n$ and it points to the location in V where the first vertex in the adjacency list for i is stored. Figure 9.24(a) gives a possible linked adjaceny list representation for the graph of Figure 9.16(a). Since $H(1) = 1$, the first vertex in the adjacency list for vertex 1 is $V(H(1))$ $= V(1) = 2$. As $LINK(1) = 0$, there is no next vertex on this list. Since, $H(2) = 0$, the adjacency list for vertex 2 is empty (i.e., there are no vertices adjacent from vertex 2). $H(3) = 4$. So, the first vertex on the adjacency list for vertex 3 is $V(4) = 2$. There is no next vertex on this list as $LINK(4) = 0$. The first vertex on the list for vertex 4 is $V(H(4)) = 3$. The next vertex is at position $LINK(H(4)) = LINK(2) = 5$. So, $V(5) = 2$ is the next vertex on this list. The third vertex on the adjacency list for vertex 4 is $V(LINK(5)) = 1$. As, $LINK(3) = 0$, there are no other vertices on this list.

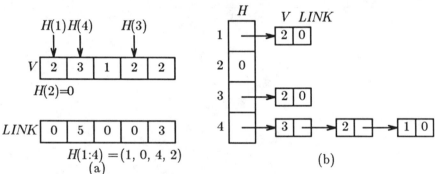

Figure 9.24 Linked adjacency lists.

When linked adjacency lists are used, we prefer to draw the representation as in Figure 9.24(b). This shows the adjacency lists explicitly. Each pair $(V(i), LINK(i))$ is called a *node*. Figures 9.25(a), (b), and (c) respectively give the linked adjacency list representation for the graphs of Figures 9.11(a), 9.13(a), and 9.14. The total number of bits needed for the linked adjacency lists representation of a graph is at most $(n+x)\lceil \log x \rceil + x \lceil \log n \rceil$ where $x = 2e$ if G is undirected and $x = e$ if G is directed. In either case, the number of bits needed is $O((n+e)\log n)$.

Linked adjacency lists permit easy addition and deletion of edges. The time needed to determine the number of vertices on an adjacency list is proportional to the number of vertices on that list.

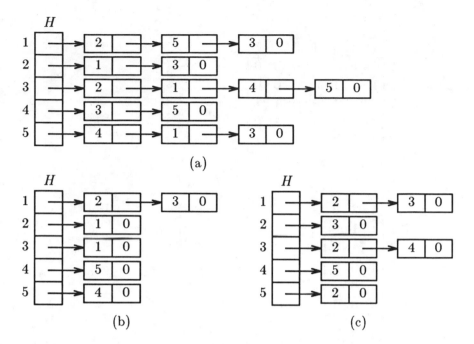

Figure 9.25 Linked adjacency lists.

The *incidence matrix* of a graph $G = (V,E)$ with $|V| = n$ and $|E| = e$ is an $n \times e$ matrix $IN(1:n, 1:e)$. We shall assume that the edges of E are labeled 1, 2, ..., e. If G is undirected, then IN is defined as:

$$IN(i,j) = \begin{cases} 1 & \textit{if edge } j \textit{ is incident on vertex } i \\ 0 & \text{otherwise} \end{cases}$$

If G is a directed graph, then IN is defined as:

$$IN(i,j) = \begin{cases} -1 & \textit{if edge } j \textit{ is incident from vertex } i \\ 1 & \textit{if edge } j \textit{ is incident to vertex } i \\ 0 & \text{otherwise} \end{cases}$$

The incidence matrices for the graphs of Figures 9.11(a), 9.13(a), 9.14, and 9.16(a) are given in Figures 9.26(a) to (d) respectively. Each figure also gives the edge labelling scheme used. If G is an undirected graph, then ne bits are needed to store IN. For directed graphs, $2ne$ bits are needed. Packed incidence lists and linked incidence lists may be defined in a manner analogous to that used for adjacency lists. Since incidence matrices and lists are seldom used, we shall not dwell on these further.

$$
\begin{array}{c}
\begin{array}{ccccccc}1&2&3&4&5&6&7\end{array}\\
\begin{array}{c}1\\2\\3\\4\\5\end{array}
\begin{bmatrix}
1&1&0&1&0&0&0\\
1&0&1&0&0&0&0\\
0&1&1&0&0&1&1\\
0&0&0&0&1&1&0\\
0&0&0&1&1&0&1
\end{bmatrix}
\end{array}
$$

edge	number
(1,2)	1
(1,3)	2
(2,3)	3
(1,5)	4
(5,4)	5
(3,4)	6
(3,5)	7

(a)

$$
\begin{array}{c}
\begin{array}{ccc}1&2&3\end{array}\\
\begin{array}{c}1\\2\\3\\4\\5\end{array}
\begin{bmatrix}
1&1&0\\
0&1&0\\
1&0&0\\
0&0&1\\
0&0&1
\end{bmatrix}
\end{array}
$$

edge	number
(1,3)	1
(1,2)	2
(5,4)	3

(b)

$$
\begin{array}{c}
\begin{array}{ccccccc}1&2&3&4&5&6&7\end{array}\\
\begin{array}{c}1\\2\\3\\4\\5\end{array}
\begin{bmatrix}
-1&0&0&-1&0&0&0\\
1&-1&1&0&1&0&0\\
0&1&-1&1&0&0&-1\\
0&0&0&0&0&-1&1\\
0&0&0&0&-1&1&0
\end{bmatrix}
\end{array}
$$

edge	number
$\langle 1,2 \rangle$	1
$\langle 2,3 \rangle$	2
$\langle 3,2 \rangle$	3
$\langle 1,3 \rangle$	4
$\langle 5,2 \rangle$	5
$\langle 4,5 \rangle$	6
$\langle 3,4 \rangle$	7

(c)

$$
\begin{array}{c}
\begin{array}{ccccc}1&2&3&4&5\end{array}\\
\begin{array}{c}1\\2\\3\\4\end{array}
\begin{bmatrix}
-1&0&0&0&1\\
1&1&1&0&0\\
0&-1&0&1&0\\
0&0&-1&-1&-1
\end{bmatrix}
\end{array}
$$

edge	number
$\langle 1,2 \rangle$	1
$\langle 3,2 \rangle$	2
$\langle 4,2 \rangle$	3
$\langle 4,3 \rangle$	4
$\langle 4,1 \rangle$	5

(d)

Figure 9.26

9.2 SPANNING TREES AND CONNECTIVITY

9.2.1 Breadth First Search

Let $G = (V,E)$ be a graph (either directed or undirected). Let $i \in V$ and $j \in V$ be two distinct vertices. Vertex j is *reachable* from i iff there is a (directed) path from i to j in G. Consider the directed graph of Figure 9.27. One way to determine all the vertices reachable from vertex 1 is to first determine the set of vertices adjacent from 1. This set is $\{2,3,4\}$. Next, we determine the set of new vertices (i.e., vertices not yet reached) that are adjacent from vertices in $\{2,3,4\}$. This set is $\{5,6,7\}$. The set of new vertices adjacent from vertices in $\{5,6,7\}$ is $\{8,9\}$. There are no new vertices adjacent from a vertex in $\{8,9\}$. So, $\{1,2,3,4,5,6,7,8,9\}$ is the set of vertices reachable from vertex 1.

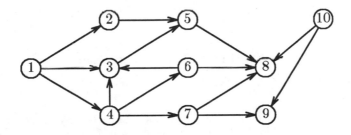

Figure 9.27.

This method of starting at some vertex in a graph and searching it for all vertices reachable from that vertex is called *breadth first search*. A formal specification of this search method is provided by procedure *BFS* (Algorithm 9.1). In this procedure, *REACHED*(1:n) is an array that is initialized to zero. When the algorithm terminates, *REACHED*(i) = 1 iff vertex i is reachable from the start vertex v. *QUEUE* is a first-in-first-out (FIFO) list.

If procedure *BFS* is used on the graph of Figure 9.27 with $v = 1$, then in the loop of lines 5 to 10, vertices 2, 3, and 4 will get added to *QUEUE* (assume that they get added in this order). In line 12, 2 is removed from *QUEUE* (by definition, vertices are removed from a queue in the order they were added to it) and the next iteration of the **for** loop results in vertex 5 getting added to *QUEUE*. Next, 3 is deleted from *QUEUE* and no new vertices are added. Then, 4 is deleted and 6 and 7 added. 5 is deleted next and 8 added. Then 6 is deleted and nothing added. Next, 7 is deleted and 9 added. Finally, 8 and 9 are deleted and *QUEUE* becomes empty. The algorithm terminates and vertices 1 through 9 have been marked as reached.

```
line  procedure BFS(v)
        //breadth first search of an n vertex graph G//
  0     global G, n, REACHED(1:n); queue QUEUE
  1     x ← v
  2     REACHED (v) ← 1
  3     initialize QUEUE to be an empty queue
  4     loop
  5       for all vertices w adjacent from x do
  6         if REACHED(w) = 0 then
  7                         REACHED(w) ← 1
  8                         add w to QUEUE
  9                     endif
 10       end for
 11       if QUEUE is empty then return endif
 12       delete a vertex x from QUEUE
 13     repeat
 14   end BFS
```

Algorithm 9.1 Breadth first search

Theorem 9.6: Let G be an arbitrary graph. Let v be any vertex of G. Procedure BFS sets $REACHED(i) = 1$ for all vertices that are reachable from vertex v (including vertex v).

Proof: Define the distance $d(v,w)$ to be the length of a shortest (directed) path from v to w in G. $d(v,v) = 0$. Clearly, for every vertex w reachable from v, $d(v,w) < n$ where n is the number of vertices in the graph being searched. We shall prove the theorem by contradiction. Suppose that when BFS terminates, $REACHED(i) = 0$ for some vertex i for which $d(v,i) < n$. Let j be a vertex for which $REACHED(j) = 0$ and $d(v,j) = \min\{d(v,i) : REACHED(i) = 0$ and $d(v,j) < n\}$. We may assume that $d(v,j) > 0$ as $d(v,j) = 0$ only if $j = v$ and $REACHED(v)$ is set to 1 in line 2.

Since $d(v,j) > 0$, there must exist a vertex t adjacent to j for which $d(v,t) = d(v,j) - 1$. By assumption, $REACHED(t) = 1$ when BFS terminates. So, t must have been added to $QUEUE$ at some time (line 8). Since $QUEUE$ is empty at termination, t must have been deleted from $QUEUE$ at some time. This deletion must have been made on some iteration of line 12. Immediately following this deletion, the **for** loop gets executed with $x = t$. At this time, j must have been marked reachable in line 7. This contradicts the assumption that $REACHED(j) = 0$ at termination (observe that $REACHED(i)$ is never changed to 0 in BFS). $\square$

Note that Theorem 9.6 is valid for both digraphs and undirected graphs. Observe that if G is a connected undirected graph, then all vertices in G are reachable from all others. So, *BFS* will mark all vertices no matter what the start vertex is. So, G is connected iff a call to *BFS* results in $REACHED(i) = 1$ for all $v \in V$. Hence, *BFS* can be used to determine if an undirected graph is connected.

The time complexity of *BFS* depends on whether adjacency matrices or lists are used to represent G. If an adjacency matrix is used and if G has n vertices, then $\Theta(n)$ time is needed to determine all the vertices adjacent from any vertex x (line 5). If a total of p vertices get marked, then x takes on p different values and $\Theta(pn)$ time is spent just to determine the vertices adjacent from all the x's. Since all vertices other than v that get marked get added to $QUEUE$ (line 8), $\Theta(p-1)$ time is spent in the **for** loop (assuming that each addition to $QUEUE$ takes $\Theta(1)$ time). Line 11 is executed p times and line 12 $p-1$ times. So, the total time spent in *BFS* is $\Theta(pn)$. In the worst case, $p = n$ and the time complexity becomes $O(n^2)$.

When adjacency lists (whether packed or linked) are used, the time needed to determine all the verices adjacent from x is $\Theta(d_x^{out})$ in the case of a directed graph and $\Theta(d_x)$ in the case of an undirected graph. Thus the total time needed for the search is $\Theta(\sum_x (1 + d_x^{out}))$ or $\Theta(\sum_x (1 + d_x))$ where the sum is taken over all values assigned to x in *BFS*. In the worst case, x takes on all values 1, 2, 3, ..., n and the time complexity of *BFS* is therefore $O(n+e)$ where e is the number of edges in G.

As far as the space complexity is concerned, n bits of space are needed for *REACHED*. If p vertices get marked, then at most $p-1$ vertices can be in $QUEUE$ at any one time. So, the space needed for $QUEUE$ is $O(p) = O(n)$. In addition, space is needed to store the graph G.

We remarked earlier that *BFS* can be used to determine if an undirected graph is connected. *BFS* can also be used to determine the connected components of an undirected graph. We shall illustrate the process by means of an example. Consider the undirected graph of Figure 9.28. If *BFS* is called with $v = 6$, then vertices 4, 5, and 6 will get marked. Since no other vertices are reachable from 6, vertices 4, 5, and 6 together with all edges connecting them form a connected component.

Now, suppose we call *BFS* with $v = 7$. This time only vertex 7 is marked and so it forms a component by itself. We are left with three unmarked vertices 1, 2, and 3. If *BFS* is now called with v set to 1, 2, or 3, then all three vertices get marked. Hence these three vertices together with all edges connecting them form another component of the graph. Since all vertices have been marked, the graph of Figure 9.28 has no other components.

The modifications needed to *BFS* so that it identifies all the components of an undirected graph are left as an exercise. It should not be too difficult to see that the modified algorithm will run in $\Theta(n^2)$ time if an adjacency matrix is used and in $\Theta(n+e)$ time if adjacency lists are used.

Figure 9.28 Graph with three components.

Let $G = (V,E)$ be an undirected graph. A subgraph $G_1 = (V_1,E_1)$ of G is a *spanning tree* of G iff $V_1 = V$ and G_1 is a tree. Since a tree is connected (Theorem 9.4), only connected graphs have spanning trees. Figure 9.29 shows a connected graph (Figure 9.29(a)) and some of its spanning trees (Figure 9.29(b), (c), and (d)).

If a breadth first search is carried out starting from any vertex in a connected graph with $|V| = n$, then from Theorem 9.6 we know that all vertices will get marked. Exactly $n-1$ of these will get marked in line 7 of *BFS*. When a vertex w is marked in line 7, the edge used to reach this previously unreached vertex is (x,w). The set T of edges used in this way is such that $|T| = n-1$. Since this set of edges contains a path from v to every other vertex in the graph, it defines a connected subgraph of G. As $|T| = n-1$, T defines a spanning tree of G (Theorem 9.4).

Consider the graph of Figure 9.29(a). If a breadth first search is started at vertex 1, then $T = \{(1,2), (1,3), (1,4), (2,5), (4,6), (4,7), (5,8)\}$. This set of edges corresponds to the spanning tree of Figure 9.29(b).

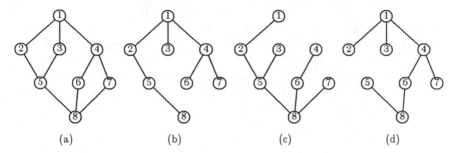

(a)   (b)   (c)   (d)

Figure 9.29 A graph and some of its spanning trees.

A *breadth first spanning tree* is any spanning tree obtained in the manner described above from a breadth first search. One may verify that the spanning trees of Figures 9.29(b), (c), and (d) are all breadth first spanning trees of the graph of Figure 9.29(a) ((c) and (d) are, respectively, obtained by starting at vertices 8 and 6).

9.2.2 Depth First Search

Depth first search is an alternate to breadth first search. Starting at a vertex v, a depth first search proceeds as follows. First, the vertex v is marked as reached. Next, an unreached vertex w_1 adjacent from v is selected. If such a vertex does not exist, the search terminates. Assume that a w_1 as described exists. A depth first search from w_1 is now initiated. When this search is completed, we select another unreached vertex adjacent from v. If there is no such vertex, then the search terminates. If such a vertex exists, a depth first search is initiated from this vertex, and so on. Algorithm 9.2 is a formal recursive specification of depth first search. As in the case of breadth first search, it is assumed that $REACHED(i) = 0, 1 \leqslant i \leqslant n$ at the time of initial call to *DFS*.

```
line    procedure DFS(v)
            //depth first search of an n vertex graph G//
0          global G, n, REACHED(1:n); integer w,v
1          REACHED(v) ← 1
2          for all unreached vertices w adjacent from v do
3              call DFS(w)
4          end for
5      end DFS
```

Algorithm 9.2 Depth first search

Let us try out procedure *DFS* on the graph of Figure 9.27. If $v = 1$, then vertices 2, 3, and 4 are the candidates for the first choice of *w* in line 2 of *DFS*. Suppose that the first value assigned to *w* is 2. The edge used to get to 2 is $<1,2>$. A depth first search from 2 is now initiated. Vertex 2 is marked as reached. The only candidate for *w* this time is vertex 5. The edge $<2,5>$ is used to get to 5. A depth first search from 5 is initiated. 5 is marked as reached. Using the edge $<5,8>$, vertex 8 is reached and marked. From 8 there are no unreached adjacent vertices. So, the algorithm backs up to vertex 5. There are no new candidates for *w* here. So, we back up to 2 and then to 1.

At this point, there are two candidates for *w*. These are vertices 3 and 4. Assume that 4 is selected. Hence, edge $<1,4>$ is used. A depth first search from 4 is initiated. 4 is marked as reached. 3, 6, and 7 are now the candidates for *w*. Assume that vertex 6 is selected. When $v = 6$, vertex 3 is the only candidate for *w*. Edge $<6,3>$ is used to get to 3. A depth first search from 3 is initiated and vertex 3 gets marked. No new vertices are adjacent from 3 and we back up to vertex 6. No new vertices are adjacent from here. So, we back up to 4. From here a depth first search with $w = 7$ is initiated. Next, vertex 9 is reached. From 9 there are no new adjacent vertices. This time, we back up all the way to 1. As there are no new vertices adjacent from 1, the algotithm terminates.

During the above depth first search, vertices 1, 2, 3, ..., and 9 are marked. Figure 9.30 shows the subgraph consisting of only those edges that were used to reach a new vertex. The numbers outside each vertex give the order in which the vertices were reached.

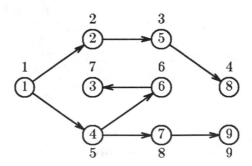

Figure 9.30.

For procedure *DFS*, one may prove a theorem analogous to Theorem 9.6. I.e., *DFS* marks vertex *v* and all vertices reachable from *v*.

Theorem 9.7: Let G be an arbitrary graph. Let v be any vertex of G. Procedure *DFS* sets *REACHED* $(i) = 1$ for all vertices (including v) that are reachable from vertex v.

Proof: Left as an exercise. □

One may also verify that *DFS* has the same time and space complexities as does *BFS* (see the exercises). However, the graphs for which *DFS* takes maximum space (i.e., stack space for the recursion) are the ones on which *BFS* takes minimum space (i.e., *QUEUE* space). The graphs for which *BFS* takes maximum space are the ones for which *DFS* takes minimum space. Figure 9.31 gives the best case and worst case graphs for *DFS* and *BFS*. One may readily verify the truth of the statements just made.

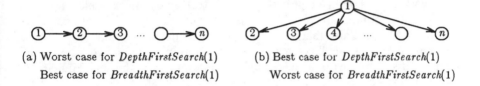

(a) Worst case for *DepthFirstSearch*(1)
Best case for *BreadthFirstSearch*(1)

(b) Best case for *DepthFirstSearch*(1)
Worst case for *BreadthFirstSearch*(1)

Figure 9.31 Worst case and best case space complexity graphs.

For undirected graphs, *DFS* has the same properties as does *BFS*. All vertices are marked reachable when a *DFS* is initiated from any vertex in the graph G iff G is connected. The connected components of G can be obtained by making successive calls to *DFS*, each time initiating a *DFS* from an as yet unreached vertex. If G is connected then exactly $n-1$ edges are used to reach new vertices. The subgraph formed by these edges is a spanning tree of G. A spanning tree obtained in this manner from a depth first search is called a *depth first spanning tree*. Figure 9.32 shows some of the depth first spanning trees of the graph of Figure 9.29(a).

In Figure 9.32 the numbers outside the vertices give the order in which the vertices were reached in the depth first search. The depth first number of a vertex is defined relative to a depth first search. The *depth first number* of vertex v is i iff v is the i'th vertex reached during the depth first search. Note that the depth first number of a vertex is not unique. It depends on the depth first search. One may verify the existence of depth first searches of the graph of Figure 9.29(a) for which the numbers outside the vertices in each of Figures 9.32(a), (b), and (c) are valid depth first numbers.

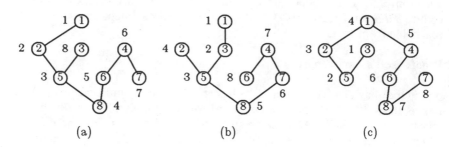

(a)   (b)   (c)

Figure 9.32 Some depth first spanning trees of Figure 9.29(a).

9.2.3 Minimum Cost Spanning Trees

Let $G = (V,E)$ be a connected undirected graph with edge costs (i.e., G is a network or weighted graph). Figure 9.33(a) gives an example of such a graph. The number on each edge is the cost of that edge. So, the cost of the edge (1,6) is 10, and that of (5,7) is 24. This network could, for instance, represent the feasible links in a proposed communication network between 7 cities. If so, the cost of an edge could be interpreted as the actual dollar cost of building (or leasing) that link. The cost of building all the links in the network is the sum of the individual edge costs. For the network of Figure 9.33(a), this figure is 167. If the link (7,8) is not built, the network remains connected and it is still possible to communicate between every pair of cities. The construction cost, however, drops to 149.

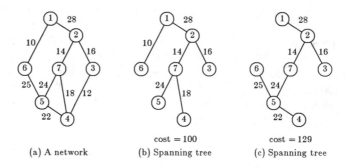

(a) A network   (b) Spanning tree   (c) Spanning tree

cost = 100   cost = 129

Figure 9.33 A network and two of its spanning trees.

Suppose that we are interested in building a least cost communication network and that we are given a weighted graph, G, that contains an edge for each feasible link. The network, T, to be built is a subgraph of G. Further, this subgraph must be connected and must include every vertex in G. We may assume that every edge in G has a positive cost. From this assumption and Theorem 9.2, it follows that a least cost network contains no cycles. To see this, observe that if T is connected and contains a cycle, then any one of the edges on any one of the cycles of T may be deleted from T. The resulting graph is connected and is of lesser cost than T. So, if T is a least cost network it must be acyclic. Since it must also be connected and include every vertex of G, T is a spanning tree of G. Different spanning trees of G have different costs (see Figures 9.33(b) and (c)). We are interested in finding a minimum cost spanning tree of G.

Is the minimum cost spanning tree of a connected weighted graph unique? The answer to this question is no. To see this, just consider any connected n vertex graph in which all edges have the same cost. Assume that this graph is not acyclic. All spanning trees of this graph have a cost of $n-1$. It can be shown that if all edge costs are distinct, then the minimum cost spanning tree is unique (see the exercises).

Given a weighted connected graph G, how can we find one of its minimum cost spanning trees? A first approach might be to generate all the spanning trees of G; sum up the edge costs of each of these spanning trees and select one with the least cost. While this approach is guaranteed to result in a minimum cost spanning tree, it is impractical because of the number of spanning trees that have to be generated. The graph K_n has at least $2^{n-1} - 1$ spanning trees. So, the worst case complexity of any algorithm based on this approach will be $\Omega(2^n)$.

Another simple approach is to view the problem as one of selecting exactly $n-1$ of the edges in G such that these n edges form a minimum cost spanning tree of G. If the edge selection is done one at a time, we could begin by first selecting an edge with least cost. Then, from the set of remaining edges we can select another least cost edge, and so on. This edge selection step is repeated until $n-1$ edges have been selected. Each time an edge is picked, we determine whether or not the inclusion of this edge into the spanning tree being constructed creates a cycle. If it does, then this edge must be discarded. If no cycle is created, this edge is included in the spanning tree. When $n-1$ edges have been selected, we must have a spanning tree as the $n-1$ edges do not contain a cycle (see Theorem 9.4). Before attempting to prove that the method just described does indeed generate a minimum cost spanning tree of G, let us look at an example.

Consider the network of Figure 9.33(a). We begin with no edges selected. Figure 9.34(a) shows the current state of affairs. Edge (1,6) is the first edge picked. It is included into the tree being built. This yields the

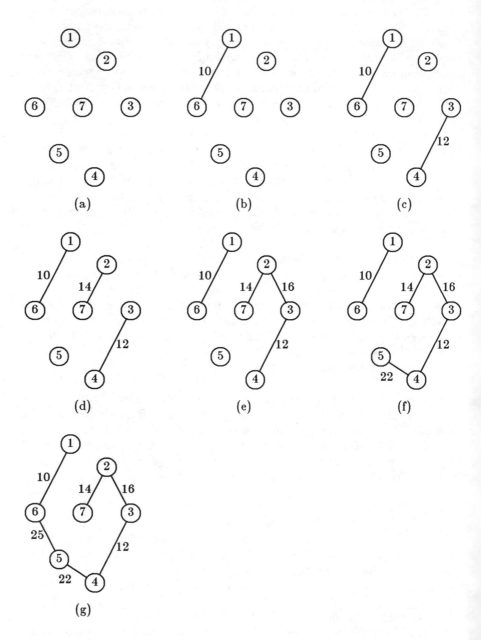

Figure 9.34 Building a minimum cost spanning tree.

graph of Figure 9.34(b). Next, the edge (3,4) is selected and included into the tree (Figure 9.34(c)). The next edge to be considered is (2,7). Its inclusion into the tree being built does not create a cycle. So, we get the graph of Figure 9.34(d). Edge (2,3) is considered next and included into the tree (Figure 9.34(e)). Of the edges not yet considered, (7,4) has the least cost. It is considered next. Its inclusion into the tree results in a cycle. So, this edge is discarded. Edge (5,4) is the next edge to be added to the tree being built. The next edge to be considered is the edge (7,5). It is discarded as its inclusion creates a cycle. Finally, edge (6,5) is considered and included into the tree being built. This completes the spanning tree. The resulting tree has cost 99.

Procedure *KRUSKAL* (Algorithm 9.3) is a more formal statement of the spanning tree generation method described above. This algorithm takes into account the possibility that G may not be a connected network. T is the set of edges to be included in the spanning tree.

```
line  procedure KRUSKAL (G,T,n)
        //G is an n vertex weighted graph with edge set E//
  1       T ← ∅; i ← 0 //i = |T|//
  2       while E ≠ ∅ and i ≠ n−1 do
  3         let (u,w) be a least cost edge in E
  4         E ← E - {(u,w)}
  5         if (u,w) does not create a cycle in T
  6                         then
  7                             T ← T ∪ {(u,w)}
  8                             i ← i+1
  9                         endif
 10       end while
 11       if i ≠ n−1 then print ("G is not connected")
 12       endif
 13    end KRUSKAL
```

Algorithm 9.3 Kruskal's spanning tree algorithm

Algorithm 9.3 is due to Kruskal. It can be implemented to run in $O(e \log e)$ time where e is the number of edges in G (see *Fundamentals of Data Structures*, by E. Horowitz and S. Sahni).

Theorem 9.8: Kruskal's algorithm generates a minimum cost spanning tree for every weighted connected undirected graph G.

Proof: The proof is in two parts. First, we shall show that procedure *KRUSKAL* always succeeds in generating a spanning tree when *G* is initially connected. Then, we shall show that the spanning tree generated is of minimum cost.

(i) Let $G = (V,A)$ be the initial graph. Initially, $E = A$. Every edge (u,w) that is not included into *T* is such that $T \cup \{(u,w)\}$ contains a cycle. From Theorem 9.2, it follows that the graph $H = (V, E \cup T)$ is connected at the start and end of each iteration of the **while** loop of procedure *KRUSKAL*. When this loop terminates, *T* must define a spanning tree of *G*. To see this, observe that if $E = \emptyset$ at this time then *T* must be a spanning tree as $H = (V, E \cup T)$ is connected, $E = \emptyset$ and *T* contains no cycles. If $E \neq \emptyset$, then *T* contains $n-1$ edges and no cycles. So, *T* is a spanning tree.

(ii) Let $T_k = (V, E_k)$ be the spanning tree constructed be Kruskal's algorithm. Let $T_{opt} = (V, E_{opt})$ be a minimum cost spanning tree of $G = (V,E)$. If $E_k = E_{opt}$, then Kruskal's algorithm has clearly generated a minimum cost spanning tree of *G*. Assume that $E_k \neq E_{opt}$. Let *e* be a least cost edge in $E_k - E_{opt}$. From Theorem 9.4, it follows that $H = (V, E_{opt} \cup \{e\})$ contains a cycle *C*. This cycle must contain an edge *e'* such that $e' \in E_{opt} - E_k$ as otherwise, this cycle is also present in T_{opt}. From Theorem 9.2 it follows that $T_1 = (V, E_{opt} \cup \{e\} - \{e'\})$ is connected. Since T_1 is connected and contains $n-1$ edges, it is a tree (Theorem 9.4).

Let $c(e)$ and $c(e')$ respectively denote the cost of the edges *e* and *e'*. Let $E' = \{(u,v) : (u,v) \in E_k$ and cost of (u,v) is less than $c(e)\}$. From the selection of *e*, it follows that $E' \subset E_{opt}$. If $c(e') < c(e)$, then Kruskal's algorithm considers edge *e'* before *e*. At the time *e'* is considered, the edges in *T* (line 5) are a subset of *E'*. $T \cup \{e'\}$ cannot contain a cycle as $T \subseteq E' \subset E_{opt}$ and $E' \cup \{e'\} \subseteq E_{opt}$ contains no cycles. So, *e'* should have been added to *T*. But, *e'* was not (as $e' \notin E_k$). Thus, the assumption that $c(e') < c(e)$ leads to a contradiction. Consequently, $c(e') \geqslant c(e)$.

Hence, the spanning tree $T_1 = (V, E_1)$ where $E_1 = E_{opt} \cup \{e\} - \{e'\}$ costs no more than does T_{opt} and is therefore also a minimum cost spanning tree. Observe that $|E_k \cap E_1| = |E_k \cap E_{opt}| + 1$. If $E_k \neq E_1$, then we may repeat the above edge replacement procedure and transform T_1 into another minimum cost spanning tree T_2 such that $|E_k \cap E_2| = |E_k \cap E_1| + 1$. Hence, T_{opt} may be succesively transformed into the min cost trees T_1, T_2, ... After at most $n-1-|E_k \cap E_{opt}|$ transformations, a min cost spanning tree T_p with $E_k = E_p$ will be obtained. This tree is identical to T_k. Hence, T_k is also a minimum cost spanning tree of *G*. $\square$

9.2.4 Cycle Basis

Let $G = (V,E)$ be an undirected graph. Let C_1, C_2, C_3, ..., C_p be the cycles of G. Assume that $p \geqslant 1$. Let E_i be the set of edges on the cycle C_i. Figure 9.35 shows a graph and six of its cycles. $C_1 = \{e_1,e_6,e_7\}$; $C_2 = \{e_2,e_7,e_8\}$; $C_3 = \{e_1,e_2,e_6,e_8\}$; etc. Notice that $C_3 = C_1 \oplus C_2$ and that $C_6 = C_4 \oplus C_5$ (recall from Chapter 3 that $\oplus$ denotes exclusive or). The cycle $C_7 = \{e_1,e_2,e_3,e_4,e_5,e_6\}$ is $C_3 \oplus C_6$.

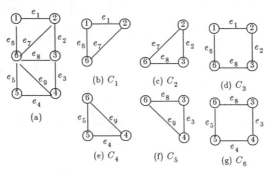

Figure 9.35 A graph and some of its cycles.

A set $S = \{C_{i_1},C_{i_2},...,C_{i_k}\}$ of cycles is a *set of independent cycles* iff no cycle $C_{i_j} \in S$ is the exclusive or of some subset of $S - \{C_{i_j}\}$. The cycle set $\{C_1,C_2\}$ of Figure 9.35 is independent. But, $\{C_1,C_2,C_3\}$ is not independent as $C_3 = C_1 \oplus C_2$. $\{C_1,C_2,C_4,C_5\}$ is an independent set of cycles. However, $\{C_1,C_2,C_4,C_5,C_6\}$ is not. The set $\{C_1,C_2,C_4,C_5,C_7\}$ is not independent either.

A set S of independent cycles is a *cycle basis* for G iff every cycle of G is the exclusive or of the cycles in some subset of S. One may verify that $S = \{C_1,C_2,C_4,C_5\}$ is a cycle basis of the graph of Figure 9.35(a). The cycle basis of a graph finds application in the analysis of electrical circuits, flow circuits, etc. In these applications, one can arrive at an independent set of circuit equations by writing equations for each of the cycles in a cycle basis.

Let G be any undirected graph. A cycle basis of G can be found in the following way. First, obtain a spanning tree for each of the components of G. This can be done, for example, by using either *DFS* or *BFS* on each component of G. Let T be the set of edges included in these spanning trees. Figure 9.36(a) shows a spanning tree for the graph of Figure 9.35(a). For this tree, $T = \{e_5,e_6,e_7,e_8,e_9\}$. It should be easy to see that G has no cycles iff $E - T = \varnothing$. Assume that $E - T \neq \varnothing$. Let (i,j) be one of the edges in $E - T$. The addition of edge (i,j) to T creates a unique cycle (see the exercises and

Figure 9.36(b)). Let $S = \{C_1, C_2, C_3, ..., C_k\}$ be the set of cycles obtained in this way by adding to T each of the edges in $E - T$ one at a time. Note that $|E - T| = k$. For the graph of Figure 9.35(a) and the spanning tree of Figure 9.36(a), the set S of cycles obtained in this way is $\{C_1, C_2, C_3, C_4\}$.

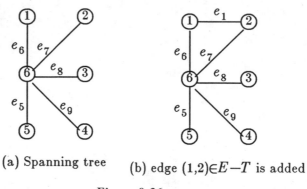

(a) Spanning tree   (b) edge (1,2)$\in E - T$ is added

Figure 9.36.

Figure 9.37(a) shows a graph that is not connected. A set of spanning trees for its components is given in Figure 9.37(b). For this set of spanning trees, $T = \{e_1, e_2, e_3, e_5, e_6, e_7, e_8, e_9, e_{12}\}$. $E - T = \{e_4, e_{10}, e_{11}\}$. Adding the edge e_4 to T creates the cycle $C_1 = \{e_1, e_2, e_4\}$. Adding e_{10} to T creates the cycle $C_2 = \{e_8, e_9, e_{10}\}$, and adding E_{11} to T creates the cycle $C_3 = \{e_8, e_9, e_{11}, e_{12}\}$.

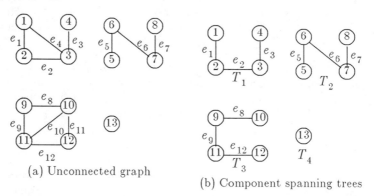

(a) Unconnected graph

(b) Component spanning trees

Figure 9.37 A graph and component spanning trees.

Let the edges in $E - T$ be labeled $a_1, a_2, ..., a_k$. Let C_i be the cycle that results when a_i is added to T. Let $S = \{C_1, C_2, C_3, ..., C_k\}$. Since each cycle C_i contains an edge (a_i) that is not in any of the cycles in $S - \{C_i\}$, S is a set of independent cycles. Is it a cycle basis for G? S is a cycle basis of G iff every cycle of G is the exclusive or of some of the cycles in S. For the case of the graphs of Figures 9.35(a) and 9.37(a), one may verify that the cycle set S is indeed a cycle basis of the respective graphs. The next theorem establishes that the cycle set S constructed as above is always a cycle basis.

Theorem 9.9: Let T be the set of edges in any spanning forest of the graph $G = (V,E)$. Let $E-T = \{a_1, ..., a_k\}$ and let C_i be the unique cycle created when edge a_i is added to T. The set $S = \{C_1, ..., C_k\}$ is a cycle basis of G.

Proof: If $k = 0$, then G has no cycles and $S = \varnothing$ is trivially a cycle basis of G. If $k > 0$, then G contains some cycles. As remarked earlier, these cycles are independent. Let C be any cycle of G. Let $D = C \cap \{E\text{-}T\} = \{a_{i_1}, ..., a_{i_r}\}$. $D \neq \varnothing$ as otherwise $D \subseteq T$. But, T contains no cycles. We leave it as an exercise to show that $C = C_{i_1} \oplus C_{i_2} \oplus ... \oplus C_{i_r}$. $\square$

The *cycle rank* of a graph $G = (V,E)$ is the number of cycles in its cycle basis. If G has k components, then the set T contains exactly $n-k$ edges. So, $|E-T| = e-n+k$ where $|E| = e$. The cycle rank of G is therefore $e-n+k$.

9.2.5 Connectivity

In this sub-section we shall be dealing with undirected graphs only. Hence, whenever we use the term graph, it should be understood to mean an undirected graph. A set S, $S \subseteq V$, of vertices is a *disconnecting vertex set* of the undirected graph $G = (V,E)$ iff the graph $H = (V-S, E-\{(i,j) \mid i \in S$ and $j \in S$ and $(i,j) \in E\})$ contains at least two components. As an example, consider the graph of Figure 9.38(a). The removal of the vertices 2 and 6 together with the edges (1,6), (1,2), (2,3), and (6,5) leaves behind the graph of Figure 9.38(b). This graph has two components. So, $\{2,6\}$ is a disconnecting vertex set of the graph of Figure 9.38(a). $\{1,2\}$ is not a disconnecting vertex set of this graph as the removal of the vertices 1 and 2 and the edges (1,2), (1,6), and (2,3) leaves us with the graph of Figure 9.38(c). This graph has only one component. The complete graphs K_1, K_2, ..., have no disconnecting vertex sets. If G is not connected then $S = \varnothing$ is a disconnecting vertex set.

If $|V| = n$ and G has no disconnecting vertex set then its *connectivity* is n. If G has a disconnecting vertex set, then its *connectivity* is the cardinality of its smallest disconnecting vertex set. So, the connectivity of the complete graph K_i is i. The graph of Figure 9.38(a) has no disconnecting vertex set of size 0 or 1. But, $\{2,6\}$ is a disconnecting vertex set. So, its connectivity is 2. The connectivity of the graph of Figure 9.39 is 1 as the removal of vertex 2 and the edges incident on 2 disconnects the graph. Observe that the connectivity of a graph G is 0 iff G is not connected.

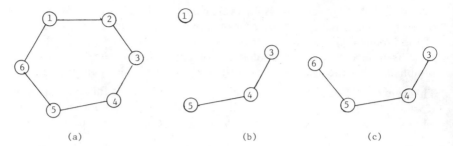

Figure 9.38 A graph and graphs that result from the deletion of some vertices and edges.

A graph G is *k-connected* iff its connectivity is at least k. All graphs are 0-connected. Every connected graph is 1-connected. All graphs with a connectivity at least 2 are 2-connected and so on. 2-connected graphs are called *biconnected graphs* and 3-connected graphs are called *triconnected graphs*. The graph of Figure 9.38(a) is biconnected while that of Figure 9.39   is only connected (or 1-connected).

We have already seen how to determine if an undirected graph is connected. We also know that the smallest (i.e., the one with the fewest number of edges) connected n vertex graph has $n-1$ edges (and so it is a tree). One may prove that the smallest n vertex biconnected graph is a cycle (and so it has exactly n edges).

Let us now consider how to determine if a graph is biconnected. To begin with, if G is not connected, it cannot be biconnected. So, let us assume that G is connected. We may also assume that $|V| > 1$. If G is not biconnected, then G must have at least three vertices as there is only one connected graph with 2 vertices and this graph is biconnected.

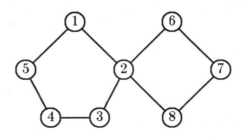

Figure 9.39 A 1-connected graph.

Let G be a connected graph with at least 3 vertices. If G is not bicon-nected then it must contain a disconnecting vertex set of size 1. A vertex v such that $\{v\}$ is a disconnecting vertex set is called an *articulation point*. Ver-tex 2 is an articulation point of the graph of Figure 9.39. By definition, a connected graph is biconnected iff it contains no articulation points. Hence, we can determine if G is biconnected by simply determining if any of its ver-tices is an articulation point.

There is an elegant way to determine if a connected graph contains any articulation points. This involves the use of a depth first spanning tree of the graph. Figure 9.40(a) shows a graph together with one of its depth first span-ning trees. This spanning tree is shown in dark edges. In Figure 9.40(b), this spanning tree has been redrawn with a depth first number assigned to each vertex. The solid edges in Figure 9.40(b) are edges in the spanning tree. The remaining edges of G are shown as broken lines.

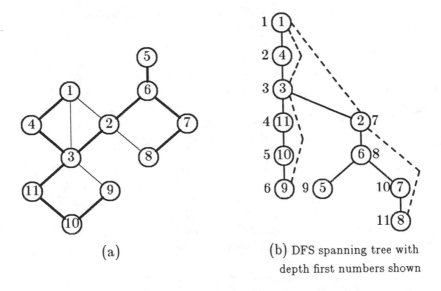

(a)

(b) DFS spanning tree with
depth first numbers shown

Figure 9.40.

The solid lines will be called *tree edges* and the broken ones *back edges*. The set T of tree edges defines the spanning tree. The vertex at which the depth first search starts has depth first number (*DFN*) 1. This vertex is called the *root* of the tree. Vertex i is a *child* of vertex j in a depth first span-ning tree T iff $(j,i) \in T$ and $DFN(j) < DFN(i)$. If i is a child of j then j is the *parent* of i. In the tree of Figure 9.40(b), 4 is a child of 1; 9 is a child of 10; 6 is a child of 2; 3 is the parent of 2; 11 is the parent of 10; etc. Note that

2 is not a child of 4 (it is a grandchild though). 4 is not the parent of 2, but it is its grandparent. Vertex j is an *ancestor* of vertex i iff j is either the parent of i, or is its grandparent, or its great grandparent, etc. Vertex i is a *descendent* of vertex j iff j is an ancestor of i. In the tree of Figure 9.40(b), 1, 4, 3, 11, and 10 are the ancestors of vertex 9. 5, 7, and 8 are the descendents of 6.

An edge $(i,j) \in E$ is a *cross edge* relative to the spanning tree T iff i is neither an ancestor nor a descendent of j. Note that none of the edges in T are cross edges. As can be seen from Figure 9.40(b), none of the edges of the graph of Figure 9.40(a) are cross edges relative to the depth first spanning tree shown in Figure 9.40(b).

Theorem 9.10: Let $G = (V,E)$ be any connected graph and let T be a depth first spanning tree of G. G contains no cross edges relative to T.

Proof: Let (i,j) be any edge in E. If $(i,j) \in T$, then either i is the parent of j or is a child of j. So, (i,j) cannot be a cross edge. Assume that $(i,j) \notin T$. Without loss of generality, we may assume that $DFN(i) < DFN(j)$. So, the depth first search that generated tree T reached vertex i before j. When vertex i is reached, a depth first search from i is initiated (see procedure *DFS*). Since j is reachable from i and not yet marked, it must get marked before the search from i is completed. In other words, j must be a descendent of i in T. Hence, (i,j) is not a cross edge. Thus, none of the edges in E are cross edges relative to the depth first spanning tree T. □

Theorem 9.10 gives us a useful tool to detect the presence of articulation points in a graph G. Let $C_1, C_2, ..., C_k$ be the children of the root of T. The root of T is an articulation point iff $k > 1$. To see this, observe that the deletion from G of the root of T together with all edges incident on the root leaves behind exactly k components. C_i and C_j, $i \neq j$ cannot be in the same component of the resulting graph as G contains no cross edges relative to T.

We can determine if any of the remaining vertices of G are articulation points in the following way. For all vertices i, $i \in V$, define $L(i)$ to be the least DFN reachable from i by using a path consisting of zero or more edges from T followed by at most one back edge (i.e., an edge in $E-T$). Since edges in T can be used only to move from a vertex to one of its children, $L(i)$ is given by the recursive formula:

$$L(i) = \min\{DFN(i), \min\{DFN(j) : (i,j) \in E-T\}, \tag{9.3}$$
$$\min\{L(j) : j \text{ is a child of } i\}\}$$

For the graph of Figure 9.40(b), $L(6) = L(7) = L(8) = 7$ ($DFN(2)$ $= 7$); $L(11) = L(10) = L(9) = 3$; $L(5) = 9$; and $L(1) = L(2) = L(3) = L(4) = 1$ ($DFN(1) = 1$).

A non root vertex i is an articulation point iff it has at least one child j for which $L(j) \geq DFN(i)$. To see this, note that if i is not the root and if it does have a child j with $L(j) \geq DFN(i)$ then j is disconnected from the parent of i when i together with all edges incident on i are removed from the graph. On the other hand if $L(j) < DFN(i)$ for each child j of i, then none of i's children gets disconnected from the remainder of the graph when i and its incident edges are removed. This is so because when $L(j) < DFN(i)$, there is a path from j to an ancestor of i that does not go through vertex i.

Our discussion on articulation points is summarised in the following theorem.

Theorem 9.11: Let $G = (V,E)$ be any connected undirected graph. Let T be a depth first spanning tree of G and let L be as defined in (9.3). Vertex i is an articulation point of G iff either:
  (a) i is the root of T and has at least two children
  or
  (b) i is not the root and has a child j for which $L(j) \geq DFN(i)$. □

The definition (9.3) of L directly leads to procedure ART (Algorithm 9.4) which computes both L and DFN. Procedure ART is a simple modification of procedure DFS. ART has two parameters u and v. v is the parent of u in the depth first spanning tree that is being generated implicitly. The initial call to ART is **call** $ART(i,0)$ where i is any vertex in G. i becomes the root of the spanning tree and is assigned the depth first number 1. It is assumed that at the time of initial call, $DFN(1:n) = 0$ and $num = 1$.

Once the depth first spanning tree and the L values (which are of course defined relative to this tree) are known, Theorem 9.11 can be used to determine which (if any) of the vertices of G are articulation points. This can also be done in-line by adding some more code to ART. One readily observes that the asymptotic complexity of ART is the same as that of DFS.

Edge Connectivity

A set A of edges is a *disconnecting edge set* of the undirected graph $G = (V,E)$ iff the graph $H = (V,E-A)$ is not connected. Let e be any of the edges in the graph of Figure 9.38(a). $\{e\}$ is not a disconnecting edge set of this graph. Let $\{e,f\}$ be a set consisting of any two of the edges in the graph of Figure 9.38(a). One may verify that $\{e,f\}$ is a disconnecting edge set. The graph of

procedure $ART(u,v)$
  //depth first search from u to determine L values//
  //initially, $DFN(1:n) = 0$ and $num = 1$//
  global $n, L(n), DFN(n), G, num$
  $DFN(u) \leftarrow L(u) \leftarrow num; \; num \leftarrow num+1$
  for each vertex w adjacent from u**do**
    if $DFN(w) = 0$ **then** //new vertex//
               call $ART(w,u)$
               $L(u) \leftarrow \min\{L(u), L(w)\}$
          else //(u,w) is a back edge//
             if $w \neq v$ **then**
                //$(u,w) \in E-T$//
                $L(u) \leftarrow \min\{L(u), DFN(w)\}$
          endif
    endif
  end for
end ART

Algorithm 9.4 Determine L values.

Figure 9.38(b) is not connected. So, $A = \varnothing$ is one of its disconnecting edge sets. The graph of Figure 9.39 has no disconnecting edge sets of size less than 2. $\{(2,6),(2,8)\}$ is one of the disconnecting edge sets of this graph.

The *edge connectivity* of an undirected graph $G = (V,E)$ with $|V| > 1$ is the size of its smallest disconnecting edge set. The edge connectivity of the graph of Figure 9.38(b) is 0. The edge connectivity of the graph of Figure 9.39 is 2. It is easy to see that the edge connectivity of a graph is no more than the degree of the least degree vertex in the graph. The edge connectivity of the complete graph K_i is $i-1$, $i > 1$.

An edge (i,j) is a *bridge* of the graph $G = (V,E)$ iff $H = (V, E-\{(i,j)\})$ has more components than does G. Clearly, G has a bridge iff one of its components has an edge connectivity of 1. The graphs of Figures 9.38(a) and 9.39 have no bridges. $(5,4)$, $(2,9)$, $(3,6)$, and $(6,7)$ are the only bridges in the graph of Figure 9.41.

The presence of bridges in a graph can be detected using a depth first search. Let $C_i = (V_i,E_i)$, $1 \leq i \leq k$ be the components of the graph $G = (V,E)$. Let T_i be a depth first spanning tree of C_i. No edge $(u,v) \in E_i - T_i$ can be a bridge. To see this, observe that since T_i contains a u to v path, the deletion of the edge (u,v) cannot increase the number of components in G. So, only the edges in the trees T_i, $1 \leq i \leq k$ can be bridges. Let $(u,v) \in T_i$ for some i. Without loss of generality, we may assume that $DFN(u) <$

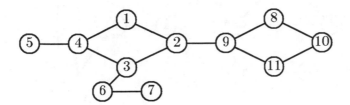

Figure 9.41.

$DFN(v)$. It is readily seen that the removal of (u,v) increases the number of components in G iff $L(v) \geqslant DFN(v)$.

9.3 PATHS

9.3.1 Euler Paths And Circuits

A path $P = i_0,i_1,...,i_e$ is an i_0 to i_e *Euler path* in the graph $G = (V,E), |E| = e$, iff P uses each edge in E exactly once. If G is directed then P is required to be a directed path. An Euler path P with $i_0 = i_e$ is an *Euler circuit*. Every graph which is a cycle has an Euler path which is an Euler circuit. For instance, 1,2,3,4,5,6,1; 2,3,4,5,6,1,2; and 3,2,1,6,5,4,3 are Euler circuits in the graph of Figure 9.38(a).

Consider the graphs of Figure 9.42. The path 1,2,3,1,5,3,4,5 is an Euler path in the graph of Figure 9.42(a). 1,3,5,4,3,2,1,5 is another Euler path in this graph. This graph has no Euler circuits. The directed path 1,2,3,5,3,4,5,2 is an Euler path in the graph of Figure 9.42(b). This graph, too, has no Euler circuits. The graph of Figure 9.42(c) does have an Euler circuit. One such circuit is 1,2,4,3,2,6,5,4,6,1. However, this graph contains no Euler paths that are not Euler circuits.

One application of Euler paths is to the postman problem. In this problem, the edges of the graph represent the streets on which the postman has to deliver mail. If there are one way streets then the graph is directed. Otherwise it is undirected. The postman has to traverse every street (i.e. edge) in the graph delivering mail. In an ideal situation the postman will have to traverse each street exactly once. He can do this iff the graph has an Euler path.

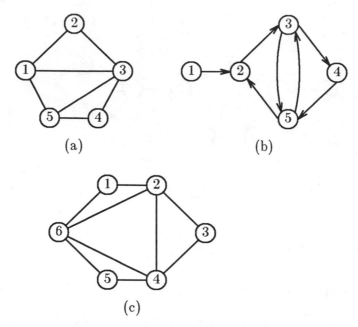

Figure 9.42

Is there any simple characterization of graphs that contain Euler paths? More insight into the existence problem for Euler paths can be obtained by examining an Euler path. Let $P = i_0, i_1, ..., i_e$ be an Euler path in G. For every i_j, $j < e$, the path P is said to *leave i_j and go to i_{j+1}*. For every i_j, $j > 0$, the path *enters i_j from i_{j-1}*. If $i_0 = i_e$, then every vertex is entered and left the same number of times. So, every vertex must be of even degree if G is undirected. If G is directed, then the in-degree of every vertex must equal its out-degree. If $i_0 \neq i_e$, then vertex i_0 is left one more time than it is entered and vertex i_e is entered one more time than it is left. So, if G is an undirected graph, then all vertices other than i_0 and i_e have even degree. Vertices i_0 and i_e have odd degree. If G is a directed graph, then all vertices other than i_0 and i_e have their in-degree equal to their out-degree. Vertex i_0 has an in-degree that is one less than its out-degree. Vertex i_e has an out-degree that is one less than its in-degree. It can be shown that every (weakly) connected graph that satisfies the degree requirements just stated has an Euler path.

Theorem 9.12: Let $G = (V, E)$ be a connected undirected graph. Assume that $|V| > 1$. G has an Euler path iff it contains either zero or exactly two vertices with odd degree. G contains an Euler circuit iff all vertices have even degree.

Proof: The discussion leading to the statement of the theorem established that if a graph G has an Euler path, then either all vertices have even degree or exactly two have odd degree. It also established that if a graph G had an Euler circuit, then all vertices are of even degree. So, we need only show that every graph G with exactly two odd degree vertices or exactly 0 odd degree vertices has an Euler path. In the latter case, the path is a circuit. The proof is by construction.

If G has exactly two vertices of odd degree, then let i_0 and i_e be these two vertices. If G has no vertex of odd degree then let i_0 be any vertex of G. In this case let $i_e = i_0$. We shall construct an i_0 to i_e Euler path. Use any edge incident on i_0 to get to an adjacent vertex i_1. Use any new edge incident on i_1 to get to an adjacent vertex i_2. Keep using previously unused edges in this way until a vertex i_k is reached such that all edges incident on i_k have been used. The path constructed in this way is $i_0, i_1, i_2, ... i_k$. For example, starting with $i_0 = 1$ and the graph of Figure 9.42(c), we may build the path 1,2,6,5,4,6,1. In this case, $i_k = 1$ as all edges incident on 1 are contained in this path.

Let n_j^{in} and n_j^{out}, respectively, be the number of times the path $i_0, i_2, ..., i_k$ enters and leaves vertex j of G. If $i_0 = i_k$, then $n_j^{in} = n_j^{out}$, $0 \leqslant j \leqslant n$ ($n = |V|$). In this case the path has used an even number of edges incident on each vertex (0 is even). Since there are no unused edges incident on $i_k = i_0$, i_0 must be of even degree. If $i_0 \neq i_k$, then $n_j^{in} = n_j^{out}$, $0 \leqslant j \leqslant n$ and $j \notin \{i_0, i_k\}$; $n_{i_0}^{in} = n_{i_0}^{out} - 1$ and $n_{i_k}^{in} = n_{i_k}^{out} + 1$. Since there are no unused edges incident on i_k, i_k is of odd degree. Hence, $i_k = i_e$ (recall that i_0 and i_e are the only vertices of odd degree). Consequently, in both cases, the graph, G_1, consisting of all vertices in V and all unused edges contains no vertex of odd degree.

If G_1 contains no edges, then an Euler path has been obtained and the construction terminates. If G_1 contains edges, then we observe that every component of G_1 must have a vertex in common with the path $i_0, i_2, ..., i_k$ (as the original graph is connected).

Let $C_i = (V_i, E_i)$ be a component of G_1 with $|V_i| > 1$. Start at a vertex u which is in the multiset $\{i_0, i_1, ..., i_k\}$ and build a path u, $u_1, u_2, ..., u_p$ as before. Path building stops at a vertex $u_p \in V_i$ such that C_i has no unused edges incident on u_p. Since every vertex in V_i is of even degree in C_i, $u_p = u$. The circuit u, $u_1, u_2, ..., u_{p-1}, u_p$ can be combined with the path $i_0, i_2, ..., i_k$ to obtain a new path $P_2 = i_0, i_2, ..., i_j, u_1, u_2, ..., u_{p-1}, u_p, i_{j+1}, ..., i_k$ where $i_j = u = u_p$ by choice of u.

The graph $G_2 = (V, E_2)$ where E_2 is the set of edges not on the path P_2 is such that every vertex is of even degree. If $E_2 \neq \varnothing$, then we may consider a component $C_j = (V_j, E_j)$ of G_2 with $|V_j| > 1$ as before and expand the path P_2 to another path P_3. By repeating this path expansion several times, an Euler path will be obtained.

This construction also shows that there is an Euler circuit iff G is connected and contains no vertex of odd degree. □

Continuing with the example begun in the proof of the preceding theorem, the removal of the edges used in the path $P_1 = 1,2,6,5,4,6,1$ gives us the graph G_1 of Figure 9.43. This graph contains only one component that has more than one vertex. There are two possible start vertices u (i.e., 2 and 4). Let us start at $u = 4$. The new path built could either be 4,3,2,4 or 4,2,3,4. Assume that the former is the case. Combining this with P_1 yields $P_2 = 1,2,6,5,4,3,2,4,6,1$. Since there are no unused edges remaining in G (Figure 9.42(c)), an Euler circuit has been constructed.

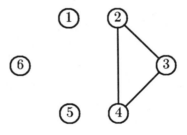

Figure 9.43

For the case of directed graphs, the following theorem may be established.

Theorem 9.13: Let $G = (V, E)$ be a weakly connected directed graph. Assume that $|V| > 1$. Let d_i^{in} and d_i^{out} respectively be the in-degree and out-degree of vertex i, $1 \leqslant i \leqslant |V|$. G contains a u to v Euler path iff G is weakly connected and either
  (a) $u = v$ and $d_i^{in} = d_i^{out}$ for every $i \in V$
or
  (b) $u \neq v$, $d_i^{in} = d_i^{out}$ for every $i \in V - \{u, v\}$, $d_u^{in} = d_u^{out} - 1$, and $d_v^{in} = d_v^{out} + 1$.

Proof: The proof is similar to that of Theorem 9.12 and is left as an exercise. □

9.3.2 Hamiltonian Paths And Circuits

In 1859 Sir William Hamilton invented a game called "Around the World". In this game the twenty vertices of a dodecahedron are labeled with the names of cities around the world. The objective of the game is to find a path starting at any city, going along the edges of the dodecahedron, passing through every city exactly once and ultimately returning to the start city. The game can be represented by a graph as in Figure 9.44. The vertices of the dodecahedron correspond to the vertices of the graph. The edges of the dodecahedron correspond to the the edges of the graph too. The path 1,2,3,...,20,1 is a path that goes through each vertex exactly once and returns to the start vertex. So, this path is a solution to Hamilton's game.

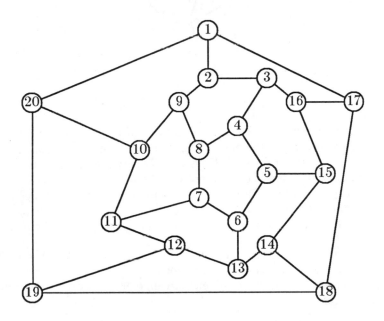

Figure 9.44

Let $G = (V,E)$ be a graph and let $P = i_1, i_2, ..., i_k$ be a path in G. P is an i_1 to i_k *Hamiltonian path* iff P is a simple path containing every vertex in V. P is a **Hamiltonian cycle** iff $i_1 = i_k$ and P is a Hamiltonian path. When G is a digraph, the terms directed Hamiltonian path and directed Hamiltonian cycle are used.

For the graph of Figure 9.44, the path 1,2,...,20 is a Hamiltonian path that is not a Hamiltonian cycle. However, 1, 2, ..., 20, 1 is a Hamiltonian cycle. The path 1, 2, 3, 4, 5, 6, 1 is a Hamiltonian cycle in the graph of Figure 9.35(a). The graph of Figure 9.39 contains no Hamiltonian path. The directed path 1, 2, 3, 4, 5 is a directed Hamiltonian path in the digraph of Figure 9.45. This digraph contains no directed Hamiltonian cycle.

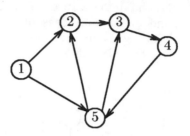

Figure 9.45 A digraph with no Hamiltonian cycle.

Suppose we have to route a postal van to pick up mail from n mail boxes and then return to the post office. The situation can be represented by an $n+1$ vertex digraph $G = (V,E)$ in which one vertex represents the post office and the remaining vertices represent the n mail boxes. Edge $<i,j>$ is in E iff it is possible to go directly from vertex i to vertex j. The postal van route is a directed Hamiltonian cycle in G. In a more realistic situation, edge $<i,j>$ will have a length associated with it (i.e., the distance between i and

j) and we will be interested in determining a shortest directed Hamiltonian cycle in the weighted digraph G. The problem of determining a shortest directed Hamiltonian cycle in a weighted digraph G is called the *traveling salesperson problem*.

As another example, suppose we wish to use a robot arm to tighten the nuts on some piece of machinery. The position of the nuts can be represented by the vertices of a digraph G. One of these vertices represents the starting position of the robot arm. The directed edge $<i,j>$ can be assigned a cost equal to the time taken to move the arm from position i to position j. The arm is required to return to the start position after tightening all the nuts. Since each nut is to be visited exactly once, the path of the arm is a directed Hamiltonian cycle in G. Because of the presence of edge costs, all cycles are not equally desirable. We are interested in determining a least cost directed Hamiltonian cycle.

Determining whether or not a graph G contains a Hamiltonian cycle is a computationally difficult problem. In fact, the fastest algorithms known have a worst case time complexity of $\Theta(n^2 2^n)$ where $n = |V|$. We shall content ourselves here by merely considering some of the known results concerning the existence of Hamiltonian cycles. To begin with, it is easy to see that only biconnected undirected graphs and strongly connected digraphs contain Hamiltonian cycles. Also, every complete graph (whether directed or undirected) contains a Hamiltonian cycle. The following Theorems provide some sufficient (but not necessary conditions) for an undirected graph to contain a Hamiltonian path or cycle.

Theorem 9.14: Let $G = (V,E)$ be an undirected graph. Let $|V| = n$ and let d_i be the degree of vertex i. If $d_i + d_j \geqslant n-1$ for every pair of distinct vertices i and j in V, then G contains a Hamiltonian path. $\square$

Theorem 9.15: Let $G = (V,E)$ be an undirected graph. Let $|V| = n$, $n \geqslant 3$. If for every i, $1 \leqslant i \leqslant \lfloor (n-1)/2 \rfloor$ G contains fewer than i vertices with degree at most i, then G contains a Hamiltonian cycle. This requirement can be relaxed when $i = (n-1)/2$ (for this n must be odd). In this case a Hamiltonian cycle exists even if G contains $(n-1)/2$ vertices of degree $(n-1)/2$. $\square$

Theorem 9.16: Let G, n, and d_i be as in Theorem 9.14. Assume $n > 3$. If for every $(i,j) \notin E$, $i \neq j$, $d_i + d_j \geqslant n$, then G contains a Hamiltonian cycle. $\square$

One may verify the existence of undirected graphs that do not satisfy the conditions of the above theorems but do contain Hamiltonian paths and cycles.

9.3.3 Shortest Paths And Transitive Closure

The *cost* of the directed path $i_1, i_2, ..., i_k$ in the weighted digraph $G = (V,E)$ is $\sum_{1 \leqslant j < k} c(i_j, i_{j+1})$ where $c(u,v)$ is the length (or weight or cost) of the edge $<u,v>$. $i_2, ..., i_{k-1}$ are *intermediate vertices*. An i to j path in G is a *shortest i to j path* in G iff G contains no i to j path of lesser length. The path 1,2,3 (Figure 9.46) has a cost of 15. It is not a shortest 1 to 3 path in this graph as the path 1,2,5,3 has a length of 14. 1,2,5,3 is a shortest 1 to 3 path in this graph.

In this section, we shall be concerned with determining a length matrix C such that $C(i,j)$ is the length of a shortest i to j path in G. If there is no i to j path in G then $C(i,j) = \infty$. We shall make the simplifying assumption that G contains no directed cycles of negative cost. With this assumption, the matrix C can be computed by an algorithm very similar to the one devised in

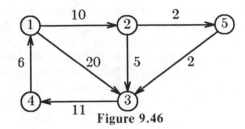

Figure 9.46

Chapter 4 to compute the transitive closure of a binary relation.

Define the $n \times n$ matrix C^k, $k \geqslant 0$ to be such that $C^k(i,j)$ is the length of a shortest i to j path in G that goes through no intermediate vertex of index greater than k (i.e., the only allowable paths are of the type i, u_1, u_2, ..., u_p, j such that $u_a \leqslant k$, $1 \leqslant a \leqslant p$. The length matrix we wish to compute is C^n. C^0 is known and is given by:

$$C^0(i,j) = \begin{cases} 0 & i=j \\ \infty & <i,j> \notin E \text{ and } i \neq j \\ c(i,j) & <i,j> \in E \end{cases} \qquad (9.4)$$

Using the same reasoning as we used to arrive at (4.1), one may verify the correctness of the recurrence:

$$C^k(i,j) = \min\{C^{k-1}(i,j), C^{k-1}(i,k) + C^{k-1}(k, j)\}, k > 0 \qquad (9.5)$$

As in the case of Algorithm 4.3, the C^ks can be computed in place as in Algorithm 9.5. When Algorithm 9.5 terminates, $C(i,j)$ is the length of a shortest i to j path. The time complexity of this algorithm is readily seen to be $\Theta(n^3)$.

procedure *SHORTEST*(C,n)
  //On entry, C is as given by (9.4). On exit, C is the//
  //shortest path length matrix of the given graph//
  declare $C(n,n)$; **integer** i,j,k,n
  for $k \leftarrow 1$ **to** n **do** //compute C^k//
    for $i \leftarrow 1$ **to** n **do**
      for $j \leftarrow 1$ **to** n **do** //compute $C^k(i,j)$//
        $C(i,j) \leftarrow \min \{C(i,j), C(i,k) + C(k,j)\}$
      end for
    end for
  end for
end *SHORTEST*

Algorithm 9.5 Computation of shortest path matrix.

The *transistive closure matrix* of a graph $G = (V,E)$ with $|V| = n$ is an $n \times n$ matrix A^+ such that $A^+(i,j) = 1$ iff G contains an i to j path containing at least one edge (directed path in case G is a digraph). $A^+(i,j) = 0$ otherwise. The *reflexive transitive closure matrix* A^* of a graph G is an $n \times n$ matrix such that $A^*(i,j) = 1$ iff G contains a (directed) path from i to j. $A^*(i,j) = 0$ otherwise. Figure 9.47 shows a digraph together with its adjacency matrix A and the matrices A^+ and A^*.

The matrix A^* can be obtained by starting with a modified adjacency matrix A' where $A'(i,j) = A(i,j)$ when $i \neq j$ and $A'(i,j) = 1$ whenever $i = j$. It should be clear that Algorithm 4.3 invoked by **call** *CLOSURE* (A', A^*, n) will compute A^* correctly. In Chapter 4, it was shown that $A^+ = A*A^*$. So, A^+ can be easily obtained from A^*.

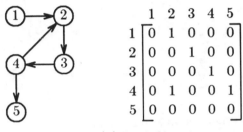

$$\begin{array}{c c} & \begin{array}{c c c c c} 1 & 2 & 3 & 4 & 5 \end{array} \\ \begin{array}{c} 1 \\ 2 \\ 3 \\ 4 \\ 5 \end{array} & \begin{bmatrix} 0 & 1 & 0 & 0 & 0 \\ 0 & 0 & 1 & 0 & 0 \\ 0 & 0 & 0 & 1 & 0 \\ 0 & 1 & 0 & 0 & 1 \\ 0 & 0 & 0 & 0 & 0 \end{bmatrix} \end{array}$$

(a) A graph   (b) Its adjacency matrix

$$\begin{array}{c c} & \begin{array}{c c c c c} 1 & 2 & 3 & 4 & 5 \end{array} \\ \begin{array}{c} 1 \\ 2 \\ 3 \\ 4 \\ 5 \end{array} & \begin{bmatrix} 0 & 1 & 1 & 1 & 1 \\ 0 & 1 & 1 & 1 & 1 \\ 0 & 1 & 1 & 1 & 1 \\ 0 & 1 & 1 & 1 & 1 \\ 0 & 0 & 0 & 0 & 0 \end{bmatrix} \end{array}$$

$$\begin{array}{c c} & \begin{array}{c c c c c} 1 & 2 & 3 & 4 & 5 \end{array} \\ \begin{array}{c} 1 \\ 2 \\ 3 \\ 4 \\ 5 \end{array} & \begin{bmatrix} 1 & 1 & 1 & 1 & 1 \\ 0 & 1 & 1 & 1 & 1 \\ 0 & 1 & 1 & 1 & 1 \\ 0 & 1 & 1 & 1 & 1 \\ 0 & 0 & 0 & 0 & 1 \end{bmatrix} \end{array}$$

(c) A^+ (d) A^*

Figure 9.47

9.4 ROOTED TREES

9.4.1 Terminology

In Section 9.1.2 we defined a rooted tree to be a directed tree in which exactly one vertex has an in-degree of 0 and the remaining vertices have an in-degree of exactly 1. The vertex with in-degree 0 is called the root. There is a standard way in which to draw a rooted tree. We start with the root at the top. All vertices adjacent from the root are drawn below the root. These vertices are called the *children* of the root; the root is the *parent* of its children. All vertices adjacent from these children are drawn next and so on. Since each vertex (other than the root) is adjacent from exactly one other vertex, the drawing procedure just described will not try to place a vertex at two different places. Figure 9.48 shows some rooted trees drawn as described above. Since all edges point downwards in the standard drawing of a tree, we may omit the arrowheads and leave the edge orientation implicit (see Figure 9.49).

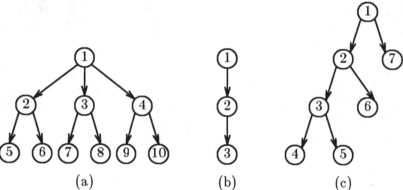

(a) (b) (c)

Figure 9.48 Some rooted trees.

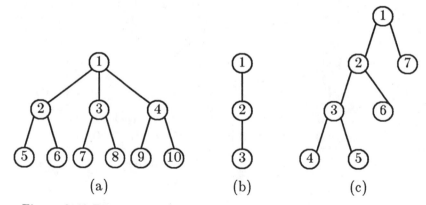

(a) (b) (c)

Figure 9.49 The trees of Figure 9.48 with implicit edge orientation.

In the remainder of this section we shall refer to a rooted tree with implicit edge orientation simply as a tree. The vertices in a tree will be called *nodes*. If node i is adjacent from node j, then we shall say that i is a child of j and j is the parent of i. Note that every node except the root has exactly one parent. The root has no parent. Nodes that have no children are called *leaf nodes*. Nodes 4, 5, 6, and 7 are the leaf nodes of the tree of Figure 9.49(c).

Node 3 is the only leaf node in the tree of Figure 9.49(b). Nodes that are children of the same parent are *siblings*. Thus, 2 and 7 are siblings in the tree of Figure 9.49(c). So also are nodes 4 and 5; and 3 and 6.

All nodes on the path from the root to a node *j* are the *ancestors* of *j*. Nodes 3, 2, and 1 are the ancestors of 4 in the tree of Figure 9.49(c). The set of *descendents* of a node may be defined recursively. If node *j* has no children then it has no descendents. Otherwise, the children of *j* together with all descendents of *j*'s children form the set of descendents of *j*. The descendents of node 1 (Figure 9.49(c)) are 2, 3, 4, 5, 6, and 7. The descendents of node 3 are 4 and 5 while 3, 4, 5, and 6 are the descendents of node 2.

For any tree $T = (V,E)$, by the subtree with root *j*, *j*∈ *V*, we mean the tree consisting of node *j* together with all its descendents plus edges in *E* connecting vertices in {*j*} ∪ {*i*|*i* is a descendent of *j* in *T*}. Figure 9.50 shows some of the rooted subtrees of the tree of Figure 9.49(a).

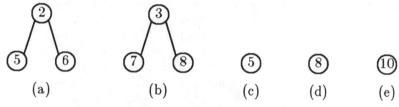

Figure 9.50 Some of the subtrees of Figure 9.49(a).

The root node is defined to be on *level one* of the tree. If node *i* is on level *j* then its children (if any) are on level *j* +1. The *height (or depth)* of a tree is the maximum level at which there are nodes. In the tree of Figure 9.49(a), nodes 2, 3, and 4 are on level 2 while nodes 5 through 10 are on level 3. This tree has a height (or depth) of 3.

The *degree* of a node is defined to be the number of children the node has. Node 1 of Figure 9.49(a) has degree 3; node 4 has degree 2; and node 7 is of degree 0. The *degree* of a tree is the degree of the node with the largest number of children. The tree of Figure 9.49(a) has degree 3 while that of Figure 9.49(c) has degree 2.

A tree of height *h* is a *full tree* of degree *k* iff every node other than those on level *h* have degree *k* (nodes on level *h* must of course be leaf nodes). The tree of Figure 9.49(b) is a full tree of degree 1 and height 3. The trees of Figures 9.49(a) and (c) are not full trees. Figures 9.51(a) and (b) show full trees of height 3 and degree 2 and 3 respectively.

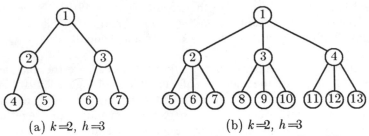

(a) $k=2$, $h=3$ (b) $k=2$, $h=3$

Figure 9.51 Full trees of height 3 and degrees 2 and 3.

The definition of a tree is usually extended to include an *empty tree* (i.e. a tree with no nodes). The empty tree is the only tree with no root node. It has a height of zero. In a tree, no ordering on the children of a node is specified. So, all the trees of Figure 9.52 are equivalent. When an ordering of the children is specified, the tree is called an *ordered tree*. If the trees of Figure 9.52 are viewed as ordered trees, then they are different. An ordered tree of degree k is called a *k-ary tree*. When $k = 2$, the ordered tree is a *binary tree*. When $k = 3$, it is a *ternary tree*. In the case of a binary tree we speak of the *left* and *right* children of a node. For example, if Figure 9.52(a) represents a binary tree, then 2 is the left child of 1 and 3 is the right child of 1.

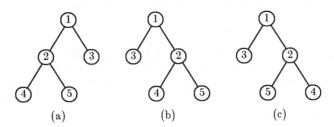

(a)   (b)   (c)

Figure 9.52 Some equivalent trees.

9.4.2 Some Properties

Theorem 9.17: The number of nodes at level i of any tree T of degree d is at most d^{i-1}, $i \geqslant 1$.

Proof: The proof is by induction on i. Let T be any tree of degree d. Only the root node may be at level one. Hence, the number of nodes on level 1 is never more than $1 = d^0$ (note that the empty tree has no nodes on level 1). Assume the theorem is true for some value of i, say $i = m$, $m \geqslant 1$. We shall show it is true when $i = m+1$. From the induction hypothesis, it follows that T has at most d^{m-1} nodes on level m. Each of the nodes on level m can have at most d children. So, there can't be more than $d*d^{m-1} = d^m$ nodes on level $m+1$. $\square$

Corollory 9.1: The total number of nodes in a tree of degree d and height h is at most $(d^h-1)/(d-1)$, $h \geqslant 1$.

Proof: Let T be any tree of degree d and height h. Since there cannot be more than d^{i-1} nodes on level i (Theorem 9.17), there cannot be more than

$$\sum_{j=0}^{h-1} d^j = (d^h-1)/(d-1) \quad \text{nodes in the tree.} \quad \square$$

Corollory 9.2: Let T be a degree d tree containing n nodes. The height of T is at least $\lfloor \log_d(n(d-1)+1) \rfloor$.

Proof: Follows directly from Corollory 9.1. $\square$

When $d = 2$, the maximum number of nodes on level i is 2^{i-1}; the maximum number of nodes in the tree is 2^h-1; and if there are n nodes in the tree, then its height is at least $\lfloor \log_2(n+1) \rfloor$.

Theorem 9.18: Let T be any n node tree of degree 2. Let n_i be the number of nodes in T of degree i, $0 \leqslant i \leqslant 2$. $n_0 = n_2+1$.

Proof: The number of edges in any n node tree is $n-1$. Also, since a node of degree i has exactly i edges emanating from it, the number of edges in T is n_1+2n_2. So, $n = n_1+2n_2+1 = n_0+n_1+n_2$. Subtracting n_1+n_2 from each side of this equality yields $n_0 = n_2+1$. $\square$

Search Trees

A *k-way search tree* is defined as follows:
   (i) An empty tree is a k-way search tree.
   (ii) If T is not empty, then T is a k-way search tree iff
      (a) The root node contains n identifiers $K_1,K_2,...,K_n$ (in this order) where $1 \leqslant n < k$ and $K_1 < K_2 < ... < K_n$;
      (b) Let the subtrees of the root be $T_0,T_1,...,T_{k-1}$. Subtrees $T_{n+1},...,T_{k-1}$ must be empty. Define $K_0 = -\infty$ and $K_{n+1} = \infty$. No identifiers in subtree T_i should be greater than or equal to K_{i+1} or less than or equal to K_i, $0 \leqslant i \leqslant n$.
      (c) The subtrees $T_0,T_1,...,T_n$ are also k-way search trees.

When $k = 2$, the k-ary search tree is called a binary search tree. When $k = 3$, it is called a ternary search tree. Figure 9.53 shows some binary and ternary search trees. The node labels have been placed outside the nodes (rather than inside). The numbers inside each node are identifiers. An immediate consequence of the definitions is that all identifiers in a k-way search tree are distinct.

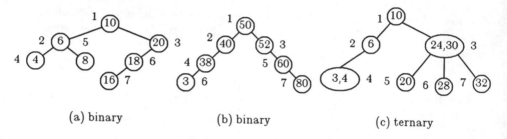

| (a) binary | (b) binary | (c) ternary |

Figure 9.53 k-ary search trees.

As suggested by the name, k-ary search trees are useful for searching. Suppose we wish to determine if the tree of Figure 9.53(a) contains the identifier $X = 16$. We may start by comparing X with the identifier in the root. Since $X > 10$, it cannot be in the left subtree. So, we move to the right subtree. Since $X < 20$, it cannot be in the right subtree of node 3. So, we move to the left subtree. As $X < 18$, it cannot be in the right subtree of node G. So, we move to node 7. Now, $X = 16$ and we have determined that X is in the tree.

Let us search the binary search tree of Figure 9.53(b) for $X = 45$. As $X < 50$, the search moves to node 2. As $X > 40$, the search moves to the right subtree of node 2. This subtree is empty and we conclude that X is not in the tree. If we wish to insert X into the search tree, then X is to be inserted as the right child of node 2.

As a final example, consider searching the ternary search tree of Figure 9.53(c) for $X = 29$. $X > 10$ and the search proceeds to node 3. $24 < X < 30$ and the search proceeds to node 6. $X > 28$ and we move to the second subtree of node 6. This subtree is empty. So X is not in the tree. X may be inserted into node 6 as this node only contains one identifier. If node 6 had contained the identifiers 25 and 28, then the search would have moved to the third subtree of this node. This subtree is also empty. But now, X can be inserted into the tree as the third child of node 6.

Procedure *KSEARCH* (Algorithm 9.6) formally describes how to search a k-way search tree. If X is in the tree, this procedure returns the node and location within the node of X. Otherwise it returns the last node used and location 0. As can be seen, the maximum number of iterations of the **while** loop is equal to the height of the search tree.

line	
	procedure $KSEARCH(T,X,k)$
	//search the k-way search tree//
	//with root T for X//
0	**integer** k; **pointer** $T,P,Q,T_0,...,T_{k-1}$;
1	**identifier** $K_0,...,K_k$
2	$P \leftarrow T;\ Q \leftarrow 0;$ //Q is the parent of P//
3	$K_0 \leftarrow \infty$
4	**while** $P \neq 0$ **do**
5	Let the identifiers in node P be
	$K_1,K_2,...,K_{n+1},\ 1 \leqslant n < k$.
6	Let the roots of the subtrees be
	$T_0,T_1,...,T_{k-1}$
7	$K_{n+1} \leftarrow \infty$
8	Let i be such that $K_i \leqslant X < K_{i+1}$
9	**if** $X = K_i$ **then return**(P,i) //X found//**endif**
10	$Q \leftarrow P;\ P \leftarrow T_i$ //Move to a subtree//
11	**end while**
12	**return**$(Q,0)$ //X not in tree//
13	**end** $KSEARCH$

Algorithm 9.6 Search of a k-way search tree.

We can distinguish between two kinds of searches in a search tree: successful and unsuccessful. *Successful* searches terminate at line 9 of procedure *KSEARCH* while *unsuccessful* searches terminate at line 12. Let us call the nodes in a k-ary tree *internal nodes*. These nodes shall be drawn as circles or ovals. Whenever a successful search is carried out, P (procedure *KSEARCH*) is left pointing at an internal node. Let N be any internal node of a k-way search tree. Let $K_1,...,K_n$ be the identifiers in this node and let $T_0, T_1, ..., T_n$ be the subtrees. If some of $T_0, T_1, ..., T_n$ are empty, replace the empty subtrees by square nodes (as in Figure 9.54). These square nodes are called *external nodes*. The resulting k-ary tree is called an *extended k-ary tree*. Note that all unsuccessful searches leave P pointing to an external node. One may readily verify that the number of external nodes in an extended k-ary search tree with b identifiers is $b+1$.

Let T be an extended binary search tree. The *path length* from the root of T to any node on level h is $h-1$. The *internal path length* of a binary tree is the sum of the path lengths of all internal nodes. The internal path length of

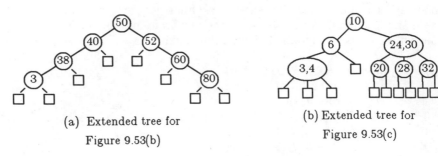

(a) Extended tree for
Figure 9.53(b)

(b) Extended tree for
Figure 9.53(c)

Figure 9.54 Extended trees.

the tree of Figure 9.54(a) is $0+2*1+2*2+2*3 = 12$. The *external path length* is the sum of the path lengths of all external nodes. The tree of Figure 9.54(a) has an external path length of $2*2+2*3+4*4 = 26$.

Theorem 9.19: Let T be any extended binary tree with external path length E and internal path length I. Let n be the number of internal nodes in T. $E = I+2n$.

Proof: The proof is by induction on n. When $n = 1$, there is only one extended binary tree (Figure 9.55(a)). For this tree, $I = 0$ and $E = 2$. So, $E = I+2n$. Assume $E = I+2n$ for some arbitrary value m of n, $m \geqslant 1$. Consider any $m+1$ internal node extended binary tree. This tree must contain at least one internal node, v, both of whose children are external nodes. Replacing this subtree by an external node gives us a new extended binary tree T' (see Figure 9.55(b)). For T', $E' = I'+2m$ (from the induction hypothesis). If node v is on level h of T, then $I = I'+h-1$ and $E = E'-(h-1)+2h = E'+h+1$. Hence, $E = I+2(m+1)$. $\square$

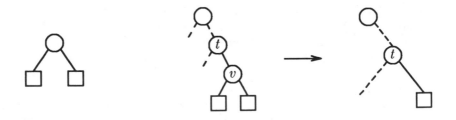

Figure 9.55

When searching a binary search tree T, the *number of identifier comparisons* made (line 8 of procedure *KSEARCH*) is h if the search is successful and X is at level h or if the search is unsuccessful and the external node reached is on level $h+1$. Let s be the average number of comparisons made for a successful search (assuming each identifier is searched for with equal probability) and let u be the average number of comparisons for an unsuccessful search. If T has n internal nodes (and hence n identifiers), then $s = (1+\frac{1}{n})u-1$, $n\geqslant 1$. For Figure 9.54(a), $s = \frac{19}{7}$, $u = \frac{26}{8}$, and $\frac{19}{7} = (1 + \frac{1}{7})*\frac{26}{8} - 1$.

Theorem 9.20: Let s, u, and n be as above. $s = (1+\frac{1}{n})u-1$, $n\geqslant 1$.

Proof: Let I and E respectively be the internal and external path lengths of the extended binary search tree. $s = (I+n)/n$ and $u = E/(n+1)$ (note that $n+1$ is the number of external nodes). From Theorem 9.19, we know that $E = I+2n$. Substituting for E and I in this equality yields $s = (1+\frac{1}{n})u-1$. $\square$

So, we see that there is a very close relationship between the average performance of a binary search tree for successful and unsuccessful searches. From Theorem 9.17, we know that the n node binary tree with minimum internal path length has 1 node on level 1, 2 on level 2, 4 on level 3, ..., and $n-2^d+1$ on level $d+1$ where $d = \lfloor \log_2 n \rfloor$. Thus the smallest value for I is

$$0 + 2*1 + 4*2 + 8*3 + ...$$

which is $O(n \log n)$.

9.4.3 The Number Of Binary Trees

How many different n node binary trees are there? First, let us be precise about when two binary trees are different. We shall say that two binary trees T and U are *equivalent* iff there is a one-to-one correspondence f between the nodes of T and U such that for every node i in T, it is the case that:

(a) i has a left child j iff $f(i)$ has a left child k such that $f(j) = k$.

and

(b) i has a right child j iff $f(i)$ has a right child k such that $f(j) = k$.

The labels on nodes are not used in determining equivalence (so, this definition is different from our earlier definition). T and U are *different* iff they are not equivalent. The binary trees of Figures 9.56 (a) and (b) are equivalent while those of Figures 9.56 (c) and (d) are not. In Figure 9.56 (c) we see that the left subtree of the root is empty while the right subtree of the root of Figure 9.56 (d) is not empty.

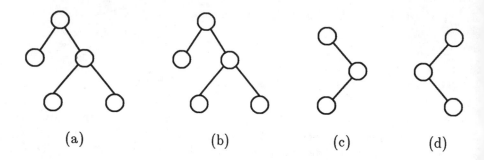

(a)   (b)   (c)   (d)

Figure 9.56

Let b_n be the number of different binary trees with n nodes. It is easy to see that $b_0 = b_1 = 1$ and $b_2 = 2$. For the case of an arbitrary n, the left subtree of the root can have i nodes and the right subtree can have $n-i-1$, $0 \leq i < n$. So, b_n is given by the recurrence:

$$b_n = \sum_{0 \leq i < n} b_i * b_{n-i-1}, \quad n \geq 1 \qquad (9.6)$$

and

$$b_0 = 1.$$

(9.6) yields $b_2 = 2$, $b_3 = 5$, $b_4 = 14$, etc. Before attempting to solve (9.6) for general n, let us see some applications of b_n. Every n node binary tree is a spanning tree of the complete graph K_n. So K_n has at least as many different spanning trees as it has different binary spanning trees.

The number of different binary search trees for a set of n distinct identifiers equals b_n.

As a third example, consider multiplying n matrices $M_1,...,M_n$. There are several ways to compute $M_1*M_2*...*M_n$ when $n > 2$. For example, if $n = 5$ we could compute $M_1*M_2*M_3*M_4*M_5$ using any of the following factorizations:

$$(((M_1*M_2)*M_3)*M_4)*M_5$$
$$((M_1*M_2)*(M_3*M_4))*M_5$$
$$(M_1*(M_2*M_3))*(M_4*M_5)$$
$$((M_1*M_2)*M_3)*(M_4*M_5)$$
$$(M_1*M_2)*(M_3*(M_4*M_5))$$
$$(M_1*M_2)*(*M_3*M_4)*M_5)$$
$$(M_1*(M_2*(M_3*(M_4*M_5))))$$
$$(M_1*((M_2*M_3)*(M_4*M_5)))$$

Let m_n be the number of different ways to compute $M_1*M_2*...*M_n$. Clearly, $m_1 = m_2 = 1$. When $n>1$, the last product formed is of the type $M_{1,k}*M_{k+1,n}$ where $M_{1,k} = M_1*...*M_k$, $M_{k+1,n} = M_{k+1}*...*M_n$, and k is in the range $1 \leqslant k < n$. There are m_k ways to compute $M_{1,k}$ and m_{n-k} ways to compute $M_{k+1,n}$. So, we get the recurrence:

$$m_n = \begin{cases} 1 & n=1 \\ \sum_{1 \leqslant k < n} m_k * m_{n-k} & n>1 \end{cases} \tag{9.7}$$

Solving this, we obtain $m_2 = 1$, $m_3 = 2$, $m_4 = 5$, $m_5 = 14$, etc. One may verify that $m_n = b_{n-1}$, $n \geqslant 1$.

Let us proceed to solve the recurrence (9.6). We shall use the generating function method. Let $G(z)$ be the generating function for $b_0, b_1, ...$. Hence,

$$G(z) = \sum_{i \geqslant 0} b_i z^i$$

From (7.39), we see that $G^2(z)$ is the generating function for $e_0, e_1 ...$ where

$$e_i = \sum_{j=0}^{i} b_j b_{i-j}, \quad i \geqslant 0 \tag{9.8}$$

Comparing (9.6) and (9.8), we see that $e_i = b_{i+1}$, $i \geqslant 0$. So,

$$G^2(z) = \sum_{i \geqslant 0} e_i z^i = \sum_{i \geqslant 0} b_{i+1} z^i$$

Hence,

$$zG^2(z) = \sum_{i \geqslant 0} b_{i+1} z^{i+1}$$

$$= \sum_{i \geqslant 0} b_i z^i - b_0$$

$$= G(z) - 1$$

or

$$zG^2(z) - G(z) + 1 = 0$$

Solving this quadratic for $G(z)$, we obtain the solutions:

$$\text{(a)} \, G(z) = \frac{1 + \sqrt{1-4z}}{2z} \tag{9.9}$$

$$\text{(b)} \, G(z) = \frac{1 - \sqrt{1-4z}}{2z}$$

From the binomial theorem (8.5), we get:

$$(1-4z)^{1/2} = \sum_{m \geqslant 0} \binom{1/2}{m} (-4z)^m$$

$$= \sum_{m \geqslant 0} \binom{1/2}{m} (-1)^m 2^{2m} z^m$$

Substituting this into (9.9), we obtain:

$$G(z) = \frac{1}{2z} \left(1 \pm \left(\sum_{m \geqslant 0} \binom{1/2}{m} (-1)^m 2^{2m} z^m \right) \right) \tag{9.10}$$

Since, from (9.6), $b_0 = 1$, $G(0)$ is defined and equals 1. From (9.10) we see that only (9.9(a)) satisfies this. So,

$$G(z) = \frac{1}{2z} \left(1 - \sum_{m \geqslant 0} \binom{1/2}{m} (-1)^m 2^{2m} z^m \right)$$

$$= \frac{1}{2z} \sum_{m \geqslant 1} \binom{1/2}{m} (-1)^{m-1} 2^{2m} z^m$$

$$= \sum_{m \geqslant 1} \binom{1/2}{m} (-1)^{m+1} 2^{2m-1} z^{m-1}$$

$$= \sum_{m \ge 0} \binom{1/2}{m+1} (-1)^m 2^{2m+1} z^m$$

Hence,

$$b_n = (-1)^n \binom{1/2}{n+1} 2^{2n+1}$$

$$= (-1)^n \frac{(1/2)(1/2-1)(1/2-2)...(1/2-n)}{(n+1)!} 2^{2n+1} \text{ (Eq (8.1))}$$

$$= \frac{1*3*5*...*(2n-1)}{(n+1)!} 2^n$$

$$= \frac{(2n)!}{2*4*6*...*2n*(n+1)!} 2^n$$

$$= \frac{(2n)!}{n!(n+1)!}$$

$$= \frac{1}{n+1} \binom{2n}{n}$$

Using Stirling's approximation (8.3), we obtain:

$$b_n = \frac{1}{n+1} \frac{(2n)!}{n!n!}$$

$$\approx \frac{1}{n+1} \frac{4\pi n (2n/e)^{2n}}{2\pi n (n/e)^{2n}}$$

$$= O\left[\frac{4^n}{n^{3/2}}\right]$$

9.5 MISCELLANEOUS TOPICS

9.5.1 Planar Graphs

There are several diagrammatic representations that correspond to a given undirected graph $G = (V,E)$ that is given in set notation. Figure 9.57 gives some of the diagrams that correspond to the graph $V = \{1,2,3,4\}$, and $E = \{(i,j) | i \in V, j \in V, i<j\}$.

We shall say that two edges *cross* each other iff they meet at a point other than a vertex to which both are incident. In the diagram of Figure 9.57(a) the edges (1,3) and (2,4) cross. No other pairs of edges cross in this diagram. In the diagrams of Figures 9.57 (b), (c), and (d), no edges of G cross. A graph G is *planar* iff it can be drawn in the plane in such a way that no two edges cross.

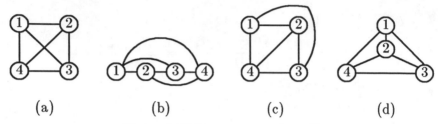

(a)   (b)   (c)   (d)

Figure 9.57 Some ways to draw K_4.

When a planar graph is drawn such that no two edges cross, the plane is partitioned into *regions*. The regions corresponding to the drawings of Figures 9.57(b), (c), and (d) are shown in Figure 9.58. In each case, there are exactly four regions. The regions of a planar drawing may be obtained by mentally cutting up the plane along the edges of the graph. Each separate piece obtained in this way is a region of the graph. So, if we cut up the plane along the edges of Figure 9.59, we shall obtain the regions shown in this figure.

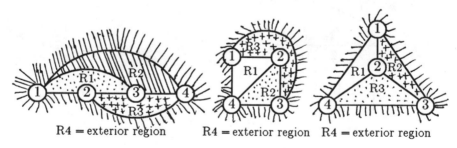

R4 = exterior region   R4 = exterior region   R4 = exterior region

Figure 9.58 Regions of a planar graph.

Theorem 9.21: Let $G = (V,E)$ be any connected planar undirected graph. Consider any planar drawing of G. Let $|V| = n$, $n \geqslant 1$, $|E| = e$, and let r be the number of regions. The following are true:

   (i) $n - e + r = 2$

   (ii) $e \leqslant 3n - 6$, if $e \geqslant 2$

Proof: (i) The proof of this part is by induction on e. When $e = 0$, there is only one connected graph. This graph has $n = 1$ and every planar drawing

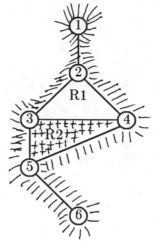

R 3 = exterior
Figure 9.59

of this graph has $r = 1$. So, $n - e + r = 2$. For the induction hypothesis, assume that $n - e + r = 2$ for every n and r and for some arbitrary value m of e, $m \geqslant 1$. We shall show that $n - e + r = 2$ for every value of n and r when $e = m + 1$.

Let $G = (V, E)$ be any connected planar undirected graph with $e = m + 1$. Consider any planar drawing of G. Let r be the number of regions in this drawing. If G contains a vertex u with degree 1, then let (u, v) be the sole edge incident on u. Clearly, the removal of (u, v) and u from the drawing of G leaves behind a connected planar graph with $n - 1$ vertices, m edges, and r regions. From the induction hypothesis, it follows that $n - 1 - m + r = 2$ or $n - (m + 1) + r = n - e + r = 2$.

If G contains no vertex of degree 1, then it must contain a cycle (as $\sum_{1}^{n} d_i \geqslant 2n$, G cannot be a tree). Hence, $r > 1$. Consider a region R of G other than the exterior region. The edges on this region form a cycle. From Theorem 9.2 we know that if any of the edges on this region are deleted, the remaining graph is connected. Also, by deleting one of these edges, R is no longer a region in the remaining graph. Remove one of the edges on R. The resulting graph has $r - 1$ regions, n vertices and m edges. From the induction hypothesis, it follows that, $n - m + (r - 1) = n - (m + 1) + r = n - e + r = 2$.

(ii) Assume that $e \geqslant 2$. If $r = 1$, then from (i) we obtain $e = n - 1$. Since $e \geqslant 2$, $n \geqslant 3$. So, $e \leqslant 3n - 6$. If $r > 1$, then every region is bounded by at least 3 edges, and every edge is on the boundary of at most 2 regions. So,

$r \leqslant 2e/3$. Substituting into (i), we obtain $e \leqslant 3n-6$. □

Corollorary 9.3: Every planar drawing of a connected planar graph has the same number of regions.

Proof: From Theorem 9.21 we know that $r = 2-n+e$. Every planar drawing of the same graph has the same number of vertices and edges. Hence every planar drawing of a planar graph has the same r. □

Corollary 9.4: Let $G = (V,E), |V| \geqslant 1$, be a planar graph with k connected components. Let $n = |V|$, $e = |E|$, and let r be the number of regions in any planar drawing of G. $n-e+r = k+1$. □

The complete graph K_5 has $n = 5$ and $e = 10$. Since, $e > 3n-6$, K_5 is not planar. The complete bipartite graph $B_{3,3}$ (Figure 9.60) has $n = 6$ and $e = 9$. $e < 3n-6$ but this graph is not planar. There is no way to draw $B_{3,3}$ in a plane and not have two edges cross. This does not violate Theorem 9.21 as this theorem only states that $e \leqslant 3n-6$ is a necessary condition for planarity (but not a sufficient condition). The non planar graphs K_5 and $B_{3,3}$ play an important role in characterizing all non planar graphs.

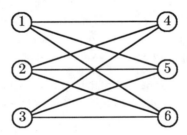

Figure 9.60 The complete bipartite graph $B_{3,3}$.

A graph G is an *elementary contraction* of a graph H iff G can be obtained from H by replacing two distinct vertices i and j in H by a single vertex k. All edges incident on either i or j (but not both) in H are now incident on k (i.e., (i,p) or (j,p) are in H iff (k,p) is in G). The graph of Figure 9.61 (b) is an elementary contraction of the graph of Figure 9.61 (a). In this case, $i = 1, j = 5$, and $k = 7$.

A graph H is *contractible* to the graph G iff G can be obtained from H by a series of elementary contractions.

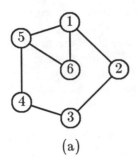 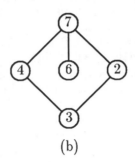

$$\text{(a)} \qquad\qquad\qquad\qquad \text{(b)}$$

Figure 9.61

Theorem 9.22: An unlabeled undirected graph G is planar iff it does not contain a subgraph that is contractible to $B_{3,3}$ or K_5 (both $B_{3,3}$ and K_5 are assumed unlabeled). ☐

Let G and H be two unlabeled undirected graphs. Graphs G and H are *homeomorphic* iff they can both be obtained from the same graph F by adding vertices onto some of the edges of F. Figures 9.62(a) and (b) show two graphs that are homeomorphic. Both can be obtained from the graph of Figure 9.62(c) by adding vertices onto the edges of this graph. In fact, all three graphs of this figure are homeomorphic. One may verify that all cycles are homeomorphic.

Theorem 9.23: An unlabeled undirected graph G is planar iff no subgraph of G is homeomorphic to K_5 or $B_{3,3}$. ☐

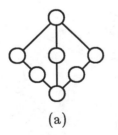 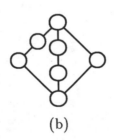 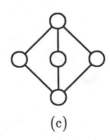

$$\text{(a)} \qquad\qquad\qquad \text{(b)} \qquad\qquad\qquad \text{(c)}$$

Figure 9.62 Homeomorphic graphs.

9.5.2 Matchings

Let $G = (V, E)$ be an undirected graph. A *matching* in G is a subset A, $A \subseteq E$ such that no two edges in A are incident on the same vertex. The cardinality of A is the *size* of the matching. If $(i,j) \in A$ then i is *matched* with j. Observe that no vertex can be matched to two or more vertices.

$\{(1,6), (5,4), (2,3)\}$ is a matching of size 3 in the graph of Figure 9.61(a). $\{(1,5), (4,3)\}$ is a matching of size 2 in the same graph. $\{(1,4), (3,5)\}$ is a matching in the bipartite graph of Figure 9.5(b). $\{(1,4), (4,6), (5,2)\}$ cannot be a matching in any graph as this edge set contains two edges incident on vertex 4.

A is a *maximum matching* in G iff G has no matching B such that $|B| > |A|$. $\{(1,5), (4,3)\}$ is not a maximum matching in the graph of Figure 9.61(a) but $\{(1,6), (5,4), (2,3)\}$ is.

If G is an undirected bipartite graph, then V can be partitioned into the sets R and S such that no edge in E connects two vertices in R or two in S. G can clearly have no matching of size larger than $\min\{|R|, |S|\}$. A matching A in the bipartite graph G is a *complete matching* iff $|A| = \min\{|R|, |S|\}$. All bipartite graphs do not have a complete matching. For example, $\{(4,7), (2,6), (1,5)\}$ is a maximum matching in the bipartite graph of Figure 9.63(a). This matching is not a complete matching. $\{(2,6), (3,7), (4,8), (5,9)\}$ is a complete matching for the graph of Figure 9.63(b).

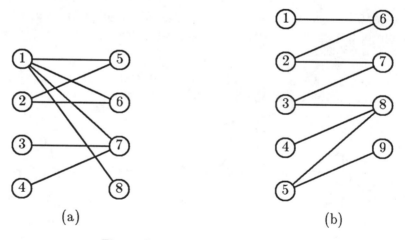

Figure 9.63 Two bipartite graphs.

The following theorem gives a necessary and sufficient condition for an undirected bipartite graph to have a complete matching.

Theorem 9.24: [Hall] Let $G = (V,E)$ be an undirected bipartite graph. Let R and S be a partitioning of V such that no edge in E connects two vertices in

R or two in S. Assume that $|R| \leqslant |S|$. For any $W \subseteq R$ let $ADJ(W) = \{i \mid i$ is adjacent to a vertex in $W\}$. G has a complete matching iff $|W| \leqslant |ADJ(W)|$ for every subset W of R. $\square$

9.5.3 Cliques And Independent Sets

A subgraph $G_1 = (V_1, E_1)$ is a *maximal complete subgraph* of the undirected graph $G = (V, E)$ iff G_1 is a complete graph and G contains no complete subgraph $G_2 = (V_2, E_2)$ such that $V_1 \subset V_2$. G_1 is a *clique* of G iff G_1 is a maximal complete subgraph of G. The *size* of the clique $G_1 = (V_1, E_1)$ is $|V_1|$. $G_1 = (V_1, E_1)$ is a *max-clique* of G iff G_1 is a clique of G and G contains no clique of size larger than $|V_1|$.

Every edge in a bipartite graph is a clique of size 2. In fact, every edge in a bipartite graph is a max-clique of that graph. The complete graph K_n contains only one clique. This subgraph is itself K_n. The graph of Figure 9.11(a) contains three cliques. Each of these is of size 3. These cliques are shown in Figure 9.64. Figure 9.65 shows another graph together with its cliques.

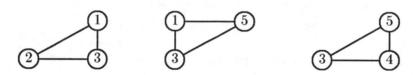

Figure 9.64 The cliques of the graph of Figure 9.11(a).

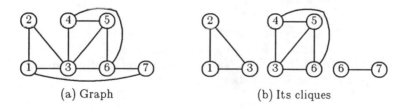

(a) Graph   (b) Its cliques

Figure 9.65 A graph and its cliques.

A set R of vertices in the undirected graph $G = (V, E)$ is *independent* iff no two vertices in R are adjacent in G. The *size* of the independent set R is $|R|$. R is a *maximum independent set* iff R is an independent set and G contains no independent set S such that $|S| > |R|$. $\{1,4\}$, $\{2,4\}$, $\{2,4,7\}$, and $\{2,5,7\}$ are some of the independent sets of the graph of Figure 9.65(a).

These sets are of size 2, 2, 3, and 3 respectively. {2,4,7} and {2,5,7} are maximum independent sets.

Define the *complement* of the undirected graph $G = (V,E)$ to be the undirected graph $\overline{G} = (V,\overline{E})$ such that (i,j), $i \neq j$, is an edge in $\overline{E}$ iff (i,j) is not an edge in E. So, $E \cap \overline{E} = \varnothing$ and $|E \cup \overline{E}| = n(n-1)/2$ where $n = |V|$.

The complement of the graph of Figure 9.65 is drawn in Figure 9.66. From the definitions, one readily sees that R is an independent set in G iff the subgraph $G_1 = (R,E_1)$ where $E_1 = \{(i,j) \mid i \in R, j \in R,$ and (i,j) is an edge in $\overline{E}\}$ is a complete graph. Hence, R is a maximum independent set of G iff G_1 is a max-clique in $\overline{G}$.

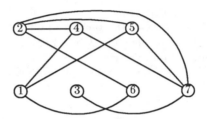

Figure 9.66 Complement of Figure 9.65(a)

9.5.4 Colorings And Chromatic Number

A *coloring* of a graph G is an assignment of colors to the vertices of a graph such that no two adjacent vertices are assigned the same color. The *size* of a coloring is the number of different colors used. The *chromatic number* of a graph is the smallest number of colors needed to color the graph.

Clearly, every graph with $|V| > 0$ and $|E| = 0$ can be colored with exactly 1 color. Figure 9.67 shows two graphs together with a coloring for each. In the graph of Figure 9.67(a) only two colors are used while in that of Figure 9.67(b) four colors are used.

A graph is *k-colorable* iff it can be colored with k colors. The chromatic number of a graph is the minimum k for which the graph is k-colorable. One may verify that every cycle of even length is two colorable and that every cycle of odd length has a chromatic number of 3. It is also known that every planar graph is four colorable (the truth of this statement had been conjectured for many years; eventually a computer generated proof examining many cases was found). One may verify that the chromatic number of the complete graph K_i is i.

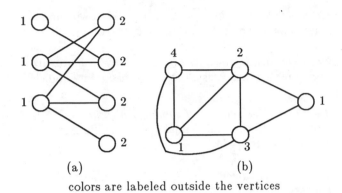

colors are labeled outside the vertices

Figure 9.67 Colored graphs.

The vertices of a k-colored graph, G, may be partitioned into k disjoint sets, $V_1, V_2, ..., V_k$. Each set contains all vertices of the same color. From the definition of a k-coloring, it follows that G contains no edges joining two vertices in the same partition V_i. Since every bipartite graph G is partitionable into two vertex sets V_1 and V_2 such that G contains no edges joining two vertices in V_1 or two in V_2, every bipartite graph is two colorable. Since no cycle of odd length is 2 colorable, it follows that no bipartite graph contains a cycle of odd length. This leads to the following characterization of a bipartite graph.

Theorem 9.25: An undirected graph $G = (V,E)$ is bipartite iff every cycle in G is of even length.

Proof: We have already seen that if G is bipartite then G contains no cycles of odd length. So, we need only show that if G contains no cycles of odd length then G is bipartite. This part of the proof is left as an exercise. □

REFERENCES AND SELECTED READINGS

The text

> *Graph theory*, by F. Harary, Addison-Wesley Publishing Co., Reading, Massachusetts, 1972.

is a good general reference for this chapter. Several interesting graph algorithms can be found in the texts:

Fundamentals of data structures, by E. Horowitz and S. Sahni, Computer Science Press, Inc., Potomac, Maryland, 1976.

Fundamentals of computer algorithms, by E. Horowitz and S. Sahni, Computer Science Press, Inc., Potomac, Maryland, 1978.

The design and analysis of computer algorithms, by A. Aho, J. Hopcroft, and J. Ullman, Addison-Wesley Publishing Co., Reading, Massachusetts, 1974.

Combinatorial algorithms: Theory and Practice, by E. Reingold, J. Nievergelt, and N. Deo, Prentice-Hall, Englewood Cliffs, New Jersey, 1977.

and

Graph algorithms, by S. Even, Computer Science Press, Inc., Potomac, Maryland, 1979.

The subject of trees is covered in more detail in the texts:

The art of computer programming: Fundamental algorithms, Vol.1, 2nd Edition, by D. Knuth, Addison-Wesley Publishing Co., Reading, Massachusetts, 1973.

The art of computer programming: Sorting and searching, Vol.3, by D. Knuth, Addison-Wesley Publishing Co., Reading, Massachusetts, 1973.

and

Fundamentals of data structures, by E. Horowitz and S. Sahni, Computer Science Press, Inc., Potomac, Maryland, 1976.

A proof of Theorem 9.14 can be found in the text:

Elements of discrete mathematics, by C. Liu, McGraw-Hill Book Co., New York, 1977.

Theorem 9.15 is proved in:

Introductory combinatorics, by R. Brualdi, North-Holland Publishing Co., 1977.

Theorem 9.22 is due to F. Harary and W. Tutte. A proof can be found in the paper:

A dual form of Kuratowski's theorems, Canadian Math. Bull., Vol.8, pp.17-20, 1965.

Theorem 9.23 is due to K. Kuratowski. A proof appears in the text:

Graph theory, by F. Harary, Addison-Wesley Publishing Co., Reading Massachusetts, 1972.

A proof of Hall's theorem (Theorem 9.26) can be found in the text:

Introduction to combinatorial mathematics, by C. Liu, McGraw-Hill Book Co., New York, 1968.

EXERCISES

1. Obtain the set representation for the graphs given in the following figures:
   (a) Figure 9.6(b)
   (b) Figure 9.7(a)
   (c) Figures 9.9(a) and (b)
   (d) Figure 9.10(a)
   (e) Figure 9.16(a)

2. For the graph of Figure 9.10(a), determine the following:
   (a) The set of edges incident on vertex 1.
   (b) The set of edges incident of vertex 3.
   (c) The vertices adjacent to vertex 5.
   (d) All simple paths together with their lengths.
   (e) The cycles in the graph.
   (f) The degree of each vertex.

3. (a) Do problem 2 for the graph of Figure 9.11(a)
   (b) Do problem 2 for the graph of Figure 9.11(d).

4. (a) Let G be any undirected graph. Show that the number of vertices with odd degree is even.
   (b) Let G be as above. Let u and v be two arbitrary distinct vertices of G. Show that G contains a u to v path iff it contains a simple u to v path.

5. Prove Theorem 9.4.

6. Determine which of the graphs of Figure 9.68 are isomorphic.

7. For each of the graphs of Figure 9.69, determine the following:
   (a) The in-degree of each vertex.
   (b) The out-degree of each vertex.
   (c) The set of vertices adjacent from vertex 2.
   (d) The set of vertices adjacent to vertex 1.
   (e) The set of edges incident from vertex 3.
   (f) The set of edges incident to vertex 4.
   (g) All directed cycles and their lengths.
   (h) All simple semipaths from 1 to 4.
   (i) All semicycles.
   (j) The strongly connected components.
   (k) The unilaterally connected components.
   (l) The weakly connected components.

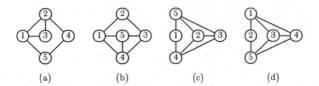

Figure 9.68 Graphs for Exercise 6.

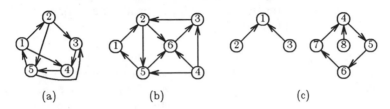

Figure 9.69 Digraphs for Exercise 7.

8. Determine which of the digraphs of Figure 9.70 are isomorphic.

9. (a) By using only the in-degree and out-degree properties of a rooted tree, show that every n vertex rooted tree contains exactly $n-1$ edges.

   (b) Use (a) and the fact that every rooted tree is weakly connected to show that no rooted tree contains a directed cycle.

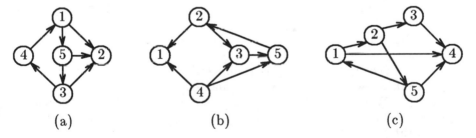

Figure 9.70 Digraphs for Exercise 8.

10. Obtain the following representations for the graphs of Exercise 1:
    (a) Adjacency matrix.
    (b) Compact adjacency lists.
    (c) Linked adjacency lists.
    (d) Incident matrix.

11. (a) Let G be an n vertex e edge undirected graph. What is the least

value of e for which the adjacency matrix representation of G uses less space than used by the compact adjacency lists representation?

(b) Do (a) for the case of a directed graph G.

12. (a) Write an algorithm to delete an edge (i,j) from the compact adjacency list representation of the undirected graph G. What is the complexity of your algorithm?

(b) Do part (a) for the case of edge insertion (i.e., a new edge (i,j) is to be added to the graph).

13. For the graph of Figure 9.42(c), do the following:
(a) Obtain a breadth first spanning tree starting at vertex 1.
(b) Obtain a breadth first spanning tree starting at vertex 3.
(c) Obtain the cycle basis relative to the spanning tree of (a).
(d) Obtain the cycle basis relative to the spanning tree of (b).
(e) What is the cycle rank of each of the cycle bases obtained?

14. Let $G = (V,E)$ be a connected undirected graph. Let T be a breadth first spanning tree with root i. Let $d^T(i,j)$ be the length of the unique i to j path in T. Let $d^G(i,j)$ be the length of the shortest i to j path in G. Show that $d^T(i,j) = d^G(i,j), 1 \leqslant j \leqslant |V|$.

15. (a) Do Exercise 13 using depth first spanning trees instead of breadth first spanning trees.

(b) Label each of the spanning tree vertices of (a) with its depth first number.

16. Prove Theorem 9.7.

17. Let $s(n)$ be the number of spanning trees in K_n.
(a) Show that $s(1) = s(2) = 1$ and that $s(n) \geqslant (n-1)s(n-1), n > 2$. Solve this recurrence using the substitution method and conclude that $s(n) \geqslant n$!

(b) Show that:

$$s(n) = \sum_{1 \leqslant i \leqslant \lceil n/2 \rceil} n(n-i) \binom{n}{i} (s(i) + s(n-i)) + (n-1)s(n-1), \ n > 2.$$

18. (a) Obtain an algorithm that determines if an undirected graph $G = (V,E)$ contains a cycle.
(b) Prove the correctness of your algorithm.
(c) Obtain the time and space complexity of your algorithm.

19. (a) Use Kruskal's method to obtain a minimum cost spanning tree of the weighted graph of Figure 9.71.
    (b) What is the cost of the spanning tree of (a)?

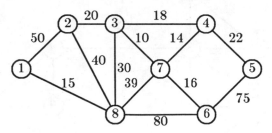

Figure 9.71 Graph for Exercise 19.

20. Let G be a weighted connected undirected graph with distinct edge costs. Use the correctness of Kruskal's method to show that G has a unique minimum cost spanning tree.

21. Let T be a tree. Show that:

    (a) The addition of edge (i,j) to T creates a unique cycle.
    (b) The removal of any edge on this cycle leaves behind a tree.

22. Prove Theorem 9.9.

23. Determine the articulation points, bridges, and biconnected components of the following graphs. Do this by first determing a depth first spanning tree for each graph and then obtaining the L values.
    (a) The graph obtained by removing edge costs in Figure 9.71.
    (b) The graph of Figure 9.72.

24. (a) Extend procedure ART (Algorithm 9.4) so that it prints out all the edges in each biconnected component.
    (b) Prove the correctness of the resulting algorithm.
    (c) Determine the time and space complexity of the new algorithm.

25. What is the fewest number of eges in any n vertex triconnected graph?

26. Determine the edge connectivity of the graph of Exercise 21(a).

27. Prove that the graph of Figure 9.72 has no Euler path.

28. Prove Theorem 9.13.

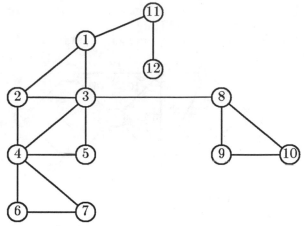

Figure 9.72 Graph for Exercise 23.

29. Find two additional Hamiltonian cycles in the graph of Figure 9.44.

30. Find an undirected graph that does not satisfy the conditions of Theorem 9.14 but contains a Hamiltonian path.

31. (a) Do Exercise 30 for Theorem 9.15.
    (b) Do Exercise 30 for Theorem 9.16.

32. (a) Find the shortest path matrix for the digraph of Figure 9.46.
    (b) Find the transitive closure matrix and the reflexive transitive closure matrix for the graphs of the following Figures:
        (i)   Figures 9.70(a) and (b).
        (ii) Figure 9.69(a).
        (iii) Figure 9.10(a).

33. Determine the internal and external path lengths of the extended binary trees of Figure 9.73.

34. Let T be a binary tree. Let u be any node in T. Let $h_{L(u)}$ and $h_{R(u)}$ be the heights of the left and right subtrees of u. T is an *AVL tree* iff $|h_{L(u)} - h_{R(u)}| \leq 1$ for every node u and T.
    (a) Which of the trees of Figure 9.73 are *AVL*?
    (b) Let N_k be the minimum number of nodes in any *AVL* tree of h.ight h. Show that $N_h = N_{h-1} + N_{h-2} + 1$, $h \geq 2$, $N_1 = 1$ and $N_0 = 0$.
    (c) Show that $N_h = F_{h+2} - 1$, $h \geq 0$ where F_h is the h'th Fibonacci number (hint: use induction on h).

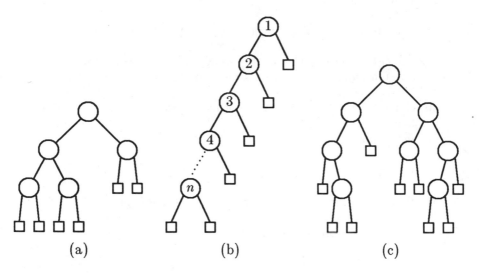

Figure 9.73 Trees for Exercise 33.

(d) Use the solution to the recurrence for F_h to obtain a closed form solution for N_h.

(e) If T contains n nodes, how large can h be?

35. A *triangulated graph* is a planar graph in which every interior region is bounded by exactly three edges. The exterior region has exactly 3 edges on its boundary. Obtain triangulated graphs for the cases $n = 6$ and $n = 10$.

36. Identify the regions in the planar graph of Figure 9.74.

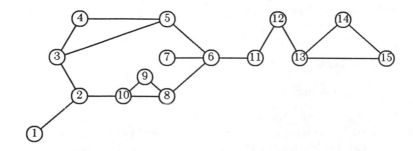

Figure 9.74 Graph for Exercise 36.

37. Prove Corollorary 9.4.

38. Show that every tree is a bipartite graph.

39. Let G be any n vertex graph. Show that if G contains no cliques of size 3 or more then G contains no more than $n^2/4$ edges.

40. Obtain a minimum coloring for the graphs given in the following figures:
    (a) Figure 9.74.
    (b) Figure 9.62(a).
    (c) Figure 9.72.

41. What is the chromatic number of K_n? Prove your answer.

42. Complete the proof of Theorem 9.25.

43. (a) Let G be an n vertex e edge undirected graph. Write an algorithm with time complexity $O(n+e)$ to determine if G is bipartitie. If G is bipartite, then your algorithm should output a partioning of the vertex set such that no edge joins two vertices in the same partition.
    (b) Prove the correctness of your algorithm.
    (c) Show that your algorithm does in fact have time complexity $O(n+e)$.
    (d) What is the space complexity of your algorithm?

44. Let $G = (V,E)$ be an undirected connected graph. Let $d(u,v)$ be the length of a shortest u to v path in G. The *eccentricity* $e(u)$, of vertex u is defined to be:

    $$e(u) = \max\{d(u,v)|v \in V\}$$

    The *radius* $r(g)$ of g is defined as:

    $$r(G) = \min\{e(u)|u \in V\}$$

    and the *diameter*, $d(G)$, of G is given by:

    $$d(G) = \max\{e(u)|u \in V\}.$$

    (a) Consider the graph of Figure 9.42(a).
        (i) Obtain the eccentricity of each vertex in this graph.
        (ii) Determine the radius and diameter of the graph.
    (b) Is $r(G) \leq d(G) \leq 2r(G)$ for every connected graph G?

CHAPTER 10

MODERN ALGEBRA

10.1 ALGEBRAS

An *algebra* is a tuple $< K, op_1, op_2, ..., op_n >$ where K is a set called the *carrier* of the algebra, and op_1, op_2, ..., op_n are operators. Each operator is a total function with domain K^i for some i and range K (i.e., $op_j: K^i \rightarrow K$). Thus, K is *closed* under each of the operations op_i. If op_j has domain K^i then op_j is an operator of *degree i*. If $i=1$, it is a unary operator; if $i=2$, it is a binary operator; it is a ternary operator if $i=3$; etc. $|K|$ is the *order* of the algebra. If K is a finite set, then the algebra is a *finite* algebra.

Example 10.1: In Chapter 1, we studied a special type of algebra called a Boolean algebra. A Boolean algebra is given by a tuple $< K, ., + >$ where . and $+$ are binary operators. Both have domain $K \times K$ and range K. The Boolean algebra $< \{\text{false, true}\}, \wedge, \vee >$ is of order 2 and is a finite algebra. $< R, *, + >$ is the algebra of reals with binary operators $*$ (multiply) and $+$ (add). This algebra is not a finite algebra. The algebra $< Z, *, + >$ is also not a finite algebra. $< R, *, +, / >$ is not an algebra as $2/0 \notin R$ (in fact, division by zero is not defined). $< Z, \|, *, + >$ is an algebra over the integers with the unary operator $\|$ ($|x|$ = absolute value of x), and the binary operators $*$ and $+$. Let O be the set of odd numbers. $< O, + >$ is not an algebra as $x+y$ is not in O for every x,y in O. In fact, $x+y \notin O$ for any x, y in O. □

Let $\odot: A \times B \rightarrow C$ be a binary operator. $0_L \in A$ is a *left zero* for $\odot$ iff $0_L \odot x = 0_L$ for every x in B. $0_R \in B$ is a *right zero* for $\odot$ iff $x \odot 0_R = 0_R$ for every x in A. 0 is a *zero* (or a two sided zero) for $\odot$ iff it is both a left zero and a right zero for $\odot$. $1_L \in A$ is a *left identity* for $\odot$ iff $1_L \odot x = x$ for every x in B. $1_R \in B$ is a *right identity* for $\odot$ iff $x \odot 1_R = x$ for every x in A. 1 is an *identity* (or a two sided identity) for $\odot$ iff it is both a left and right identity for it.

Example 10.2: Let A be a set of 2×2 matrices and let $B=C$ be the set of 2×1 matrices. Assume that the elements of the matrices in A, B, and C are from the set of real numbers. Let $*: A \times B \rightarrow C$ be the standard matrix multiplication operator restricted to the sets A and B. One may easily verify that:

$$X * \begin{bmatrix} 0 \\ 0 \end{bmatrix} = \begin{bmatrix} 0 \\ 0 \end{bmatrix} \text{ for every } X, X \in A$$

Hence, $\begin{bmatrix} 0 \\ 0 \end{bmatrix}$ is a right zero for $*$. $*$ has no left zero as $X * Y \notin A$ for any X in A.

Since, $\begin{bmatrix} 1 & 0 \\ 0 & 1 \end{bmatrix} * Y = Y$ for every Y in B, $\begin{bmatrix} 1 & 0 \\ 0 & 1 \end{bmatrix}$ is a left identity for $*$. There is no right identity for $*$. □

Example 10.3: Consider the Boolean algebra $<\{$false, true$\}, \wedge, \vee>$. false is a two sided zero for $\wedge$. $\vee$ has no left or right zero. true is an identity (two sided) for $\wedge$ and false is an identity (two sided) for $\vee$. □

Example 10.4: Consider the algebra $<R, *>$ where R is the set of real numbers and $*$ denotes multiplication. 0 and 1 are, respectively, the zero and identity for $*$. In the algebra $<R, +>$, 0 is the identity for $+$. There is no zero for $+$. □

Theorem 10.1: Let $\odot$: $A \times A \to A$ be a binary operator.
  (1) If a left zero 0_L and a right zero 0_R for $\odot$ exist, then $0_L = 0_R$ and $0 = 0_L = 0_R$ is the unique two sided zero for $\odot$.
  (2) If a left identity 1_L and a right identity 1_R for $\odot$ exist, then $1_L = 1_R$ and $1 = 1_L = 1_R$ is the unique two sided identity for $\odot$.
Proof: We shall prove (1) only. Assume that the left and right zero exist. As $0_L \in A$ and $0_R \in A$, $0_L = 0_L \odot 0_R$ and $0_R = 0_L \odot 0_R$. Hence, $0_L = 0_R$. Let $0 = 0_L = 0_R$. Suppose 0 is not a unique zero for $\odot$. Let 0' be another zero for $\odot$. We see that $0 = 0 \odot 0'$ and $0' = 0 \odot 0'$. Hence, $0 = 0'$. This contradicts the assumption that $0 \neq 0'$. Hence, 0 is a unique zero for $\odot$. (2) may be established in a similar way. □

From the preceding theorem, it follows that $\odot$: $A \times A \to A$ has a zero (identity) iff it has both a left and right zero (identity).

10.2 BINARY ALGEBRAS

A *binary algebra* is an algebra $<K, \odot>$ in which $\odot$ is a binary operator ($\odot$: $K \times K \to K$). $<R, +>$, $<Z, ->$, and $<N, +>$ are examples of binary algebras. $<N, ->$ is not a binary algebra. In fact, since $3-4 = -1 \notin N$, $<N, ->$ is not even an algebra.

A *semigroup* is a binary algebra $<K, \odot>$ for which the binary operator $\odot$ is associative. $<Z, ->$ is not a semigroup as - is not associative.

$<K, \odot>$ is a *monoid* iff $\odot$ is a binary associative operator with an identity. Thus every semigroup that has an identity element for its binary operator is a monoid. $<R, +>$, $<Z, *>$, $<\{$true, false$\}, \wedge>$, and $<\{$true, false$\}, \vee>$ are monoids.

Let $\odot$: $A \times A \to A$ be a binary operator with identity 1. $x_L^{-1} \in A$ is a *left inverse* of $x \in A$ iff $x_L^{-1} \odot x = 1$. $x_R^{-1} \in A$ is a *right inverse* of x iff $x \odot x_R^{-1} = 1$. If $x_L^{-1} = x_R^{-1}$ then x has an inverse (two sided) $x^{-1} = x_L^{-1} = x_R^{-1}$.

A monoid $<K,\odot>$ in which every element of K has an inverse is called a *group*. $<Z,+>$ is a group. To see this, note that this binary algebra is a monoid. The identity element is 0 and the inverse of $x \in Z$ is $-x$. The monoid $<\{\text{true,false}\}, \wedge>$ has identity true. However, the element false has no inverse. So, this monoid is not a group. $<E=\{0, 2, ...\}, +>$ is a monoid as $+$ is a binary associative operator and 0 is the additive identity. $<E,+>$ is not a group as 0 is the only element that has an inverse.

Theorem 10.2: Let $<K,\odot>$ be a group. Every element x, $x \in K$ has a unique inverse.

Proof: From the definition of a group, it follows that every x, $x \in K$ has an inverse. Suppose that for some x, $x \in K$, there are two inverses b and c. We shall show that $b = c$. From the definition of an inverse, it follows that $b \odot x = 1$ where 1 is the group identity. So, $(b \odot x) \odot c = 1 \odot c = c$. Since $\odot$ is associative, $c = (b \odot x) \odot c = b \odot (x \odot c) = b \odot 1 = b$. Hence, the inverse of x is unique. $\square$

From the above proof, it is clear that if $<K, \odot>$ is a monoid and if some x in K has an inverse then the inverse of x is unique.

Let $<K,\odot>$ be a binary algebra. The element c, $c \in K$, is *left cancellative* iff $\forall x,y \in K$ $[c \odot x = c \odot y \Longrightarrow x = y]$. c is *right cancellative* iff $\forall x,y \in K$ $[x \odot c = y \odot c \Longrightarrow x = y]$.

Theorem 10.3: Let $<K,\odot>$ be a group. Every c, $c \in K$, is both left and right cancellative.

Proof: Let c, x, and y be arbitrary elements of K. Assume that $c \odot x = c \odot y$. Since $<K,\odot>$ is a group, c has an inverse c^{-1}. So,

$c^{-1} \odot (c \odot x) = c^{-1} \odot (c \odot y)$

or

$(c^{-1} \odot c) \odot x = (c^{-1} \odot c) \odot y$ (associativity of $\odot$)

or

$1 \odot x = 1 \odot y$ (1 is the identity)

or

$x = y$

Hence, c is left cancellative. c may be shown right cancellative in a similar way. $\square$

$<K',\odot>$ is a *subgroup* of the group $<K,\odot>$ iff $K' \subseteq K$ and $<K',\odot>$ is a group. $<Z,+>$ is a subgroup of $<R,+>$.

An *Abelian group* (or commutative group) is a group $<K,\odot>$ in which $\odot$ is commutative (in addition to being associative). The properties of an Abelian group are:

(1) $\odot$ is an associative and commutative binary operator.
(2) Every element in K has an inverse.
(3) $\odot$ has an identity element.

$<Z,+>$ and $<\{\pm 0, \pm 2, \pm 4, \pm 6, ...\}, +>$ are examples of Abelian groups.

Let $<K,\odot>$ be a group with identity element 1. For any x, $x \in K$, and any k, $k \in Z$, we define the kth *power*, x^k, of x as below:

$$x^k = \begin{cases} 1 & k=0 \\ x \odot x^{k-1} & k>0 \\ (x^{-k})^{-1} & k<0 \end{cases}$$

So, $x^1 = x$; $x^2 = x \odot x$; $x^{-2} = (x^2)^{-1}$, etc.

Theorem 10.4: Let $<K,\odot>$ be a group. $x^a x^b = x^{a+b}$ and $(x^a)^b = x^{ab}$ for every $x \in K$, $a \in Z$, and $b \in Z$.

Proof: The theorem is easily proved using the associativity of $\odot$. The proof is left as an exercise. $\square$

A binary algebra $<K,\odot>$ is said to be *generated* by the set $G = \{a_1, a_2, \cdots \} \subseteq K$ iff every x in K can be represented as

$$x = a_{i_1} \odot a_{i_2} \odot a_{i_3}... \odot a_{i_n}$$

where $a_{i_j} \in G$ for every j and n is a natural number. The group $<Z,+>$ is generated by the set $\{+1,-1\}$. A *cyclic group* is a group that can be generated by a set of size 1. $<Z,+>$ is not a cyclic group as it cannot be generated by a set of size 1.

Example 10.5: The algebra, $<A,*_{10}>$, generated by 2 under multiplication modulo 10 ($x*_{10} y = (x*y)$ mod 10 for every x and y) is obtained by using the fact that $*$ is a binary operator with domain $A \times A$ and range A. So, $2*2$ mod 10 = 4 is in A. Hence, $4*2$ mod 10 = 8 is also in A. $8*2$ mod 10 = 6 must also be in A. One sees that $<\{2, 4, 6, 8\}, *_{10}>$ is the generated algebra. This algebra is in fact a monoid. To see this, observe that multiplication modulo 10 is associative and that 6 is the identity ($2*6 \bmod 10 = 2$; $4*6 \bmod 10 = 4$; etc.). This monoid is not a group as 4 and 8 have no inverse. The order of the monoid is $|A| = 4$. □

Example 10.6: Let $<A,*>$ be the algebra generated by x and y. Assume that $*$ is commutative and that $x^2 = y^4 = (xy)^2 = 1$. 1 is the multiplicative identity and $1*z = z*1 = z$ for all z. Further, assume that x, y, xy, y^2, and $y^3 \neq 1$. Since $u*v$ is in A for every u and v in A and since x and y are known to be in A, we get that $x*x = x^2 = 1$; xy; y^2; xy^2; y^3; and xy^3 are all also in A. Furthermore, there are no other elements in A. One may verify that every u in A has an inverse. For example, $x^{-1} = x$; $y^{-1} = y^3$; $(y^2)^{-1} = y^2$; $(xy^3)^{-1} = xy$; etc. Hence, $<A, *>$ is a group of order 8. Since $*$ is commutative, it is an Abelian group. □

Let $<K, \odot>$ be a binary algebra. Let A be a subset of K and let a be an element of K. The *left coset*, $a \odot A$, of a with respect to A is the set $\{a \odot x \mid x \in A\}$. The *right coset*, $A \odot a$, of a with respect to A is the set $\{x \odot a \mid x \in A\}$.

Example 10.7: Consider the monoid of Example 10.5. Let $A = \{2, 4, 8\}$. The coset $2*_{10}A$ is the set $\{2*_{10}2, 2*_{10}4, 2*_{10}8\} = \{4, 8, 6\}$. For the group of Example 10.6, the coset $x*\{y, 1, y^3\}$ is the set $\{xy, x, x y^3\}$. The coset $\{x, y, 1, xy^2\}*y^3$ is the set $\{xy^3, y^4, y^3, xy^5\} = \{xy^3, 1, y^3, xy\}$. □

Theorem 10.5: Let $<K,\odot>$ be a group. Let $<A,\odot>$ be a subgroup of $<K, \odot>$ and let $a \odot A$ and $b \odot A$ be two left cosets with respect to A.

(1) Either $a \odot A = b \odot A$ or $(a \odot A) \cap (b \odot A) = \emptyset$
(2) $|a \odot A| = |A|$

Proof: We shall prove (1) only. The proof of (2) is left as an exercise. Suppose that $(a \odot A) \cap (b \odot A) \neq \emptyset$. So, there exist d and e in A such that $a \odot d = b \odot e$. Since $<A, \odot>$ is a subgroup, the inverse d^{-1} of d is in A. Hence, for any f in A we have $a \odot f = b \odot e \odot d^{-1} \odot f$ which is in $b \odot A$. I.e., $a \odot A \subseteq b \odot A$. A similar proof shows that $b \odot A \subseteq a \odot A$. Therefore, $a \odot A = b \odot A$. This completes the proof of (1). □

Theorem 10.6: [Lagrange] The order of every subgroup of a finite group divides the order of the group.

Proof: Left as an exercise. □

From Lagrange's theorem, it follows that if the order of a group is a prime number then the group is cyclic.

10.3 RINGS AND FIELDS

A *ring* is an algebra $<K, \oplus, \odot>$ in which both $\oplus$ and $\odot$ are binary operators that satisfy the following properties:

(1) $\oplus$ is associative and commutative.
(2) $\odot$ is associative.
(3) Both operators have identities. $\oplus$ is called the *additive* operator and $\odot$ is the *multiplicative* operator. 0 is the additive identity and 1 is the multiplicative identity.
(4) Every x in K has an additive inverse. I.e., for every x in K, there is an x^{-1} such that $x^{-1} \oplus x = x \oplus x^{-1} = 0$.
(5) $\odot$ is left and right distributive with respect to $\oplus$.

From the above properties of a ring, it follows that $<K, \oplus>$ is an Abelian group and $<K, \odot>$ is a monoid. A *commutative ring* is a ring in which the multiplicative operator $\odot$ is commutative. $<Z, +, *>$ is an example of a commutative ring.

Theorem 10.7: Let $<K, \oplus, \odot>$ be a ring. Let 0 denote the additive identity. 0 is also the multiplicative zero.

Proof: We need to show that $0 \odot x = x \odot 0 = 0$ for every x in K. Let 1 be the multiplicative identity. From the definition of an identity, it follows that $(1 \oplus 0) \odot x = 1 \odot x = (1 \odot x) \oplus 0$. Using the distributivity of $\odot$ with respect to $\oplus$, we obtain $(1 \oplus 0) \odot x = (1 \odot x) \oplus (0 \odot x)$. Hence, $(1 \odot x) \oplus 0 = (1 \odot x) \oplus (0 \odot x)$. Using left cancellation of $(1 \odot x)$ with respect to $\oplus$, we obtain $0 = 0 \odot x$. Hence, 0 is a left zero for $\odot$. The proof that 0 is a right zero is similar. $\square$

An *integral domain* is a commutative ring $<K, \oplus, \odot>$ which satisfies the following cancellation rule:

$$\forall x, y \in K \, [c \odot x = c \odot y \text{ and } c \neq 0 \implies x = y]$$

Once again, 0 is the additive identity which by Theorem 10.7 is also the multiplicative zero. One may verify that the commutative ring $<Z, +, *>$ ia an integral domain.

Theorem 10.8: Let $<K, \oplus, \odot>$ be an integral domain. If x and y are two elements of K such that $x \neq 0$ and $y \neq 0$, then $x \odot y \neq 0$.

Proof: Left as an exercise. $\square$

A *field* is a commutative ring $<K, \oplus, \odot>$ such that $<K-0, \odot>$ is a group. Specifically, $<K, \oplus, \odot>$ is a field iff:

(1) $<K, \oplus>$ and $<K-0, \odot>$ are Abelian groups

and

(2) $\odot$ is distributive over $\oplus$

From the definition of a field, it is evident that every field is an integral domain. $<R, +, *>$ is an example of a field. The ring $<Z, +, *>$ of integers is not a field as $<Z - \{0\}, *>$ is not an Abelian group.

Example 10.8: While the ring of integers $<Z, +, *>$ is not a field, the ring $<N(p), +_p, *_p>$ of integers modulo p where $N(p) = \{0, 1, 2, ..., p-1\}$; $x +_{py} = x+y \bmod p$; $x *_p y = x*y \bmod p$; and p is a prime number is a field. While we shall not prove this here, we shall verify it for the case $p = 7$. In this case, $N(7) = \{0, 1, 2, 3, 4, 5, 6\}$. The additive identity is 0 and $+_p$ is both associative and commutative. The additive inverses of 0, 1, 2, 3, 4, 5, and 6 are, respectively, 0, 6, 5, 4, 3, 2, and 1. Hence, $<N(p), +_7>$ is an Abelian group. 1 is the multiplicative identity and $*_7$ is both associative and commutative. The multiplicative inverses of 1, 2, 3, 4, 5, and 6 are, respectively, 1, 4, 5, 2, 3, and 6. So, $<N(7) - 0, *_7>$ is also an Abelian group. One may verify that $*_7$ distributes over $+_7$. Hence, $<N(p), +_7, *_7>$ is a field. The order of this field is 7. □

10.4 HOMOMORPHISMS

Let $<K, op_1, op_2, op_3, ..., op_n>$ and $<K', op'_1, op'_2, op'_3, ..., op'_n>$ be two algebras such that the degree of op_i equals that of op'_i for every i. A *homomorphism* from $<K, op_1, ..., op_n>$ to $<K', op'_1, ..., op'_n>$ is a function $f: K \rightarrow K'$ such that for every x, y, and z in K:

(1) $f(op_i(x)) = op'_i(f(x))$ if op_i is unary
(2) $f(x \ op_i \ y) = f(x) op'_i f(y)$ if op_i is binary
(3) $f(op_i(x,y,z)) = op'_i(f(x), f(y), f(z))$ if op_i is ternary

.

.

.

etc.

$f(x)$ is called the *homomorphic image* of x. Let $f(K)$ be the set $\{x \exists y \in K \ [f(y) = x]\}$. The algebra $<f(K), op'_1, op'_2, ..., op'_n>$ is the *homomorphic image* of the algebra $<K, op_1, op_2, ..., op_n>$.

There are several types of homomorphisms possible from algebra $A = <K, op_1, op_2, ..., op_n>$ to algebra $B = <K', op'_1, op'_2, ..., op'_n>$. These are summarized in Table 10.1.

	HOMOMORPHISM	CONDITIONS
1	endomorphism	$A = B$
2	epimorphism	$f: K \rightarrow K'$ is onto
3	monomorphism	f is one to one (injective)
4	isomorphism	f is one to one onto (bijective)
5	automorphism	$A = B$ and f is an isomorphism

Table 10.1 Homomorphisms from A to B.

Example 10.9: Let $A = <Z,+>$ and $B = <N,+>$. The function $f: Z \rightarrow N$ such that $f(x) = |x|$ for all x in Z is not a homomorphism from A to B. To see this, note that $f(2+(-3))=1$ while $f(2)+f(-3)=5$.

Suppose that $A = <E,+>$ and $B = <O,\oplus>$ where E is the set of even numbers and O is the set of odd numbers. $\oplus$ is defined such that $x \oplus y = x+y+1$ for every x and y in O. The function $f: E \rightarrow O$ such that $f(x) = x+1$ for every x in E is also not a homomorphism from A to B. Observe that $f(2) = 3$; $f(2+2) = f(4) = 5$; but $f(2) \oplus f(2) = 3+3+1 = 7$. If $\oplus$ is defined as $x \oplus y = x+y-1$, then we see that $f(x+y) = x+y+1 = f(x) + f(y) - 1 = f(x) \oplus f(y)$ for every x and y. With this definition of $\oplus$, f is a homomorphism from A to B. In fact, since f is a one to one onto function, it is an isomorphism. □

Example 10.9: Consider the binary algebras $A = B = <Z,+>$. The function $f: Z \rightarrow Z$ such that $f(x) = x$ is an automorphism from A to B. The function g such that $g(x) = 2x$ is a monomorphism from A to B. It is not an automorphism as g is not bijective. □

Homomorphisms are important as the homomorphic image of an algebra A generally has the same properties as A.

Theorem 10.9: Let $A = <K,\odot>$ and $B = <K',\oplus>$ be two binary algebras. Let f be a homomorphism from A to B.

(1) If A is a semigroup then the homomorphic image $<f(K),\oplus>$ of A is

also a semigroup.

(2) If A is a monoid then $< f(K), \oplus >$ is also a monoid.

(3) If A is a group then $< f(K), \oplus >$ is also a group.

Proof: We shall prove (1) only. The proofs of (2) and (3) are left as an exercise. Assume that A is a semigroup. We need to show that $\oplus$ is associative over $f(K)$. Every x in $f(K)$ is the image of at least one element in K. Let x, y, and z be arbitrary elements of $f(K)$. Let s, t, and u be such that $f(s)=x$, $f(t)=y$, and $f(u)=z$. So, we need to show that $(f(s) \oplus f(t)) \oplus f(u) = f(s) \oplus (f(t) \oplus f(u))$. Since A is a semigroup, $\odot$ is associative. Hence,

$$\begin{aligned}
(f(s) \oplus f(t)) \oplus f(u) &= f(s \odot t) \oplus f(u) \\
&= f((s \odot t) \odot u) \\
&= f(s \odot (t \odot u)) \\
&= f(s) \oplus f(t \odot u) \\
&= f(s) \oplus (f(t) \oplus f(u)) \quad \square
\end{aligned}$$

Theorem 10.10: The homomorphic image of a Boolean algebra is also a Boolean algebra.

Proof: Left as an exercise. $\square$

REFERENCES AND SELECTED READINGS

Some good references and sources for further reading on the subject of modern algebra are:

A survey of modern algebra, by G. Birkhoff and S. MacLane, The Macmillan Co., NY, 1965.

Algebra, by S. MacLane and G. Birkhoff, The Macmillan Co., NY, 1967.

Modern applied algebra, by G. Birkhoff and T. Bartee, McGraw-Hill, NY, 1970.

and

Topics in algebra, 2nd Edition, by I. Herstein, Xerox College Publishing, Lexington, Mass., 1975.

Group theory has found considerable application in the design of error detecting and error correcting codes. A good reference for this is:

Coding and information theory, by R.W. Hamming, Prentice-Hall, Inc., Englewood Cliffs, NJ, 1980.

EXERCISES

1. Prove part (2) of Theorem 10.1.

2. Let $<K,\odot>$ be a semigroup. Let x in K be a left zero and let y in K be a right zero. Show that:
   (a) $a \odot x$ is a left zero for every a in K
   (b) $y \odot a$ is a right zero for every a in K

3. Let $N(p) = \{0, 1, ..., p-1\}$ and let $*_p$ be multiplication modulo p (i.e., $x *_p y$ is the remainder of $x*y$ divided by p). Show that $<N(p),*_p>$ is a semigroup for every natural number p, $p>0$.

4. Let $K = \{a^i \mid i \in N\}$. Assume that $<K,\odot>$ is a monoid. Monoids of this type (i.e., those generated by a single element) are called *cyclic monoids*. Show that every cyclic monoid is commutative (i.e., $\odot$ is commutative over K).

5. Is $<\{true,false\},\vee>$ a group?

6. Let $<K,\odot>$ be a finite monoid with the property that $x \odot x = 1$ for every x in K (1 is the identity element). Show that $<K,\odot>$ is an Abelian group.

7. Let $C = \{a + ib \mid a,b \in R$ and $i = \sqrt{-1}\}$ be the set of complex numbers. Show that $<C-0+i0, *>$ is an Abelian group.

8. Let $<K,\odot>$ be a group. Show that:
   (a) If $a \odot x = b$ for some x, a, and b in K, then $x - u^{-1} \odot b$.
   (b) If $x \odot a = b$ for some x, a, and b in K, then $x = b \odot a^{-1}$.
   (c) The inverse of the identity 1 is 1.
   (d) $(x \odot y)^{-1} = y^{-1} \odot x^{-1}$ for every x and y in K.

9. Let X and Y be the 2×2 matrices:

$$X = \begin{bmatrix} 0 & 1 \\ 1 & 0 \end{bmatrix} \quad \text{and} \quad Y = \begin{bmatrix} 0 & 1 \\ -1 & 0 \end{bmatrix}$$

   Let $*$ denote matrix multiplication.
   (a) What is the algebra generated by X, Y and $*$?
   (b) Show that this algebra is an Abelian group.
   (c) What is the order of the generated algebra?

10. Suppose that the algebra $<K,*>$ is generated by the elements x and y. Assume that $*$ denotes multiplication and that $x^3 = y^3 = (x*y)^2 = 1$.

(a) List the algebra generated. Show that this algebra is a group.
(b) What is the order of this group?
(c) List all the subgroups of this group.

11. Do exercise 10 with $x^3 = y^3 = (x*y)^2 = 1$ replaced by $x^3 = y^2 = (x*y)^3 = 1$.

12. Show that if $<K,\odot>$ is a group of order p where p is a prime number, then it is cyclic.

13. Let $<K,\odot>$ be a group of order n. Let a be an element of K and let 1 be the group identity. Show that $a^n = 1$.

14. Consider the group $<NP(11),+_{11}>$ (see exercise 3) where $+_{11}$ is addition modulo 11. Obtain the cosets $2 +_{11} N(11)$, $N(11) +_{11} 2$, $5 +_{11} \{0,5,8\}$, and $\{2,3,8,10\} +_{11} 7$.

15. Prove Theorem 10.4.

16. Prove part (2) of Theorem 10.5.

17. Prove Theorem 10.6.

18. Let $<K,\odot>$ be a group. The order of any x in K is defined to be the least positive integer i such that x^i equals the group identity. If no such i exists, the order of x is infinite. Show that the order of every element in a finite group is finite and divides the order of the group.

19. Prove Theorem 10.8.

20. Prove parts (2) and (3) of Theorem 10.9.

21. Prove Theorem 10.10.

22. A ring $<K,\oplus,\odot>$ is a Boolean ring iff $x \odot x = x$ for every x in K. For a Boolean ring, show that:
(a) $x \oplus x = 0$ (the additive identity) for every x in K.
(b) The ring is commutative.
(c) If the order of the ring exceeds 2 then it cannot be an integral domain.

23. Show that every finite integral domain is a field.

24. Let $<K,\oplus,\odot>$ be a field. Let 0 be the additive identity. For $x \neq 0$

define $y/x = y \odot x^{-1}$ where x^{-1} is the multiplicative inverse of x. Show that when $v \neq 0$ and $x \neq 0$:

(a) $u/v = w/x$ iff $u \odot x = w \odot v$

(b) $(u/v) \oplus (w/x) = [(u \odot x) \oplus (v \odot w)]/(v \odot x)$

(c) $(u/v) \odot (w/x) = (u \odot w)/(v \odot x)$

APPENDIX A

SYMBOLS

A.1 GREEK LETTERS

LC	UC	NAME	LC	UC	NAME
α	A	Alpha	ν	N	Nu
β	B	Beta	ξ	Ξ	Xi
γ	Γ	Gamma	o	O	Omichron
δ	Δ	Delta	π	Π	Pi
ϵ	E	Epsilon	ρ	P	Rho
ζ	Z	Zeta	σ	Σ	Sigma
η	H	Eta	τ	T	Tau
θ	Θ	Theta	υ	Υ	Upsilon
ι	I	Iota	ϕ	Φ	Phi
κ	K	Kappa	χ	X	Chi
λ	Λ	Lambda	ψ	Ψ	Psi
μ	M	Mu	ω	Ω	Omega

LC = lower case UC = upper case

Table A.1 Table of Greek letters

A.2 MATHEMATICAL SYMBOLS

LOGIC

1	$\wedge$	and	§1.1
2	$\vee$	or	§1.1
3	$\neg$	not	§1.1
4	$\bar{}$	not	§1.1
5	$\implies$	implies	§1.1
6	$\iff$	iff	§1.1
7	$\models$	inference	§1.3
8	$\forall$	for all	§1.6
9	$\exists$	there exists	§1.1

SETS

10	{}	set	§3.1
11	()	ordered set	§3.1
12	∅	empty set	§3.1
13	∈	member of	§3.1
14	ϵ	null element	§3.1
15	∪	union	§3.3
16	∩	intersection	§3.3
17	⊆	subset	§3.1
18	⊂	proper subset	§3.1
19	⊕	exclusive or	§3.3
20	⁻	complement	§3.3

OTHER

21	$\dfrac{d}{dx}$	derivative	
22	$\int$	integral	
23	$\sum$	sum	
24	$\prod$	product	
25	$\sqrt{}$	square root	
26	!	factorial	§8.2
27	$\left(\begin{matrix} n \\ x \end{matrix}\right)$	combination	§8.1
28	$\left\{\begin{matrix} n \\ x \end{matrix}\right\}$	Stirling number	Chap.8
29	$\left[\begin{matrix} n \\ x \end{matrix}\right]$	Stirling number	Chap.8
30	$\lceil x \rceil$	smallest integer $\geqslant x$	
31	$\lfloor x \rfloor$	largest integer $\leqslant x$	

APPENDIX B

ALGORITHMIC LANGUAGE

The algorithmic language used in this text is quite similar to block structured programming languages such as Pascal, PL/I, and Algol. We shall now explicitly describe the various control structures used in this text. In what follows, S, $S1$, $S2$, ... denote blocks of program statements and *cond*, *cond*1, *cond*2, ... denote Boolean expressions (i.e., they have value true or false).

while statement

The construct:

while *cond* **do**
  S
end while

is equivalent to the following block diagram:

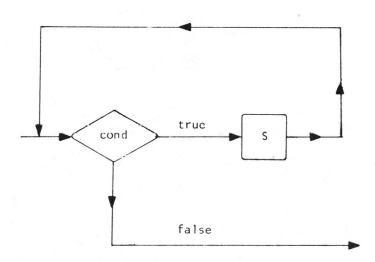

for statement

The **for** statement:

for $i \leftarrow$ *start* **to** *finish* **by** *increment* **do**
  S
end for

is equivalent to the following construct:

$i \leftarrow$ *start*
fin $\leftarrow$ *finish*
incr $\leftarrow$ *increment*
while $(i - fin) * incr \leqslant 0$ **do**
  S
  $i \leftarrow i + incr$
end while

repeat-until statement

The construct:

repeat
  S
until *cond*

corresponds to the block diagram:

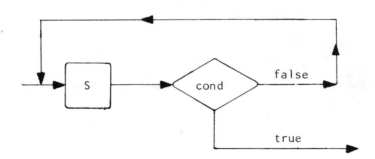

case statement

The construct:
case
  :*cond*1: $S1$
  :*cond*2: $S2$

.
.
.

:*condn*: *Sn*
:else: *Sn*+1
end case

is equivalent to the block diagram:

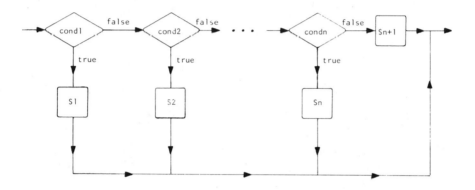

The construct:

case
  :*cond*1: *S*1
  :*cond*2: *S*2

  .

  .

  .

  :*condn*: *Sn*
end case

is equivalent to:

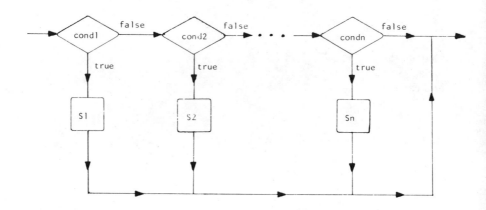

if statement

  if *cond* **then** *S* **end if**

is equivalent to:

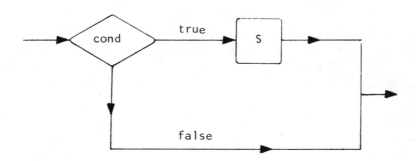

while

  if *cond* **then** *S*1
        else *S*2
  end if

is equivalent to:

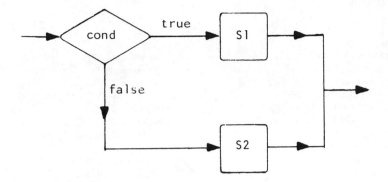

INDEX

Abelian group, 452
Ackermann's function, 76, 120, 216
Adjacency list, 386
Adjacency matrix, 385
Aho, A., 199, 286, 440
Algebra, 449-461
  binary, 450-451
  Boolean, 8-16
  field, 455
  finite, 449
  group, 451
  integral domain, 454
  monoid, 450
  ring, 454
  semigroup, 450
  subgroup, 452
Algorithms
  analysis, 251-285
  complexity, 251-273
  recursive, 151-238
AND, 5
Association, 148
Associativity, 142
Automorphism, 456

Backus Naur Form, 135
Bartee, T., 458
Beeri. C., 204
Bell, J., 51
Bentley, J., 322
Bernoulli trial, 351-353, 354
Bernstein, P., 204
Binary algebra, 450-451
Binary relation
  equivalence, 175-178
  g.l.b., 186
  l.u.b., 186
  lattice, 186
  partial order, 178-186
  topological order, 181-183
Binary search, 84-85, 87-89
Binomial coefficient, 336-339
Binomial distribution, 352
Binomial theorem, 337
Birkhoff, G., 458
Boolean algebra, 8-16
Boole's inequality, 349
Breadth first search, 391-395
Brualdi, R., 322, 361, 440

call
  by reference, 221
  by return, 221
  by value, 221
Carroll, L., 38-40, 51, 56, 60
Chang, C.L., 52
Characteristic roots, 303-318
Chebychev's inequality, 357-358
Clause, 12
Combinations, 340-341
Combinatorics, 336-341
Commutativity, 142
Complexity
  asymptotic, 273-282
  average, 270
  practical, 282-285
  space, 251, 256-273
  time, 251-256
  worst case, 270
Computability, 238-244
Conjunctive normal form, 12, 26, 30-33
Constable, R., 118
Context free grammar, 135
Context sensitive grammar, 135
Contradiction, 8
Correspondence, 148-154
Coset
  left, 453
  right, 453
Countability, 148-154
Covariance, 356

Date, C., 204
Davis, M., 30-33, 52, 56
Davis-Putnam procedure, 30-33, 56
Dekel, E., 245
DeMorgan's laws, 18
Depth first search, 395-398
Deo, N., 287, 440
Derangement, 249
Derivation, 123, 136, 137
Diagonalization, 154-158
Dijkstra, E., 93
Disjunctive normal form, 12, 26-30
Distributivity, 142
Domino, 119

Ehrlich, G., 245
Endomorphism, 456

Epimorphism, 456
Equivalence relation, 175-178
Equivalence rule, 17
Euclid's gcd algorithm, 89-90, 91
Euler circuit, 411-414
Euler path, 411-414
Even, S., 440
Event
    compound, 346
    independent, 351
    mutually exclusive, 348
    random, 347
    simple, 346
Expressions
    arithmetic, 136-139
    Boolean, 11
    logical, 6, 36
    set, 145-147

Fagin, R., 200
Feller, W., 361
Fibonacci numbers, 70
First order logic, 33-49
Functions, 205-213
    composition, 209
    domain, 205-213
    into, 209
    inverse, 210
    multivalued, 207
    one-to-one, 209
    onto, 209
    partial, 205
    range, 205
    total, 205
Friedman, A., 51

Gardner, M., 245
Garey, M., 199
Generating functions, 318-332
Grammar, 132-139
    ambiguous, 137
    BNF, 135
    context free, 135
    context sensitive, 135
    regular, 135
    type 0, 135
Graphs, 367-448
    acyclic, 378
    bipartite, 371
    cliques, 437
    coloring, 438-439
    complement, 438

    complete, 373
    components, 375, 381
    connected, 375, 405-411
    cycles, 403-405
    digraph, 368
    directed, 368
    edge, 367
    independent set, 437
    isomorphic, 374
    matching, 435-437
    multigraph, 369
    paths, 374
    planar, 431-435
    pseudograph, 369
    representations, 384-390
    shortest paths, 417-418
    spanning tree, 394, 395, 397, 398-402
    subgraph, 373
    strongly connected, 381
    transitive closure, 419
    undirected, 368
    unilaterally connected, 382
    vertex, 367
    weakly connected, 382
Gray code, 215, 245
Greatest common divisor gcd, 89-90
Gries, D., 126
Group
    Abelian, 452
    cyclic, 452
    semigroup, 450
    subgroup, 450

Haken, D., 334
Halting problem, 240
Hamiltonian circuit, 415-417
Hamiltonian path, 415-417
Hamming, R, 458
Harary, F., 439, 441
Harmonic numbers, 120
Harris, R., 52
Hennie, F., 245
Herstein, I., 458
Hill, F., 51
Homomorphisms, 455-461
Hopcroft, J., 286, 440
Horowitz, E., 199, 245, 286, 440

Identity
    left, 449
    right, 449
    two sided, 449

Incidence matrix, 389
Inclusion-exclusion theorem, 146
Induction, 67-127, 311-312
    strong, 67-77
    weak, 77-81
Inference rule, 17
Inverse
    left, 451
    right, 451
    two sided, 451
Isomorphism, 456

Knuth, D., 286, 351, 440
Kuratowski, K., 441

Lattice, 186
Law of large numbers
    strong, 359
    weak, 358
Lee, R., 52
Literal, 11
Liu, C., 332, 351, 440, 441
Logic, 1-61
    clause, 12
    conjunctive normal form, 12, 26, 30-33
    Davis-Putnam procedure, 30-33, 56
    disjunctive normal form, 12, 26-30
    equivalence rule, 17
    first order, 33-49
    inference rule, 17
    literal, 11
    maxterm, 11, 12
    minterm, 11, 12
    predicate calculus, 33-49
    proof methods, 16-26
    propositional calculus, 1-26
    tableau method, 26-30, 56
    truth table, 4
    well formed formula, 6, 36
Loop invariant, 90-93

Machover, M., 51
MacLane, S., 458
Maxterm, 11, 12
Mean, 353
Minterm, 11, 12
Moment, 354
Monoid, 450
Monomorphism, 456
Multinomial coefficient, 343-346
Multinomial theorem, 344
Multiset, 129

NAND, 52
Nassimi, D., 245
Nievergelt, J., 287, 440
NOR, 52
NOT, 3
Numbers
    even, 148
    Fibonacci, 70
    harmonic, 120
    natural, 148
    odd, 148
    rational, 149
    real, 154-156
    Stirling, 363-354

O'Donnell, M., 118
One-to-one correspondence, 148
OR, 4

Paul, W., 286
Permutations, 235-237, 341-343
Peterson, G., 51
Pigeon hole principle, 211-213
Poisson distribution, 365
Polynomial, 151-152
Predicate, 34
Predicate transformer, 93-118
Prenex normal form, 42-45
Probability, 346-360
    conditional, 349-351
    distribution, 353
    joint distribution, 355
Proof methods, 16-26
Propositional calculus, 1-26
Putnam, H., 30-33, 52, 56

Random variable, 353-357
Recurrence relation, 291-335
Recursion, 213-238
    direct, 217
    indirect, 217
    stack, 220-221
Reingold, E., 287, 440
Reflexivity, 171
Regular grammar, 135
Relation, 163-204
    binary, 164, 168-186
    closure, 172
    composition, 172
    equivalence, 175-178
    k-ary, 163, 183-191
    partial order, 178-186

recurrence, 291-335
Relational databases, 166, 186-198
   functional dependencies, 187-195
   join, 190
   lossless join, 194
   multivalued dependencies, 193-195
   normal forms, 195-198
   projection, 188
   scheme, 187
   selection, 188
   semijoin, 190
   superkey, 196

Sahni, S., 199, 245, 286, 440
Sample point, 346
Sample space, 346
Satisfiable, 8
Saxe, J., 332
Schedule, 65
Selection, 235-236
Selection sequence, 340
Semigroup, 450
Set, 129-162, 165
   cardinality, 130
   closure, 141, 142
   complement, 141
   correspondence, 148
   countability, 148-154
   diagonalization, 154-158
   difference, 141
   disjoint, 142
   discrete, 148
   empty, 130
   exclusive or, 141
   inclusion-exclusion, 146
   intersection, 140
   multiset, 129
   ordered, 130
   power set, 142
   product, 141
   proper subset, 130
   specification, 130-139
   subset, 130
   union, 140
   universal, 130, 139
   Venn diagram, 140
Smullyan, R., 52
Sorting
   insert sort, 232-233
   merge sort, 233-234, 291
   quicksort, 234-235, 292

Spanning trees
   breadth first, 395
   depth first, 397
   minimum cost, 398-402
Stack, 220
Standard deviation, 355
Sirling's approximation, 337
Stirling numbers, 363-364
Subgroup, 452
Subset, 130

Tableau method, 26-30
Tarjan, R., 118
Tautology, 8
Theorem, 8, 46
Theory, 51
Thomasian, A., 361
Topological order, 163, 181-183
Traveling salesperson problem, 416
Trees
   AVL, 446
   binary, 422
   breadth first, 395
   depth first, 397
   forest, 379
   minimum cost, 398-402
   number of, 427-431
   rooted, 383-384, 419-431
   search trees, 423-427
Trotter, H., 245
Truth table, 4
Tutte, W., 441
Type 0 grammar, 135

Ullman, J., 199, 286, 440
Universe of discourse, 36

Variance, 355
Venn diagram, 139-140

Weakest precondition, 96-118
Well formed formula, 6, 36

Zero
   left, 449
   right, 449
   two sided, 449